The fascinating secrets of

OCEANS & ISLANDS

The fascinating secrets of
OCEANS & ISLANDS

**The fascinating secrets of Oceans & Islands
was edited and designed by The Reader's Digest Association**

First Edition

Copyright © 1972 The Reader's Digest Association Limited

For full details of copyright see page 368

Printed in Great Britain

Published by The Reader's Digest Association Limited
London   Montreal   Sydney   Cape Town

MAJOR CONTRIBUTORS

William J. Cromie
William H. Amos

SCIENCE EDITOR

Durward L. Allen
Purdue University, U.S.A.

SPECIAL CONSULTANTS

Agatin T. Abbott
University of Hawaii

Stanley and Kay Breeden
naturalists/authors (Australia)

S. G. Brown
National Institute of Oceanography of Great Britain

Ronald I. Currie
Scottish Marine Biological Association

Peter David
National Institute of Oceanography of Great Britain

Raymond de Lucia
American Museum of Natural History

F. Raymond Fosberg
Smithsonian Institution

Thomas Gladwin
University of Hawaii

Richard Gould
American Museum of Natural History

Joel W. Hedgpeth
Marine Science Centre, Oregon, U.S.A.

Martin Holdgate
Central Unit on Environmental Pollution, London

J. R. Lewis
Wellcome Marine Laboratory, University of Leeds

N. B. Marshall
British Museum (Natural History)

Olin Sewall Pettingill Jr.
Cornell University, U.S.A.

Garrett Smathers
U.S. National Park Service

C. Lavett Smith
American Museum of Natural History

D. R. Stoddart
Department of Geography, Cambridge University

Hobart Van Deusen
American Museum of Natural History

CONTENTS

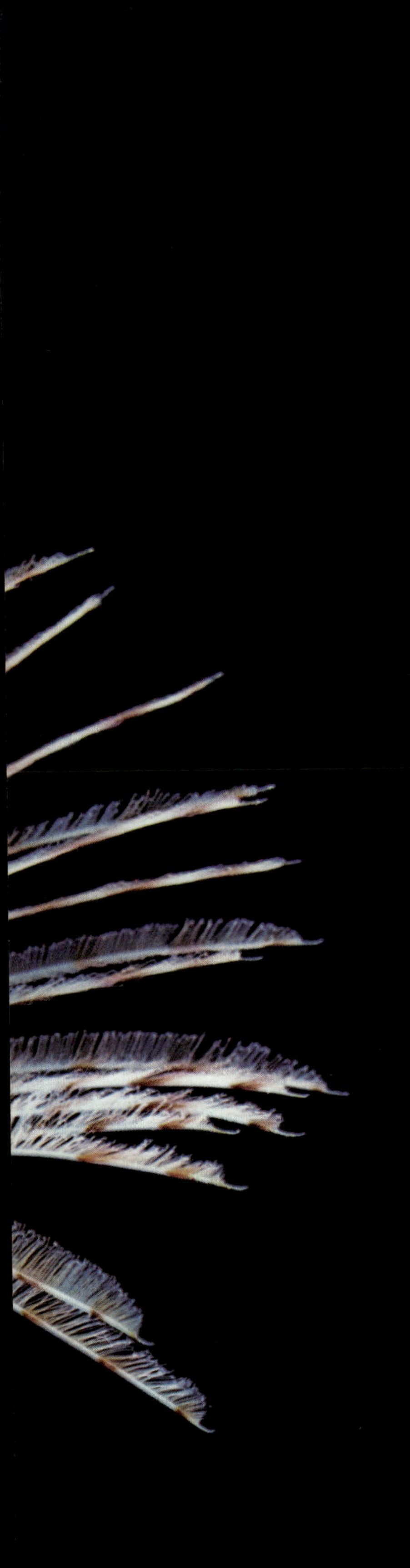

Fanworm *Sabella pavonina*

FOREWORD BY
CAPTAIN JACQUES-YVES COUSTEAU
Explorer of the ocean depths
co-inventor of the aqualung

From Captain Jacques-Yves Cousteau, aboard the Calypso

THE SEA, OUR HERITAGE

Earth is the only planet in the solar system where water exists in any appreciable quantity – covering as it does some seven-tenths of the surface of the globe. But this water, in which life first evolved and on which all living things still depend for survival, has been given to us once and for all: there will be no more.

History will look back on our age as a time of crisis when man, the source of all pollution, endangered his whole environment, including the life-sustaining sea. By allowing industry to poison the sea with its waste products, by refusing to accept strict international controls on fishing rights, we squander the birthright of generations to come.

That birthright also includes islands, places that have fascinated mankind from the beginning of time. Looming out of the sea, they have safeguarded and preserved a remarkable diversity of plant and animal life, and their isolation has made them a last refuge for many fish, corals and other sea creatures. Now these oases are themselves threatened and deserve our special concern.

Yet technology, though it is blamed for all the evils of the modern world, is in fact mankind's best weapon against the pollution that so often seems to be the price of progress; for it is not the advances of science and industry that menace man and nature, but the clumsiness and thoughtlessness with which man uses his technological achievements.

It is my hope that an increased awareness of the sea, which for thousands of years has taught man wisdom, will inspire in him once more those thoughts and actions that will preserve the balance of nature and maintain life itself.

PART ONE
Secrets of the seas

Life along the shore

Wherever land and sea meet – along rocky shores, sandy beaches, tidal marshes, mangrove thickets and river estuaries – plants and animals thrive abundantly in a turbulent world swept by the waves and tides

Nearly three-quarters of the earth is covered by the sea. It is the unique asset of our small world; as far as we know, no other star or planet has water on its surface. It was the source of life; the first living things emerged in the sea, and it was many millions of years before certain small fish with bony fins and air-breathing lungs learnt how to propel themselves over land when their mud pools dried up in the sun. This rare gift of life which the sea bestowed on the land still depends on the sea; without the water which evaporates from the oceans and falls as rain, all the plants and animals of the land would wither away, leaving only dry desert and barren rock.

The sea is man's past and present, for the blood flowing through his veins is little more than modified sea-water. It is also his future; for it is likely that only by the conservation and cultivation of these great waters can we find food to feed ourselves and so survive the crisis of our rapidly increasing numbers.

For thousands of years the open sea offered the ultimate adventure; beyond its trackless wastes lay the unknown, a world of boundless promise and danger. With incredible courage, the great explorers set out across these storm-tossed seas, knowing nothing of the shores that lay beyond. Today their work is done; most of the remote parts of the continents have yielded their secrets. But the sea itself remains unconquered. Many of its creatures and their patterns of life are still beyond our understanding. Its deepest chasms are barely within our reach, almost entirely uncharted and unknown. The ocean has become the last great frontier, the last challenge on earth to man's ingenuity and courage.

Most people, of course, are unlikely ever to venture into the perpetual night of the great abyssal deeps; many will never have the chance to swim under water, weaving among shoals of silent fish; and only a lucky few can hope to enjoy the opportunity of a trip over the multi-coloured corals of Australia's Great Barrier Reef. But to almost everyone, old or young, is given the priceless experience of viewing the ocean at its margin, where land and water meet in eternal conflict and compromise. And here on the shore, with both feet firmly planted on dry sand or solid rock, it is possible to begin the ever-rewarding study of the sea. For the animals of the tidal shore are among the most beautiful and fascinating creatures of the sea. Their world is a narrow one – the widest beaches are seldom more than a few hundred feet across – but it extends for more than a million miles around all the earth's islands and continents.

RESISTING THE WAVES *All starfish resist the pounding force of the waves by clinging to rocks, using the suction cups in their hundreds of tube-shaped feet, which they use also to prise open shellfish. When an arm breaks the starfish can regenerate a replacement part. Starfish are found throughout the world's seas*

HARSH ENVIRONMENT OF THE SHORE

The plants and animals of the tidal zone live in one of the sea's harshest environments. Unlike inhabitants of the stable world of deep water, shore dwellers endure constant change. The waves pound and pull them incessantly. The daily comings and goings of the tide submerge them in water, then expose them to air.

Yet all the basic requirements of life – light, water, oxygen and minerals – are here in abundance. An astonishing variety of living things populates this world of ebb and flow.

Ebb and flow of the tides

As the positions of the moon and sun change, their gravitational pulls create the shifting patterns of the earth's tide. The moon, being much closer to earth than the sun, exerts the stronger pull and produces two bulges in the seas at opposite sides of the globe. Because the earth rotates, most coasts are 'hit' by both bulges in a 24-hour period and so have two high tides daily.

The sun produces smaller bulges, which are usually masked by those of the moon. About twice a month, at full moon and new moon, the earth, moon and sun are in a straight line, and the gravitational tug of the sun is added to that of the moon. At such times, flood tides rise highest on the shore and ebb tides fall lowest. These are *spring tides* – referring not to the season of the year but to the 'springing up' of the water. Also about twice monthly, the three bodies form a right angle, and the sun's influence partially cancels that of the moon. This is the time of *neap tides*, when the difference between high and low tides is smallest.

Geographic factors, such as the contours of shorelines and the depth of coastal waters, play a large part in determining tidal range. In the Mediterranean, the approaches to the Baltic and through much of the western Pacific, the difference between high and low tide seldom exceeds 2 ft. In the Bay of Fundy, between New Brunswick and Nova Scotia, spring flood tides may rise 50 ft above the low tidemark, while ranges of 20–40 ft are not uncommon on parts of the coast of the British Isles and north-west France.

At the edge of the sea, the pulse of life beats to a tidal rhythm. Shore creatures require a watery environment even when the tide is at its ebb. Some simply move out with the retreating waters, or take refuge in tide pools scattered among the rocks. Others that move slowly or not at all have developed ways of retaining moisture during the drought of the ebb tide. Mussels almost shut their shells, keeping their own private 'ocean' inside. Limpets – snails with flattened, conical shells – form a watertight seal by clinging firmly to rocks.

HOW PLANTS AND ANIMALS SURVIVE THE WAVES

On any exposed shore, plants and animals must resist being swept away by the waves. These organisms – and their counterparts on rugged coastlines throughout the world — are especially equipped to cling

KELP AND SEA PALM *Members of the sea palm family have no roots; they cling by many branches*

STARFISH *Hundreds of tube feet on the underside of a starfish's arms keep it fixed firmly to its rock*

SEA URCHIN *Sheltering in hollows created by its ancestors, the sea urchin attaches itself by suction*

ACORN BARNACLES *A powerful cement secreted from their limy shells anchors barnacles permanently*

MUSSEL *A cluster of tough, sticky threads is secreted by the mussel to hold it firmly on the rocks*

LIMPET *The limpet's large, muscular foot clamps it to a rock and its streamlining resists the waves*

When the water rises, the tempo of life quickens in the inter-tidal world. Mussels open their shells and filter water through their bodies, removing anything edible in the process. Limpets move to graze on algae on the rocks.

Assault of the waves

Just as living things have modified their behaviour and anatomy to cope with alternate exposure and submergence by tides, so they have evolved defences against waves. Under normal wind conditions, four or five waves crash on a shore every minute. Between waves, the backwash exerts a drag towards the sea.

Seaweed holds its own against this pounding and pulling by gripping rocks with root-like hold-fasts. Instead of resisting the waves, the seaweed's flexible stipes (stems) and fronds (leaves) stream in and out with the rushing water. Mussels are anchored to rocks with tough threads, and oysters are cemented in place. Snails, abalones, limpets and other molluscs hold on by the suction-cup action of a muscular foot. Starfish cling to hard surfaces with hundreds of tube feet tipped with suction cups. Sea urchins wedge themselves into crevices with strong, needle-like spines.

Acorn barnacles confront the onslaught of the waves, sometimes withstanding forces that measure many tons per square foot. Their secret lies in the extraordinarily strong glue that cements them to rock surfaces, and in their sturdy, conical shells, which divert the flow of oncoming water.

24 hours. On a shore with moderately heavy surf, the most conspicuous feature of the middle zone is a white ribbon of acorn barnacles. In some places, more than 3000 may be crammed into a square foot. Among the barnacles, and also living below them, are mussels. Both are prey of voracious dog whelks, which are snails that can force open shells partly by using their powerful foot, but mainly by use of their radula, a tongue-like organ equipped with horny teeth which drills holes through the shell. Cannibals, they eat their own eggs as well as the eggs of other snails. And even while still inside egg capsules, the first young to hatch devour the remaining eggs.

Scattered among barnacles and mussels of the middle zone are cone-shaped limpets. On soft rock each limpet scrapes out a shallow, circular depression that conforms exactly to the shape of its shell. When covered by the tide, it creeps as far as 3 ft away and scrapes algae off rocks with its radula. When the tide begins to ebb, the little snail finds

LITTLE-KNOWN RESIDENT *Unknown as a British resident until 1955, the leopard-spotted goby inhabits the almost vertical rock crevices just below the low-tide mark. It is found also in Madeira, north Spain and in the Mediterranean*

its way back to its own rock scar – no mean feat considering that it lacks eyes, ears and a nose.

The characteristic plants of the middle zone are brown wracks with branching, ribbon-like fronds. On exposed shores, stunted wrack less than a foot long grows just below the white band of barnacles. But in sheltered bays and inlets, wrack 6 ft long may completely cover the inter-tidal rocks.

As the tide goes out, this rubbery seaweed collapses into a sodden, jumbled mat – an ideal place for marine animals to escape wind and sun. Periwinkles with olive-green, yellow or orange shells hide among the rockweed. Unlike those of the upper zone, they cannot breathe oxygen directly from the air and must stay wet. Entirely dependent on wracks, they eat tissue scraped from the plants and lay eggs on the fronds.

LOWER ZONE The lowest zone of a rocky coast is uncovered only by the ebbing spring tides. Here, where exposure to air is minimal, there is a marked increase in the number and variety of living things. The brown wrack of the middle zone gives way to many types of smaller algae, generally red or reddish-brown in colour and growing in the form of compact cushions, short tufts, leathery ribbons or hair-like tresses. Here are such types as 'Irish moss', 'dulse' and 'pepper dulse', but they are all overshadowed by the dense tangles of large, strap-like kelp, the seaweed which dominates the lower zone of the shoreline in all cool, temperate waters.

In some areas there is hardly a stem of Irish moss that is not encrusted with a 'sea mat', or bryozoan colony, resembling a piece of brittle lace. Under a magnifying glass, rows of minute limy cups appear, each holding a tentacled tenant.

Many kelp fronds bear what look like delicate clusters of flowers. Each cluster is a colony of hydroids – tiny tubular animals wearing crowns of thread-like tentacles. Despite the hydroids' fragile look, their gossamer tentacles are armed with poisonous stinging structures. When a small creature blunders near, it is impaled on tiny darts and injected with a paralysing venom.

In addition to fixed animals, larger, more mobile creatures live in the lower zone. Sea urchins move slowly on tube feet; spider crabs creep along on slender, jointed legs. In the Irish moss, crabs hide from watchful gulls and wait until nightfall before feeding. When the tide is in, rock eels, rock bass, gobies and other fish swim among the swaying kelp fronds. Below the low-tide mark, the influences of tides and waves diminish. Shore life blends imperceptibly into ocean life.

LIFE ON A SANDY SHORE

The zonation of life, so sharply etched on a rocky shore, is blurred and indefinite on a sandy beach. The sand offers no firm foothold for organisms such as kelp and barnacles. Beach dwellers tend to be mobile, free to move in and out with each turn of the tide, or even with each wave. Yet zonation is apparent even here.

On the upper beach, beyond reach of the highest tides, live land insects and marine creatures, such as ghost crabs and sandhoppers, that may be in the evolutionary process of leaving the sea.

On the inter-tidal beach, animals such as clams and burrowing crabs live in the sand.

On the lower, or sub-tidal, beach are animals that cannot stand much exposure to air – burrowing sea cucumbers, anemones, shrimps, swimming crabs, sand-eels and other fish, and many types of clams.

Sandhoppers, only $\frac{1}{2}$–1 in. long, live on beaches all over the world, and on European and many other temperate shores they are the only noticeable inhabitants of the upper levels. By day they lie just below the surface or under cast-up weed at the tideline; but at night they move towards the water in search of food, hopping 2–3 ft at a time.

Another and larger inhabitant of upper beaches is the sand-coloured ghost crab, found on American Atlantic shores. Its legs are useless for swimming; when it dashes into the surf to avoid an enemy, it runs along the bottom. The crab normally enters the water only to wet its gills. It still must get oxygen from water in the manner of a fish, and always carries a drop of the sea in a chamber surrounding its gills.

With few exceptions, burrowing is the common way of life on inter-tidal beaches, and the inhabitants spend most of their lives hidden from view beneath the sand. Only a few barely noticeable signs betray their presence.

In some parts of the world, a V-shaped trail may lead to half-buried, spine-covered heart urchins that plough along eating scraps of organic material mixed with the sand grains. Their sharp spines usually point backwards to make the going easier but can straighten quickly and inflict a painful wound on a hand or bare foot.

A ribbon-like track may lead to a sand dollar or burrowing starfish. Sand dollars are extremely flat, wafer-like sea urchins that move by walking on their short spines. They ingest sand grains coated with plant matter and tiny edible particles mixed with the sand. Burrowing starfish, on the other hand, eat small snails and crustaceans, swallowing them whole, then regurgitating the shells. Unlike their rocky-shore relatives, these starfish have no suction cups on the tube feet under their arms. They cannot cling to rocks, but they are excellent burrowers. When placed on the sand, they quickly sink out of sight.

A moving hill of sand may hide a necklace shell searching for clams. The shell pushes aside sand with its large foot, its passage eased by a smoothly rounded shell. When it finds a clam, the necklace shell curls its foot around the prey, drills a neat, circular hole with its radula, and rasps out the flesh. Necklace shells are known to eat more than

POISED FOR FLIGHT *A ghost crab stands ready to retreat to its burrow near the high-tide line. A mature crab of this species can scuttle at a speed of up to 5 ft per second, fast enough to elude most of its natural enemies*

a third of their weight in clams each week.

The pace of life on inter-tidal beaches is keyed to rhythms of the sea, and activity increases tremendously as the tide comes in. Parchment worms in U-shaped tubes begin moving their fan-like 'paddles' back and forth, creating a current of water from which they filter out one-celled plants. Fanworms living in tubes of fine sediment and mucus project their fan of tentacles to trap tiny food particles. Clams push up their tube-like siphons and draw in sea-water, extracting oxygen and trapping tiny organisms on the sticky mucus of their gill filaments. Whelks cruise on the bottom and locate clams through currents set up by their siphons. Crabs and predatory worms seize small clams or shrimps. Small fish search out a variety of crustaceans, worms and feebly swimming larval creatures. Large fish devour small ones. Skates and rays, seeking clams and snails, plough the bottom.

Some of the more mobile sand dwellers follow the tide back and forth. Mole crabs, for example, suddenly bubble up from the quiet sand and are carried up the beach as a wave sweeps over them. When the force of the rushing water slackens, they swiftly burrow backwards in the sand and feed only as the water flows back to the sea. Fine bristles on two long feeding antennae strain microscopic organisms from the backwash. Their oval, mole-like bodies and flattened, paw-like limbs make these crabs well suited for rapid digging.

Colourful 1-in. coquina clams, members of the cockle family, also follow the tide. As the water comes in, hundreds, sometimes even thousands, of them rise out of the sand at one time and allow waves to wash them up the beach. Before the backwash can carry them away, they use their feet and wedge-shaped shells to dig into the sand. Then they thrust up their siphons and begin feeding. As the tide goes out, the clams move further down the shore, always staying near the surf.

With the ebb tide, activity slackens on the inter-tidal beach. Clams and worms cease feeding, and large predators move offshore. A few birds wheel in the air or run up and down the beach, thrusting their bills expectantly into the sand. But most burrowers on the exposed beach have stopped feeding and await the return of the flood.

LIFE RHYTHMS FOLLOW THE TIDE
Perfect timing by shore animals

Many seashore animals synchronise not only their day-to-day feeding activities but also their reproductive cycles with the patterns of the tides.

On California beaches during high tides of spring and summer, for example, grunion, silvery fish 5–8 in. long, engage in a spawning ritual precisely keyed to tidal rhythms. From March to August on the second, third and fourth nights after a full moon, grunion begin arriving just as the tide starts to ebb.

A grunion spawning run may last for an hour, but the mating time of individuals is extremely brief. Females, each accompanied by one or more males, ride the incoming waves as far up the beach as they can, then bore into the sand with their tails. Each female deposits about 1000 eggs 2 in. below the surface. Males curl around the females, cover the eggs with milt, and immediately slip back into the water. The females free themselves and follow the males on the wash of the next wave. The entire process may take only 30 seconds.

This extraordinary performance ensures that the fertilised eggs will be hidden from predators in the moist sand and kept warm for about ten days. The lower tides that follow do not reach the eggs; when the next series of spring tides uncovers them, the embryos are fully developed, ready to leave the protection of the egg. Minutes after the waves pick up the eggs, the baby grunion hatch and drift out to deeper water.

The critically timed spawning of the grunion is a strikingly effective adaptation to conditions on tidal shores. If they spawned during lower tides, or even an hour earlier on the same tides, their eggs would be washed away and probably destroyed. If they spawned during the high tides of the dark moon, when the water rises higher than it does during the full moon, the eggs would have to wait a month to be uncovered again. Scientists have so far been unable to determine how the grunion developed this uncanny timing ability. Somehow, over millions of years, the pulse of the tides has become part of them, and it is a sense that each generation passes on to its successors.

IN TIME WITH THE TIDE *Thousands of grunion wriggle up the sand at high tide to spawn. The female in the centre of the picture (right) has buried her tail and released her eggs; the male curls round her and fertilises them*

On the east coast of North America, there is another annual spawning congregation as impressive as that of the grunion. Certainly it is far older, for the creatures involved are the only remaining members of a group that has existed for 500 million years. These are horseshoe crabs – not crabs at all, but distant relatives of scorpions and spiders, shaped like a large horse's hoof to which a long, dagger-like tail has been added.

On nights of the spring tides in May and June, the horseshoe crabs crawl ashore in huge numbers, males clinging to larger females. The females scoop out nests, and each one lays thousands of eggs. Males release their sperm, and waves carry it to the eggs. As the ebbing tide covers the eggs with sand, the parents return to the sea.

Within a few days, the greenish eggs are several times their original size. In about two weeks, those not dug up by crabs, seabirds and other predators are ready to hatch. The next spring tide uncovers the eggs, and the wave-stirred sand tears them open, freeing the young. They scramble into the surf and head for deeper water.

MATING BY THE TIDE *Spawning horseshoe crabs gather on beaches on the nights of spring tides in May and June. Each female, dragging its smaller mate, burrows several shallow nests in the sand and lays as many as 20,000 eggs*

SHELTER FROM THE WAVES *A tiny rock pool provides a temporary haven for a purple starfish and three sea anemones, one partly camouflaged with a covering of pebbles. At any hint of danger, the anemones can retract their tentacles and draw themselves into a tight mound*

SURVIVAL IN THE TIDE POOLS

The hazard of life in a miniature sea

Ebbing tides leave behind them miniature seas on the rock-rimmed coasts of the world. For the small marine plants and animals that inhabit them, these tide pools offer a precarious existence. Although the pools protect living things from exposure to the air, they also present problems in survival. As their waters are heated by the sun the oxygen in the water diminishes and the animals may suffocate. Another hazard is the changing amount of salt in the water: heavy rains can dilute the salinity, but evaporation can cause an equally dangerous increase. If a tide pool becomes too fresh, starfish, anemones, crabs and other animals may absorb too much water, become bloated and die before the tide returns. If the water becomes too briny, they can lose fluid under the heat of the sun and shrivel to death. Although these pools are occupied mainly by species typical of the middle and lower zones of the shore, species trapped by the tide can occasionally offer unexpected information about the animals that normally inhabit the rocks and crevices just below the low-water mark.

PROLIFIC PARENT *So-called because of its ear-like tentacles, the sea hare lays strings of eggs in tide pools. Some of these molluscs reproduce once and die; but scientists found one hare laid 478 million eggs in four months*

STONY PROTECTION *The limy tubes of the fanworm protect it from predators. The cilia on their feathery gills gently stir the water and strain plankton. At the slightest provocation, these colourful gills snap back into the safety of the tubes*

CO-OPERATING FOR SURVIVAL *An empty whelk shell on the Devonshire coast provides a home for a hermit crab and a place of attachment for sea anemones. The anemones' tentacles help to protect the crab; it leaves scraps that the anemones eat*

SHORELINE WATERFALL *Water cascades from one tide pool to another over blue mussels and white goose barnacles, often found together on Pacific shores. In the upper pool, sea urchins lie submerged and pink acorn barnacles encrust its rim*

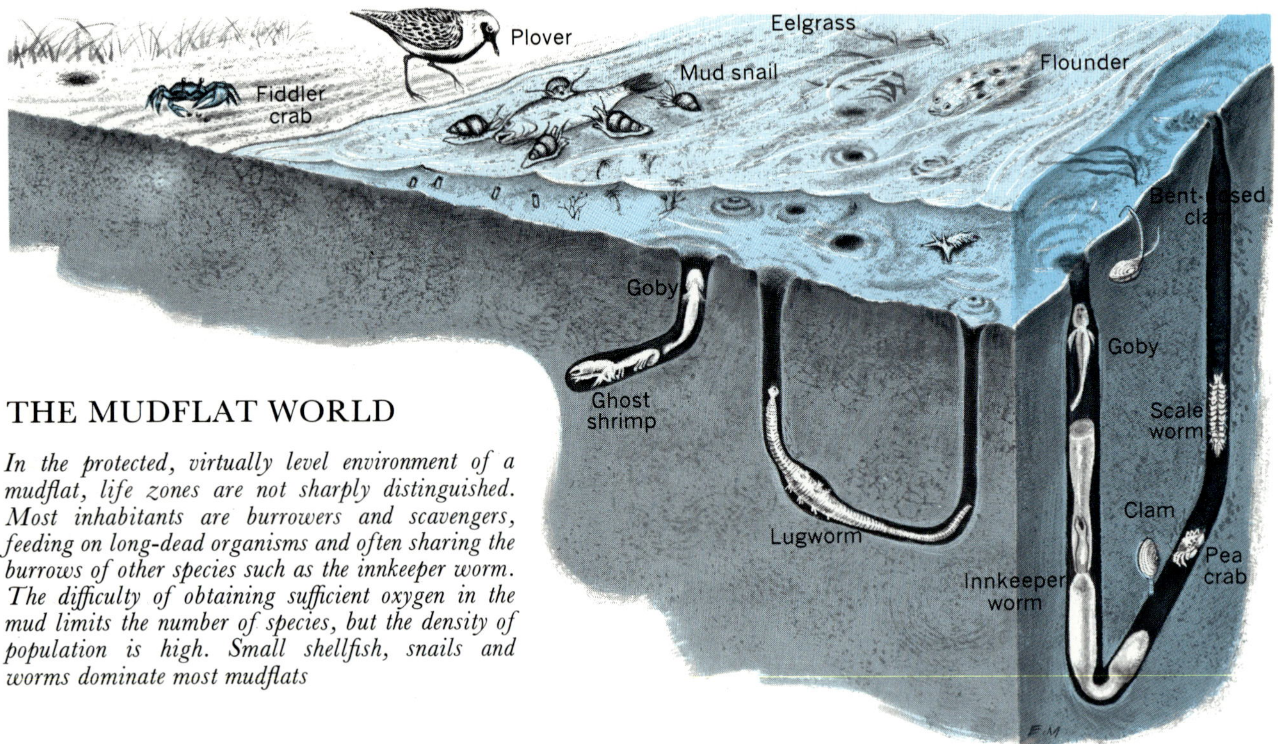

THE MUDFLAT WORLD

In the protected, virtually level environment of a mudflat, life zones are not sharply distinguished. Most inhabitants are burrowers and scavengers, feeding on long-dead organisms and often sharing the burrows of other species such as the innkeeper worm. The difficulty of obtaining sufficient oxygen in the mud limits the number of species, but the density of population is high. Small shellfish, snails and worms dominate most mudflats

LIFE ON A MUDDY SHORE

Mudflats rim gently sloping shores of bays and river mouths, wherever there is protection from the assault of the waves. Mudflats contain rich deposits of plant and animal material – remains of dead mud dwellers, debris from adjacent grassy tidal marshes, and organic matter carried in from the sea by the tides. Some people find the pungent odour of the flats unpleasant, but it is the smell of life – an indicator of vast reserves of food available to mud dwellers. Sand is often mixed with mud in varying amounts, and it is in such localities that the largest populations of animals live. Fine mud particles hold nutrients, and the leavening of larger sand particles produces a firm bottom, ideal for burrowing.

As on sandy shores, most animals live beneath the surface, burrowing down to avoid exposure to air and to predators that come from land, sea and air. A mudflat, however, is not as porous as a sandy beach, and the dense, closely packed silt inhibits the circulation of water carrying vital dissolved oxygen. Thus, burrowing animals that breathe through their skins, such as some sea urchins and starfish, seldom inhabit mudflats. Only those animals with special adaptations for obtaining oxygen can survive.

Many types of clams inhabit mudflats throughout the world, living permanently embedded in mud. To obtain oxygen and tiny food particles, water is pumped in and out of the body through a fleshy siphon extending to the surface. When the tide goes out, the buried soft-shelled clam can switch over to anaerobic, or oxygenless, respiration – a metabolic process akin to fermentation, used by many bacteria and by some animals to supplement their oxygen intake. In a laboratory, soft-shelled clams have survived for as long as eight days without oxygen.

Lugworms abound on many mudflats, sometimes reaching concentrations as high as 82,000 per acre. These rough-skinned creatures, normally about 6 in. long and 1 in. thick, live in U-shaped burrows and continually pump a current of water over their large, bushy gills. They feed in the same way as earthworms, by swallowing large quantities of mud and digesting whatever organic matter is present. Wastes from this diet form heaps around one opening of each burrow.

One of the most curious inhabitants of mudflats is the fat, foot-long innkeeper worm, found on the Pacific coast of America. This flesh-coloured, sausage-shaped creature lives in a U-shaped burrow through which it continually pumps water in much the same way as a ragworm. Instead of eating mud, however, the innkeeper spins a slimy mucus net that catches tiny food particles brought in with the current. When enough of these particles have accumulated, the innkeeper eats the net and begins to spin another.

The innkeeper is so named because its burrow is nearly always occupied by a variety of guests. A 2-in. red scale worm lives only in the inn-keeper's burrow; it stays in almost constant contact with its host, nibbling on rejected food and sometimes on the net itself. A pair of pea crabs may also occupy the burrow and fight with the scale worm over larger pieces of food. Gobies frequently shelter in the innkeeper's burrow, too. But rather than fighting with the crabs, these little fish have actually been observed bringing the crabs pieces of food that were too big to swallow.

Other mudflat burrowers are the ghost shrimps, and some of these also take in lodgers. A single ghost-shrimp burrow may shelter nine different kinds of creatures, including small clams, pea crabs, worms, several tiny shrimp-like animals and gobies. Each burrow consists of an almost-vertical shaft with a number of tunnels leading off later-ally. Tunnels branch and re-branch, widening at junctions to provide room for their owner to turn around.

To excavate its tunnels, a ghost shrimp digs out mud with its first two pairs of legs, storing the accumulated particles in a receptacle formed by fleshy appendages around its mouth. When this receptacle is full, the shrimp goes to the burrow entrance and pushes the load of mud outside.

'Grasslands' in the sea

From the Arctic to the tropics on the coasts of Europe, North America, Asia and Australasia, marine meadows of eelgrass frequently grow on the seaward sides of mudflats. Eelgrass thrives in pro-tected waters from the low-tide zone out to a depth of about 20 ft; its long, branching roots anchor in sandy and muddy bottoms that will not support seaweed holdfasts.

Eelgrass is not a true grass at all, but is related to the pondweeds. Its popular name was probably inspired by its thin, ribbon-like leaves, $\frac{1}{4}$ in. wide and often 1 yd long, which resemble blades of grass. Eelgrass invaded the sea from land, and it is one of the relatively few flowering plants that have adapted to salt water. The flowers – small, inconspicuous and short-lived – bloom in late spring and summer. The pollen grains are trans-

MANY-EYED SCALLOP *Two rows of bright blue eyes peer from between the fluted shells of a deep-sea scallop. Tentacles around its fleshy mantle are sensitive to chemicals in the water and alert the scallop to prey and predators*

ferred from plant to plant by currents, and the seeds are eaten by many birds.

Eelgrass provides shelter for a variety of animals. Large numbers of scallops – usually found only in deeper offshore waters – live in eelgrass beds on the north island of New Zealand and the east coast of the United States. These clam-like creatures swim by quickly clapping together their valves, causing a jet of water to shoot out and pro-pel them forwards. They can also move backwards by squirting water through openings on either side of the hinge. When disturbed, jackknife clams can also swim through the eelgrass beds by shoot-ing water from their siphons or from between their valves. Many different snails and worms creep on the bottom among the stalks and roots of these underwater jungles. Small snails fall prey to larger ones, and their empty shells are often occupied by hermit crabs. Eels, flounders and other fish come in from deeper water to hunt.

A microcommunity lives on the surfaces of the eelgrass leaves. Older and longer blades may be completely covered by tiny organisms such as hydroids, tubeworms and tufts of red algae. These in turn may be coated with other plants and pieces of organic matter. A variety of animals, including the delicate and brightly coloured sea-slugs, feeds on this abundant food supply.

INVADERS OF THE SEA

The boundary lines of sea and land are forever in a state of flux. Where the coast is exposed to battering waves, the sea may encroach on the land; but where the shore is sheltered, the land may inch its way forward year by year.

This gradual advance of the land into the sea is spearheaded by the few plants and trees which can survive in salt water. Among these invaders of the sea are the cordgrasses which form tidal marshlands on temperate coasts, and the three varieties of mangrove trees – red, black and white – which succeed one another in a land-building cycle on sheltered tropical shores.

Life in a tidal marsh

From the inter-tidal area to dry land, along the shoreward edge of mudflats and their twisting maze of tidal creeks, there may be meadows of cordgrass and sea rushes. The short, fine cordgrass grows no taller than 2 ft and forms a lush carpet on flats shoreward of the taller, coarser cordgrass, which may reach a height of 10 ft. Where tall cordgrass grows thickly, almost no light penetrates to the mud. Tidal currents sweep away dead plants, leaving a clear, dark floor under this dwarf forest. Because they are further back from shore, the stands of shorter cordgrass are not swept by currents. There, dead plants mat the floor, forming an ideal moist shelter.

A bed of cordgrass is often the first stage in the slow transformation of shallow, open water into a wet meadow and, eventually, into dry land. These rooted plants prevent waves and currents from carrying away sediment washed in by rivers. Gradually, the area fills with sand and silt, which becomes mud as it is enriched with decaying plants and dead animals. Small muddy islands start appearing; the cordgrass traps more sediment and the islands grow together.

As land builds up, it is flooded at high tide for shorter periods. Short cordgrass edges forward to replace tall cordgrass, only to be replaced in turn by rushes. Land plants become established, and land animals move in. As marsh becomes dry land, the different kinds of plants and animals appear in a definite order, or succession.

Cordgrass provides food and shelter for a distinct community. Black snails crawl up the leaves to feed on algae coating them. Purple marsh crabs come out of their burrows at low tide to nibble on marsh grass. Clamworms as big as small snakes feed on dead razor clams.

The marsh snail shelters under tall cordgrass at low tide but avoids the rising water by climbing the stalks. It has a lung and breathes air. If a particularly high tide submerges the snail, it can survive without breathing for about an hour – the time it takes for the tide to ebb. The marsh snail anticipates the flood tide and climbs the grass *before* the water reaches it. Apparently it possesses some kind of internal clock that sounds an alarm every 12 hours.

One of the most conspicuous animals of the salt marsh is the fiddler crab. At low tide, large groups of these 1 in. wide crustaceans flit across the mud beneath the tall cordgrass in search of edible morsels in the mud. The fiddler crab has a brightly coloured outsized claw that develops on one of its front legs. Only the male grows this claw, which he uses in disputes with other males and to attract a female at breeding time. When a female is near, the male extends his claw full length and then whips it back towards his body. The gesture is repeated until the female either departs or retires with him to his burrow.

The fiddler crab seems nearly independent of the water. Under its shell, just above the legs, the fiddler has a primitive lung cavity. As long as this stays moist, the fiddler can breathe oxygen from the air, sometimes surviving for several weeks without immersing itself in water.

Beneath the tall cordgrass the fiddler excavates a deep burrow, bringing mud out in neatly formed pellets and piling them up at the entrance. As the tide approaches, the crab constructs a tightly fitting door of mud pellets, and in the burrow's air-filled chamber awaits the departure of the water. When the tide ebbs, the fiddler again emerges and resumes the fighting, feeding and wooing activities that occupy much of its time.

By constantly 'turning over' the bottom, fiddler crabs – and other burrowing animals as well – replenish the depleted surface mud with nutrient particles from below, thus performing a vital function in the ecology of the marsh.

At high tide, predators from offshore – squids and fish – hunt in the cordgrass. At low tide, visitors from land invade the marsh: swarms of insects and spiders, animals such as foxes, otters, and deer, and a great variety of birds.

SALT MARSH INHABITANTS

SEARCHING FOR FOOD *Skimming over the waters of a salt marsh, a black-winged stilt hunts for insects and small fish. When it spots its prey, the stilt wades breast-deep on its spindly legs to make a catch*

IN DANGER OF EXTINCTION *Waddling across a marsh, a diamond-backed turtle heads for water – and a meal of snails and other small invertebrates. The reptile is itself a prize meal, and turtle fanciers have killed most of the species*

FLIGHT FROM THE WATER *Fiddler crabs, only 1 in. long, scuttle from the shelter of the cordgrass, where they seal themselves in airtight burrows against high-water floods. The males wave enlarged claws in territorial displays*

Life in a mangrove swamp

On many miles of tropical and sub-tropical coasts, ranging from equatorial Africa to northern New Zealand, there is a similar type of land-building and succession, but here mangrove trees are the pioneer plants. Mangroves drop seedlings that have already germinated and sprouted roots. Some of these anchor in the muddy waters beneath the parent; others drift out with the tide.

Like the marsh grasses, mangroves colonise a mudflat and are the first stage in a succession of plant communities that transform inter-tidal areas into dry land. Red mangroves, the first colonisers in Florida, collect soil around their roots in such large quantities that the trees eventually die from lack of oxygen. Black mangroves, however, thrive in this built-up region, for their roots contain cavities that hold air. But, finally, the black mangroves, too, are replaced – first by white mangroves and then by other trees that take root as more soil accumulates. The debris from plants and trees helps the process of land building.

Mangrove seeds grow into strong young trees buttressed by arching prop roots that penetrate deep into the bottom mud. As more and more seedlings take root on a shoal, their prop roots

TREE-DWELLING CRAB *A yellow-spotted mangrove crab scurries among the branches of the mangrove, living on its leaves. During its larval stages, the crab lives in the water among the tangled roots of the mangrove trees*

STATELY STORKS *Wood ibises, members of the stork family, nest on red mangroves on Pelican Island, Louisiana. The long-legged birds feed on small fish which they catch in their bills as they wade through mangrove swamps*

form an impenetrable tangle in which silt, sand, decaying vegetation, shells, coral fragments and other debris come to rest. The floating larvae of barnacles, mussels and oysters also settle on the prop roots, grow into adults, and soon all these forms of sedentary life are competing for space with one another. Snails feed on algae coating the mangrove roots, mussels and oysters. Crown conch snails pry open oysters with their powerful feet and devour the succulent meat. Mangrove crabs spend their larval stages in the water, then take up residence in the mangrove branches and gnaw on leaves.

Many animals common to mudflats and salt marshes, such as fiddler crabs, hermit crabs, marsh crabs, marsh snails and ghost shrimps, live among mangrove roots. And other animals that could not survive on mudflats, such as sea squirts, starfish and brittle stars, also find a home and are able to survive in the environment.

Foraging for these creatures are a variety of predators, including land animals such as racoons and various reptiles, for instance turtles, water snakes and, occasionally, a crocodile. In the upper branches, pelicans, herons, egrets, wood ibises, roseate spoonbills and other birds roost or nest and feed on the inhabitants of the swamps.

INTENT ON FOOD *A hungry racoon crouches beneath arching roots at the edge of a mangrove swamp, its nimble fingers searching along the shallow bottom for fish and shellfish. Racoons frequently visit Florida mangrove swamps*

TAKING PREY BY SURPRISE *While sunning themselves beneath red mangroves in Everglades National Park, Florida, American crocodiles seize unwary birds and mammals and hold them under water to drown. They also hunt fish under water*

LIFE AT THE RIVER MOUTH

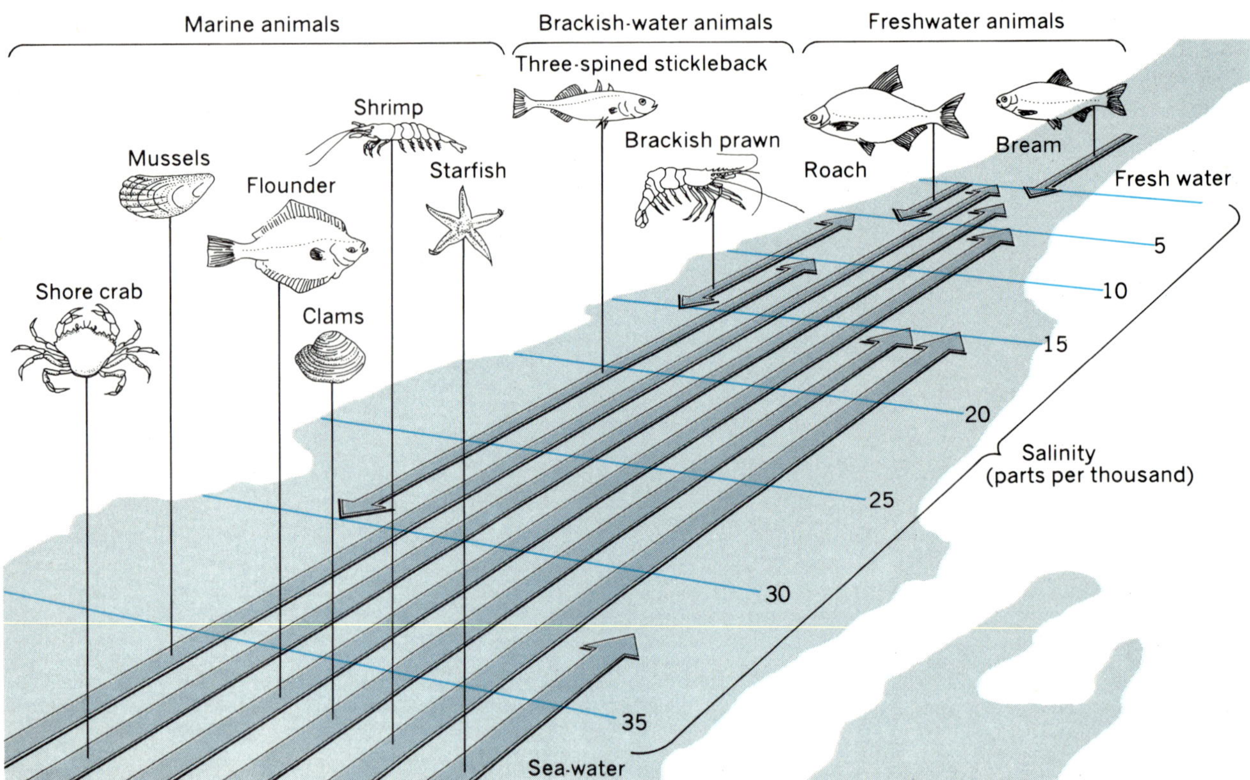

A few animals of the estuary live only in brackish waters where river and sea meet; others come downstream from the river. Most inhabitants are marine species able to withstand alterations in salt level as the tides ebb and flow

ESTUARIES – GATEWAYS TO THE SEA

Mudflats with their accompanying marshes and mangrove stands are commonly found in estuaries – the regions where rivers and streams empty into the sea. Here fresh water flowing from the land mixes with salt water.

Living conditions in estuaries can be hazardous. As a river spreads out in an estuary, it slows down, and the fine silt that it brings from land settles to the bottom in a smothering blanket. Water temperatures may rise and fall drastically from season to season and even from day to day. But by far the greatest danger to estuarine life is the abruptly changing salinity, or salt content, of the water. Salinity may fluctuate more or less regularly twice daily as tides move back and forth through the river mouth, and it may also fluctuate with shifting weather conditions as rain makes the water less salty or drought makes it more salty.

Plants and animals that live in water – whether salt or fresh – must maintain a certain concentration of dissolved salts in their body tissues. If an organism gets into water with a lower concentration of salt than that in its body fluids, water flows into its tissues and it may bloat. If it gets into water with a higher concentration of salt, water flows out of its tissues and it may shrivel.

As the brackish waters of an estuary give way further upstream to the progressively fresher waters of a river, one group of plants and animals is slowly replaced by a different one. This gradual change is a kind of zonation, but it is based on a tolerance for varying degrees of salinity rather than on an ability to withstand exposure to air. Starfish and sea urchins, for example, need the full salt concentration of the sea; they can live only at the mouths of estuaries where there are 30 to 35 parts of salt for every 1000 parts of water. Some clams require water with 35 parts of salt per 1000, while

FRESH AND SALT WATER MEET *At the mouth of a river, the tell-tale muddy outflow at low tide reveals how far fresh water has penetrated into the sea. Many rivers also carry quantities of sewage and other pollutants into the seas*

others can live in water with only 9 parts of salt per 1000. Thus, these creatures can venture further from the sea. Some marine worms, crabs and flat-fish survive in water with less than 5 parts of salt per 1000 and are found in almost fresh water.

Despite the constant stress of changing salinity, life abounds in estuaries. Indeed, estuaries support many of man's principal food animals: oysters, clams, crabs and shrimps. Estuaries also serve as nurseries or feeding grounds for a variety of commercial and game fish: flounder, fluke, mackerel and varieties of herring. In fact, about 85 per cent of the fish and shellfish sold in markets of the world come from coastal waters, and many of these spend all or part of their lives in estuaries.

Unfortunately, estuaries now reveal man's failure to conserve the earth's most valuable natural assets. From rivers which are all too often regarded as convenient natural sewers, estuaries receive large volumes of pesticides, industrial chemicals, sewage and all manner of floating rubbish, together with a more subtle but equally fatal pollutant, hot water.

Fish and shellfish in many polluted estuaries have already accumulated bacteria or toxic chemicals to levels of concentration that are dangerous to seabirds and man. Sandwich terns, found dying from convulsions on the Danish coast, had been poisoned by pesticides in herrings and sand eels, which in turn had fed on smaller organisms affected by discharges from the Rhine. In Japan between 1953 and 1960 about 100 people died from eating fish contaminated by mercury-laden effluent from a plastics factory.

Oil-tanker accidents are now all too frequent, but man's eagerness to save the beaches by using detergents to remove oil may destroy more life.

Some of the damage already done may be irreversible, but much can be undone, and all of it should be checked. Britain has made good progress in cleaning several rivers. Wherever the estuaries are choked by pollution, the task of cleaning is urgent. For beyond these gateways lie the bountiful coastal seas and the open ocean; and the harvest of these waters is probably mankind's best hope in the war against world hunger.

The bountiful coastal seas

*The shallow waters covering the submerged margins
of the continents harbour the greatest – and most
varied – concentration of living things on our planet*

Almost four-fifths of all the plants and animals on earth live in the shallow coastal seas fringing the continents. Countless minute plants drift in the waters, staining them green. Millions of tiny animals graze these floating 'pastures' and in turn are eaten by huge schools of prowling fish. Seabirds skim above the waves, scooping up prey with their bills. On the sea floor, forests of undulating seaweed cling to rocks; a host of bizarre creatures filter the water for food; a steady rain of dead and dying organisms fertilises the bottom ooze.

Only awesome numbers can describe the density of coastal-sea populations. A thimbleful of water contains thousands of the microscopic drifting plants and animals that form plankton, the first link in the ocean's chain of life. An estimated 10,000 cod live above each acre of ocean floor on the Grand Banks off Nova Scotia and Newfoundland. On some parts of the sea floor around the British Isles, 250 million brittle stars crowd into each square mile.

The coastal seas – collectively, about equal in area to Asia – make up less than one-tenth of the ocean's expanse, but they yield almost nine-tenths of the world's annual harvest of fish and shellfish. Compared to these fertile seas, the open ocean is a barren desert.

The continental shelves

Most coastal waters lie over continental shelves, extensions of land once believed to be nearly as flat and featureless as their name implies. Detailed studies have revealed, however, that they are wrinkled with gentle hills and carved with deep gorges. The gradual descent of the shelves from shore – only 5 ft in every mile – would be almost unnoticed on land.

The continental shelves average about 45 miles wide, but range from less than a mile wide off stretches of the western coasts of North and South America to over 900 miles off the Arctic coast of Siberia. At the outer edges, the sea is about 450 ft deep. Here, the slightly pitched bottom suddenly becomes a steeper slope or, in some places, a sheer wall plunging thousands of feet to the deep ocean floor. A mighty precipice 500 miles long lies in the Gulf of Mexico off the Florida coast; in one place there is an almost perpendicular drop of over a mile. Near Santiago, Chile, the narrow continental shelf ends in a steep incline that drops without interruption to a depth of 20,000 ft. These continental slopes, as they are called, form the true boundaries of the continents.

During the Ice Ages the seas were much lower

A CROWDED WORLD *The floors of the world's continental shelves harbour some of the sea's most complex communities of animal and plant life. Mussels, tentacled sea anemones and sponges feed on the plankton brought by the ocean currents; spiny sea urchins graze on slimy algae; a starfish prises open a mussel*

VALUABLE FRINGES AROUND THE WORLD'S CONTINENTS

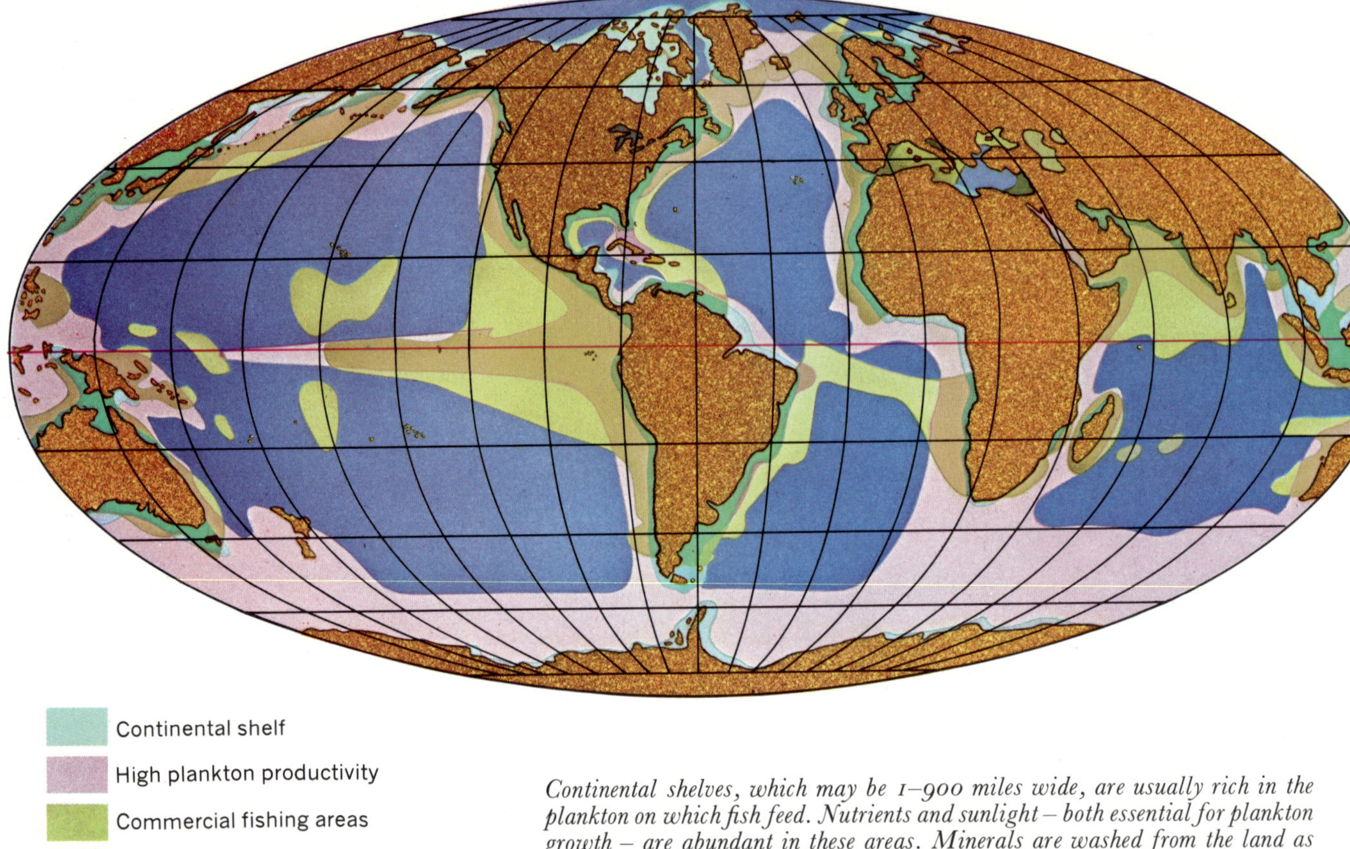

Continental shelves, which may be 1–900 miles wide, are usually rich in the plankton on which fish feed. Nutrients and sunlight – both essential for plankton growth – are abundant in these areas. Minerals are washed from the land as sediment or are brought from the ocean's bottom by upwelling currents; sunlight can penetrate the upper levels of the relatively shallow water. Most continental shelves are good commercial fishing grounds

than they are today because glaciers locked up vast amounts of water. Large areas of the continental shelves were exposed. Bones dredged up from what is now the ocean floor reveal that musk oxen, giant moose, horses, tapirs and huge ground sloths once ranged far beyond present coastlines. Brick-sized teeth of mammoths and mastodons have been discovered on the Atlantic ocean floor, far from where they could have been carried by currents and tides.

The North Sea is one example of land inundated by the rising sea level. Scientists have found plenty of evidence of land vegetation, of ancient land animals and even prehistoric men where some of Britain's most important fisheries now exist.

Glaciers have played their part in moulding the continental shelves, planing down the land like bulldozers; this process is still continuing around the glacial coasts of Antarctica.

Rivers meandered across the continental shelves when they were above sea level, and their paths can still be traced in the sea-bed today. On the floor of the North Sea, scientists have been able to map the channels carved by the great rivers of Europe when Britain was joined to the Continent across the English Channel.

Marine geologists believe that turbidity currents – torrents of silt-laden water cascading down underwater slopes – have extended and deepened some of these ancient river beds into spectacular underwater canyons. The Hudson Canyon, a continuation of the Hudson River beyond its mouth at New York City, is one of the best examples. This canyon has been traced more than 100 miles away from land and 12,000 ft down the continental slope to the floor of the abyss. The V-shaped notch formed at the edge of the shelf is 3600 ft deep and 5½ miles wide from rim to rim.

Yet this Grand Canyon of the ocean is not the largest undersea gorge. The Ameghino Canyon off Argentina claims that title, but it is possible that canyons still deeper await discovery.

In addition to the changing levels of the sea, the margins of the world's continents have been affected by the up-and-down movement of the earth's crust over many millions of years. In the Mediterranean, for instance, there are some underwater canyons which are so deep that they cannot be accounted for by changes in the level of the sea. Scientists believe that they may be drowned river gorges which have been carried deep into the sea by an enormous subsidence in the earth's crust in the Mediterranean area.

Deposits of sediment

The rivers of the world carry nearly 750 million tons of sediment into coastal waters every year. Millions of tons are deposited as silt on the shelves, and millions more are dissolved in sea-water. The minerals present in this sediment read like the list of ingredients on a bag of lawn fertiliser: phosphates, nitrates, calcium and silica. All are essential for life in the sea. Just as plants on land cannot exist without minerals in the soil, so marine plants require the same minerals in the water surrounding them.

Nutrients are also brought into coastal seas in some areas by currents called upwellings. Winds and the earth's rotation cause surface currents to flow away from the land. This water is replaced by vertical currents that bring up accumulated minerals and organic materials. Thus the coastal seas are fertilised with nutrients from both the land and the ocean floor.

Such upwelling occurs mainly on the western coasts of continents, and as a result these areas have become some of the most fertile parts of all the world's seas. Indeed in some areas – for example, off the coast of South-West Africa – so much organic matter is produced that not all of it can be eaten by the ocean inhabitants, and a great deal is deposited on the bottom of the continental shelf. As it decomposes, the oxygen in the water is gradually used up and eventually the sea in the area becomes almost devoid of oxygen. Organic matter continues to be deposited, but without oxygen its decomposition follows a different pattern. The organic layers that result are similar to the sediments laid down millions of years ago, which now form the world's oil deposits. The land adjacent to some of these areas has proved a rich source of oil.

FOUR KINDS OF SHELF

Most continental shelves are composed of recently deposited sediments covering deep layers of sediment turned to rock. Where a dam exists, the sea bottom tends to be extremely rugged, especially on the seaward side. Beyond a fault dam *– a wedge of rock thrust up from the ocean floor by volcanic or earthquake activity – the slope of the shelf floor is very steep. One of the best examples is the dam along the west coast of the United States. A* reef dam *may consist of ancient algae or coral growths or even living coral – for example off north-eastern Australia. Upward movement by an offshore salt bed forms a* salt dam, *for example off the west coast of the Gulf of Mexico. If no dam or barrier holds back the sediments, as along the coast of Europe, the shelf slopes gradually downwards*

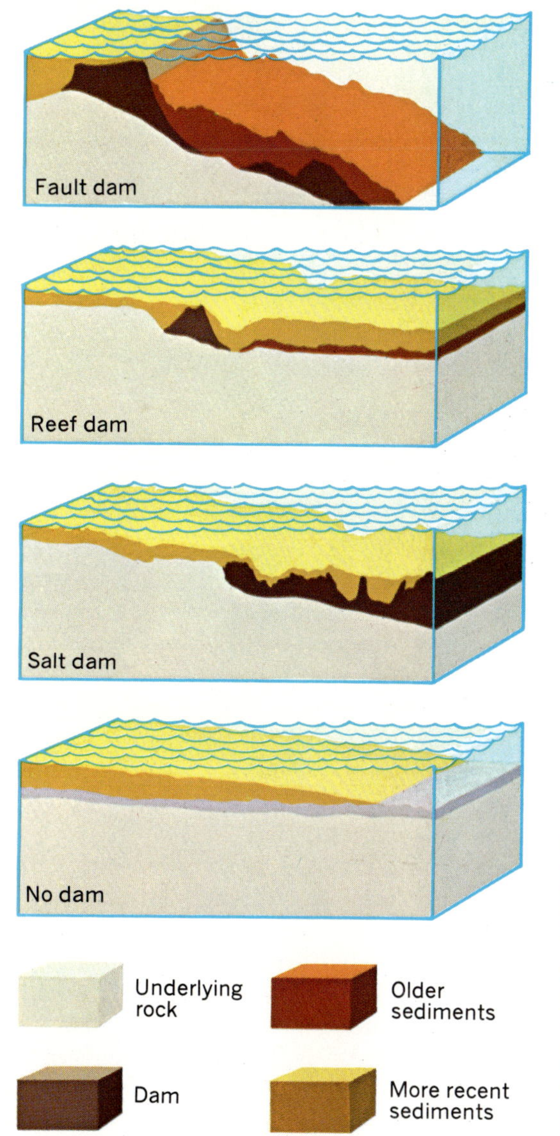

Each species of diatom secretes its own unique silica shell, here magnified 600 times. The many thousands of types, however, are basically similar in form and are the first links in many ocean food chains. The plant cell, enclosed in its silica casing, is kept afloat in the sea by natural oil droplets. These ocean drifters flourish even in cold northern waters

MOST NUMEROUS MARINE PLANT *Diatoms are so abundant in the ocean that in spring their blooms may actually stain the water a dark brownish-green. One gallon of sea-water may contain millions of these minute, drifting, one-celled plants*

PLANT PLANKTON, SOURCE OF LIFE

Nowhere else in the ocean are sunlight and nutrients – the basic elements of life – brought together in such profusion as over the continental shelves. Surface waters further out have ample sunlight but lack a plentiful supply of vital nutrients; the waters of the abyss may be rich in nutrients but receive no sunlight. In the shallow coastal waters marine plants, upon which all animal life in the sea depends, have everything they need for growth.

Like the familiar plants of gardens, grasslands and forests, plants of the sea convert inedible minerals into food through photosynthesis. They absorb water, carbon dioxide and various minerals, and use the energy of sunlight absorbed by their green pigment, chlorophyll, to create the various sugars, starches, fats and proteins of their living tissue.

During photosynthesis, plants give off more oxygen than they need, and thus supply oxygen for sea animals as well. In fact, scientists estimate that marine plants produce 70 per cent of the oxygen necessary to support all animals now living on this planet.

Drifting grasses of the seas

Along rocky coasts, the leafy growth that people associate with the word 'plants' is abundant in the form of seaweed, but seaweed supports only a very small percentage of marine animals. The great majority of sea plants are minute one-celled organisms that drift with the currents. The most abundant of these, particularly in colder seas, are the diatoms, often called the grasses of the sea.

A single quart of sea-water may contain hundreds of thousands of diatoms. Under a microscope, they look like exquisite, transparent jewel boxes, each containing a tiny blob of brownish-green living matter. The 'boxes' are actually external skeletons of silica, a mineral that

36

Diatoms reproduce by dividing in two. First the cell nucleus divides, and the rest of the cell material separates into two masses, each with its own nucleus. Then each mass forms a cover for its nucleus

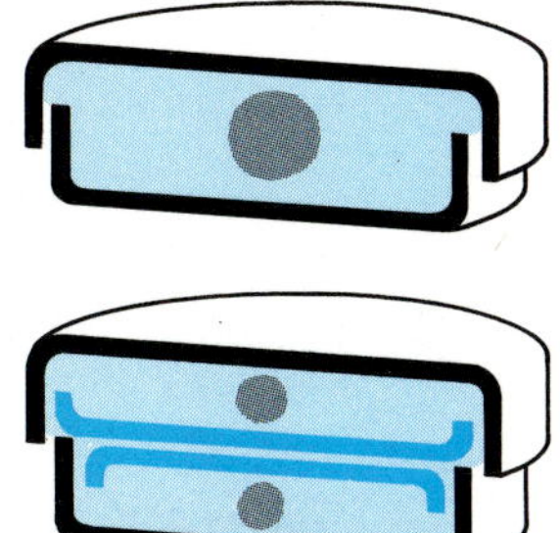

Since the new silica covers are formed inside the old protective shell, one of them must always be smaller than the other. Because of this, the average size of the diatoms that are descended from the smaller one diminishes steadily at every stage of the reproductive process. At some point, however, after many divisions, the continuation of the species is assured by the minute diatom's ability to discard its greatly diminished silica shell and secrete a new one two or three times larger. So the process begins again

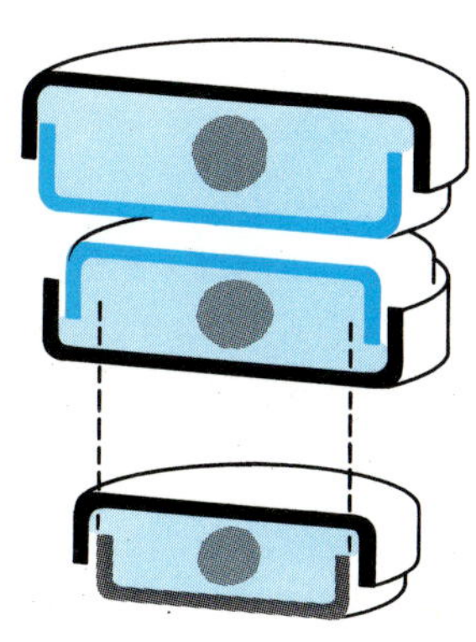

diatoms extract from sea-water and secrete around themselves. Each skeleton has two parts that fit together neatly like the top and bottom of a box. These may be round, rectangular, spindle-shaped or oval; often they are drawn out in lacy lattices and spiny designs of astonishing beauty.

Striations, pits and perforations on the surfaces of diatom skeletons form intricate patterns peculiar to each of the estimated 9200 species. The designs counteract the process of sinking by increasing friction with the water.

A diatom usually reproduces by dividing in half. One part goes with the 'lid' of the box, the other with the bottom; each daughter cell secretes the missing half of the skeleton. Some species remain attached to form long chains.

The marathon migrators

In warm waters, diatoms are frequently outnumbered by dinoflagellates, single-celled organisms possessing characteristics of both plants and animals. Like diatoms, some dinoflagellates use

ANIMAL OR PLANT?

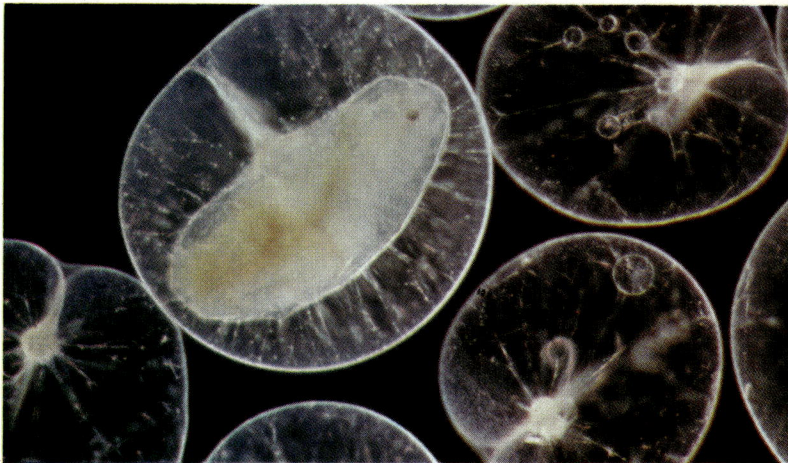

NIGHT LIGHTS *Luminescent dinoflagellates, called Noctiluca (night light), eat smaller planktonic animals. The bean-shaped pod (above) inside the Noctiluca is a copepod*

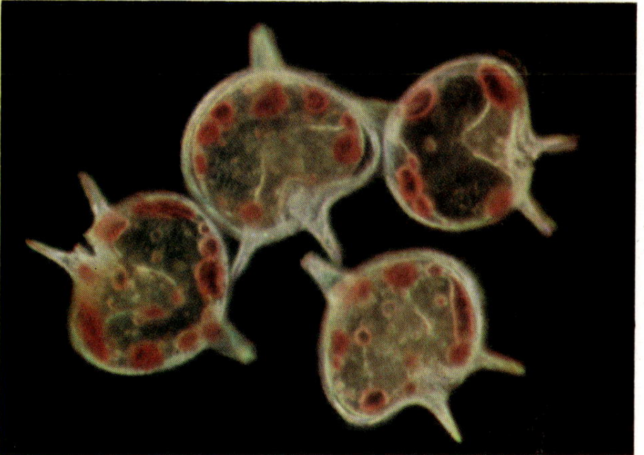

EATING AND MAKING FOOD *Some types of dinoflagellate can eat food as animals do, or produce it as plants do, depending on whether prey or light is more abundant*

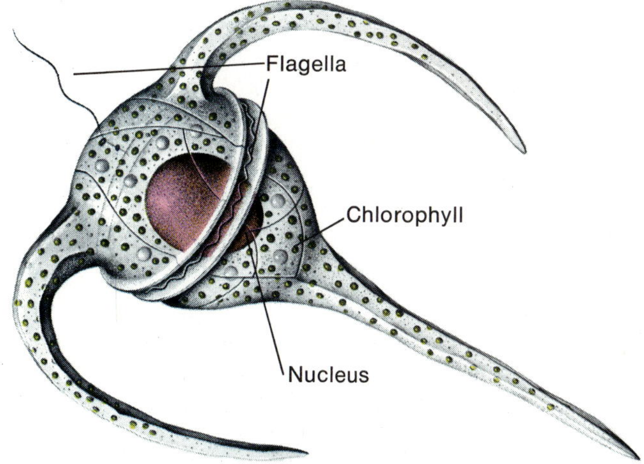

KEEPING AFLOAT *Because warm waters offer less resistance to sinking, this anchor-like dinoflagellate (above) grows its horns longer in the tropics to keep it afloat*

the sun's energy to make food from water, carbon dioxide and minerals. Others do not photosynthesise but capture food with their tiny whip-like appendages, or flagella. Many dinoflagellates are encased in cellulose, the same material that gives bulk to land-dwelling plants. Droplets of oil often provide buoyancy.

During the day, dinoflagellates usually swim towards the surface. A small red 'eyespot' senses light. Some species migrate 150 ft vertically in a 24-hour period – nearly 2 million times their own length. To match this feat proportionally, a grown man would have to swim more than 2000 miles in a day. Although dinoflagellates can control their up-and-down movements, they are swept along horizontally wherever ocean currents carry them.

Many dinoflagellates are luminescent. The same complicated chemical reaction that makes fireflies flash produces light within the bodies of dinoflagellates. Millions of them drifting near the surface at night can turn the water a blue-green. In the north, on a summer night, a splashing oar may stimulate their telltale flashes of light.

Dinoflagellates sometimes bring disaster to shelf communities. When temperatures are high, nutrients abundant and skies clear, certain species undergo population explosions, which stain the waters green, yellow, brown or red. They use up more oxygen than they produce. Then the bloom ends and the dinoflagellates – and most other living things – suffocate.

Some dinoflagellates also secrete poisons that kill fish and other marine organisms. In waters off the Atlantic and Gulf coasts of the United States, one such species less than a thousandth of an inch across periodically occurs in concentrations of 90 million per quart of water. Just one outbreak of this so-called Red Tide may kill thousands of tons of fish.

Other plant plankton

Diatoms and dinoflagellates are the two most important synthesisers of food in the ocean, but they are not the only floating plants. Bright green *Halosphaera* cells, less than $\frac{1}{25}$ in. in diameter, occur in all seas from the tropics to the polar regions. Coccolithophores, microscopic plants covered by plates of lime, sometimes turn parts of the North Sea white and may even outnumber diatoms at certain times of the year.

The composition of plant plankton changes from place to place and from season to season. As one plant is depleted, another species replaces it.

PERMANENT PLANKTON *Most species of animal plankton spend their whole lives in the plankton. The Radolarian (above), for*

TEEMING WORLD OF ANIMAL PLANKTON

The 'pastures' of floating diatoms, dinoflagellates and other marine plants are grazed by the animal plankton, untold millions of drifting creatures, scarcely bigger than the plants they feed on. They also feed voraciously on one another, and with the plant plankton they form the staple food of innumerable fish, ranging from herring to blue whales.

The animal plankton are of two types: those which spend their lives as permanent drifters of the seas, and those which are planktonic only in their larval stages.

TEMPORARY PLANKTON *Some planktonic animals spend only part of their lives as drifters. A sea snail larva (below) swims*

The most abundant of the permanent plankton are the copepods. Seldom bigger than grains of rice, they are more numerous and widespread in the sea than insects on the land, and they may well be the most numerous creatures on earth. Concentrations of up to 3000 individuals per cubic foot are not unusual in fertile coastal waters.

Many fish and invertebrates which spend their adult lives on the tidal shores, on the ocean floor or in the deep sea float freely as part of the animal plankton during their larval stages. The lives of these minute creatures are so hazardous that only a small proportion of them survive to become adult sponges, clams, snails, starfish, crabs, barnacles and fish. In their planktonic stage, they no more resemble their parents than caterpillars resemble butterflies.

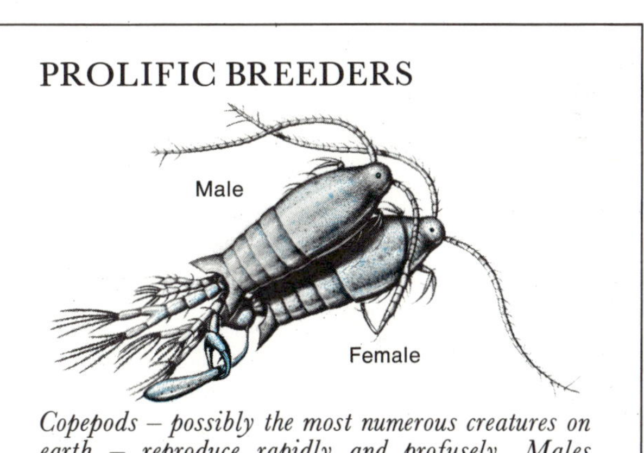

PROLIFIC BREEDERS

Copepods – possibly the most numerous creatures on earth – reproduce rapidly and profusely. Males fertilise eggs by transferring sperm to the female with a specially formed claw. Females carry the fertilised eggs for several days until they hatch. A few weeks later the new generation begins to reproduce

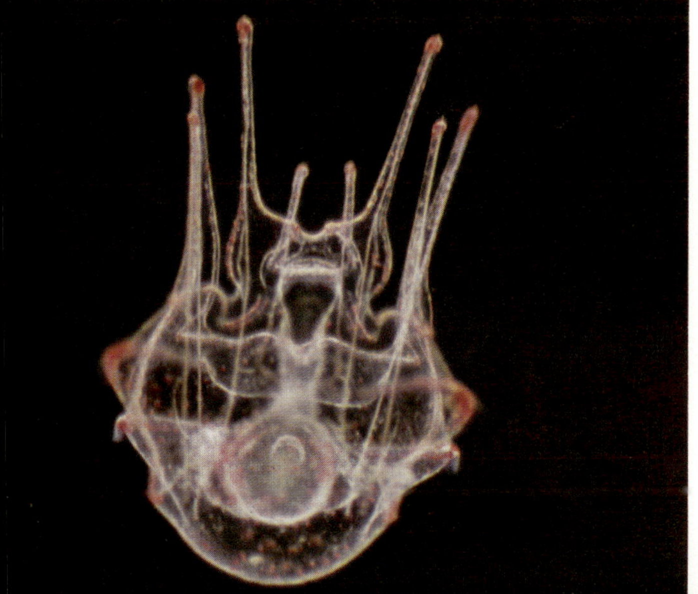

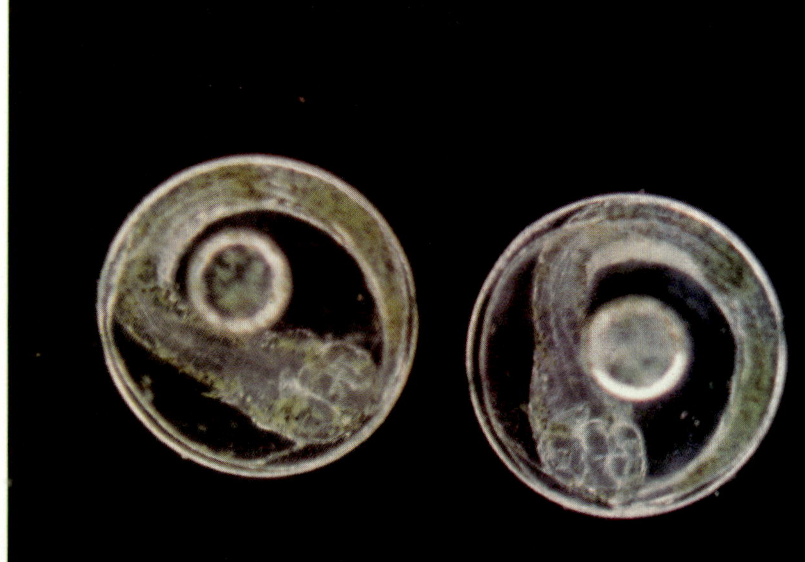

THE OCEAN CHAIN OF LIFE AND DEATH

The fate of every animal in the ocean is to eat and be eaten. Not even the killer whales, which go unchallenged during their lifetimes, are exceptions to this rule. When they die, their bodies are broken down by organisms, and the nutrients released are re-circulated along the ocean's endless interlocking food chains.

Diatoms, dinoflagellates and other marine plants form the first link in all ocean food chains. There would be no life in the sea without the plants, which convert inorganic substances into food through photosynthesis.

Copepods, the chief consumers of diatoms, are the second link in many food chains. By vibrating their antennae, they set up currents that sweep diatoms into their bristly mouth parts. Enormous numbers of diatoms are consumed by these diminutive crustaceans in just one season; a single copepod can eat 120,000 diatoms in a day, and the worldwide copepod population is estimated to be larger than that of all other multicellular animals combined.

Copepods themselves are eaten in huge quantities by herring, among the most numerous fish in the sea – and the third link in many ocean food chains. Herring have gill rakers, structures that strain food from water passing over their gills. Each time a herring swallows a copepod, it is indirectly eating thousands of diatoms.

The herring itself may be eaten, perhaps by a porpoise. And the porpoise then may be caught by a killer whale.

Clearly, something is lost at each link in a food chain; the porpoise does not convert every bit of the herring into flesh. In fact, the porpoise 'burns up' most of the herring's food value just in the process of staying alive – swimming, feeding, breathing, and so on. From diatom to copepod, copepod to herring, herring to porpoise, porpoise to killer whale – at each step in the food chain, 80–90 per cent of the food value passed on from the previous link is converted into energy and dissipated by the activities of the living.

This situation in which food is passed along to progressively larger animals may be compared to a pyramid. The base of the ocean's pyramid of life is the enormous mass of drifting plants, which use solar energy to convert raw materials into

THE PLANKTON EATERS

The plankton-eaters of the ocean – from tiny copepods to the giant baleen whales – consume their prey in remarkably varied ways. Some strain their meals from the sea through specially adapted teeth. Some suck it in through tubes. Some simply swallow the plankton as they swim. Others actively hunt prey, stinging the victim to paralyse it before eating it. Even the smallest eaters of the plankton are selective. The pictures show how they catch their prey from the flow of water (indicated by blue arrows)

Sponge

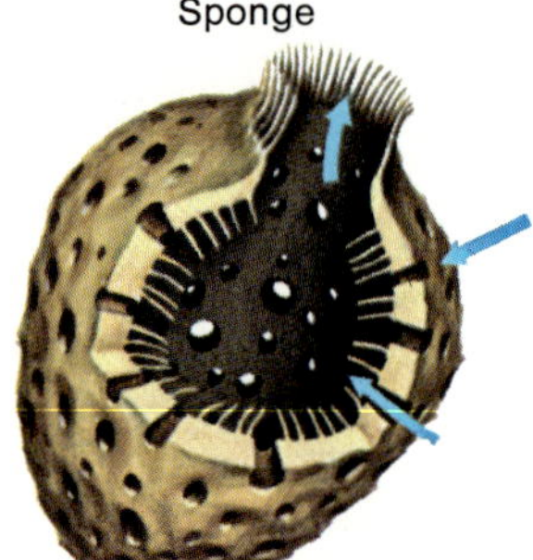

Tiny whips within a sponge's body cavity create a current of water that enters the animal's minute pores. Cells in the sponge's collar filter and digest any food particles

Copepod

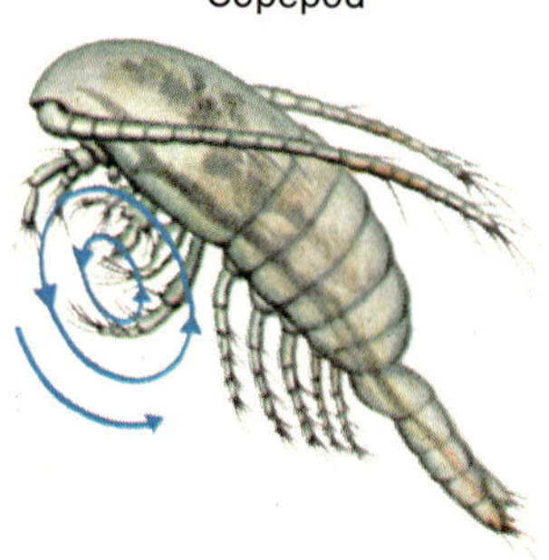

Fan-like appendages around a copepod's mouth swirl water through a sieving process. The plankton that are trapped in this way are pushed into the animal's mouth from time to time

Herring

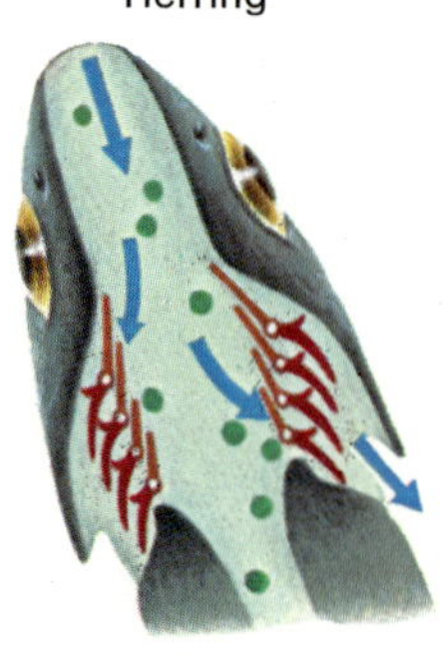

As a herring swims through the sea, water enters its mouth and pours over its stiff gill-rakers. These sieve the planktonic organisms on which the fish usually feed

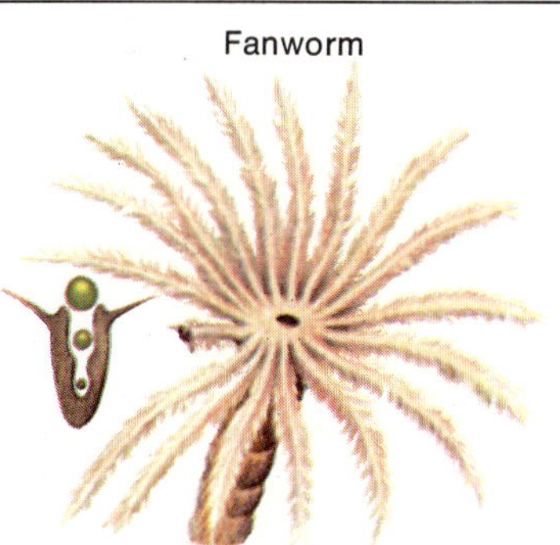

Fanworm

When the fanworm is feeding, it extends sticky, feathery gills from its tube burrow to ensnare its planktonic victims. Mucous channels carry the food to the animal's mouth

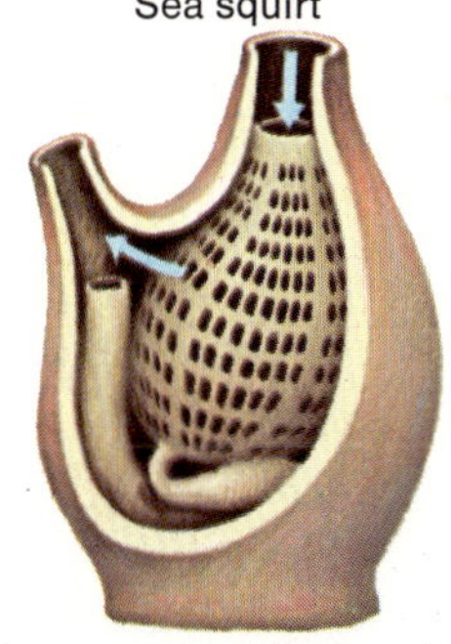

Sea squirt

Hair-like cilia drive a stream of water into a sea squirt's gill basket, where any food particles are trapped. The animal disposes of the water through an exit siphon

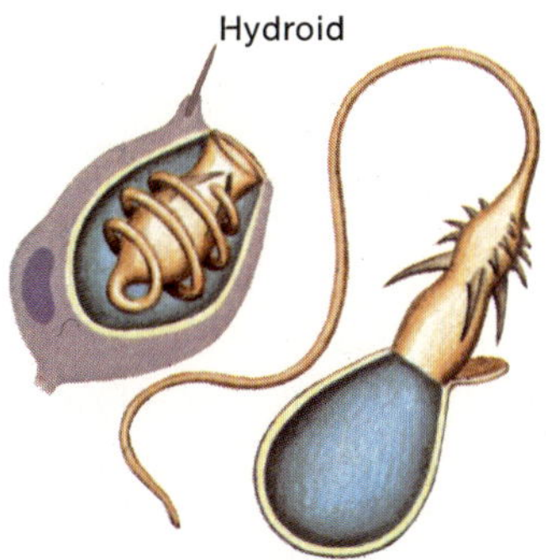

Hydroid

The hydroid stings its prey to paralyse it before it eats. When it is about to feed, the barbed stinging whip springs out of a capsule cavity where it is normally coiled

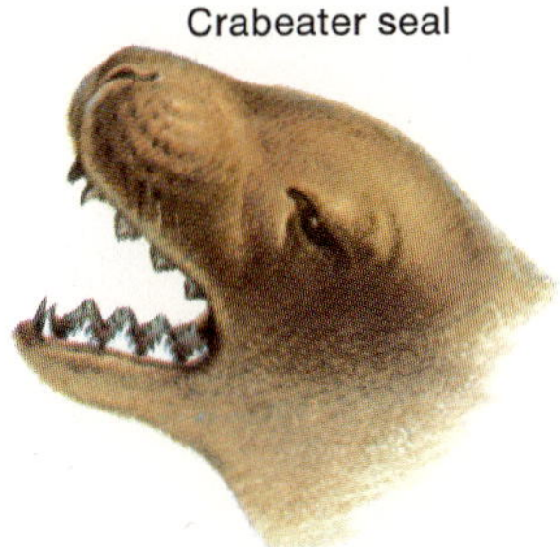

Crabeater seal

A crabeater seal gulps mouthfuls of sea-water and expels it again through tiny spaces between its teeth, selecting for consumption only the shrimp-like krill on which it lives

living tissue. On the next level are the most abundant animals in the sea, the copepods and other minute creatures that feed on the plants. They convert some of the plants into flesh, which is eaten by the next level's animals, ranging in size from herring to blue whales. At each successive level, the animals tend to be larger but fewer in number, and the total mass of animal life decreases. At the peak of the pyramid is a relatively small number of dominant carnivores: killer whales, giant squid and large fish, such as white sharks, that are rarely hunted by other animals.

Creatures of the ocean floor

The shallow sea floor of the continental shelf supports an animal community different from that of the waters above. A continual rain of food slowly falls from the sunlit surface waters to the dim sea-bed. This detritus – the bodies of planktonic organisms, large dead and dying animals, and debris swept from land – is the base of the sea-floor food pyramid, and a host of animals feeds on it. Some live permanently attached to the bottom; others burrow, crawl, hop or slither across the silt and sand.

Fixed feeders such as sponges, mussels and sea squirts feed directly from the sea, by filtering food particles from the enormous amounts of water that they pump through their bodies. Other creatures burrow through the bottom, eating embedded food particles. Bristle worms, the most important burrowers, literally eat their way through the ooze inches below the surface. Brightly coloured sea anemones hide under the mud, leaving only their mouths and tentacles exposed.

Leathery, sausage-shaped sea cucumbers gather food from the bottom with sticky tentacles around their mouths. Spiny sea urchins creep slowly over rocky shallows chewing off algae and detritus with their teeth.

More-mobile bottom dwellers prey on the filter feeders, burrowers and plant eaters. The 18-in. clamworm eats young clams and mussels. Carnivorous snails attack mussels and clams and even their own kind; rasping a hole through their victims' shells, they suck out the flesh. Carnivorous sea slugs, which are distantly related to land slugs, feed on stationary animals such as sponges. Skates and rays, flattened relatives of sharks, cruise along the bottom and with teeth like nutcrackers crush the shells of crabs, snails and other invertebrates. Crabs and lobsters scuttle across the ocean floor, eating whatever they can trap in their strong pincers.

Starfish move slowly on rows of tube feet that line the bottom of their arms. If one of their five arms is lost another will grow in its place. Each cylindrical foot has muscular walls and a sucker on the end. When a tube foot touches something solid, its muscle fibres contract, the sucker adheres, and the starfish is drawn towards the point of attachment. When a starfish finds a clam, oyster or slow-moving scallop, it wraps its arms around the victim and exerts a steady pull with its tube feet. When the valves separate slightly, the starfish inserts its stomach through the small opening and dissolves the soft body inside with powerful digestive juices.

Brittle stars, close relatives of starfish, slither across the bottom with snake-like movements of their five long arms. Some dig in the bottom with tube feet and eat decaying material. Others lie upside-down and extract food from the water. These brittle stars can pass food to their mouths by 'bouncing' it from one tube foot to another.

Many fish spend their lives on or near the bottom of the continental shelf. Large fish, such as hake, cod, haddock, pollack and ocean perch, are caught commercially in coastal waters throughout most of the Northern Hemisphere. They feed on one another as well as on their own young, on other fish, and on invertebrates such as crabs, molluscs and starfish.

Flatfish – flounders, halibuts, turbots and soles – are beautifully adapted for life on the sea-bed. A newly hatched flatfish swims upright like

HUNGRY STARFISH *A red starfish envelops a mussel and prises at its shell with five arms lined with sucker-tipped tube feet. When the shell opens, the starfish inserts its stomach to digest its victim. Whelks and sea urchins await the leftovers*

MATING CRABS *A male and female Jonah crab mate soon after the female has shed her old shell. While she waits for the new shell to harden she has the protection of the male partner against* *any predator. The female crab carries the eggs that have been fertilised until they hatch into larvae which spend their early lives in the plankton community before developing*

other fish and has an eye on each side of its head. But in the first months of life, it begins leaning over to one side. Gradually the eye on the underside of the fish moves across or through the head until both eyes are on the upper side. Most flatfish turn their left side towards the bottom. This eyeless underside becomes white. The upper side remains pigmented and can change colour to match the pattern of the sea floor.

Essential scavengers

A diver observing the concentration of life on the floor of a shallow coastal sea would not see one of the most important links in any food chain – the link connecting bottom and surface dwellers, the living and the dead. Countless bacteria work in the bottom sediments decomposing dead carcasses and scraps from the bottom and surface. Without these microscopic scavengers, the nutrients which plants require would stay locked in the remains of dead organisms. Bacteria, the principal organisms of decay in the sea as well as on land, break down animal and plant tissues into basic components – carbon dioxide, phosphates, nitrates and other nutrients. Upwelling currents eventually carry these vital compounds back to surface waters, where they can be used by plant plankton.

The community living on the sea floor of the continental shelves is one of the most complex in the ocean, and all the links in its tangled food chains have not yet been traced.

SEA ANEMONE UNDER ATTACK *Sea slugs appear to be immune to the stinging capsules of a sea anemone's tentacles. The curling of the anemone tentacles may be a defence mechanism, but it is more usual for the anemone to retract fully into its tube*

EAGLE RAY *Gliding over the sea bottom, this predator of the Great Barrier Reef pounces on crustaceans and clams and crushes their shells with powerful flat tooth plates. Rays, most abundant in the tropics, may grow to a fin span of 7 ft*

STRANGE FISH OF THE OCEAN FLOOR
Camouflaged and stealthy, they watch for passing prey

Bottom-dwelling fish feed voraciously on invertebrates and each other. Modifications such as camouflaged and flattened bodies, retractable fishing lures, and sense organs developed from fins enable these species to live and hunt successfully on the ocean floor. Many of them spend long periods motionless on the sea-bed, and because slow-moving creatures are more vulnerable to attack, they have developed an extraordinary range of camouflage adaptations. One shark, the tasselled wobbegong or carpet shark, is fringed with untidy barbels giving it the appearance of a weed-covered rock as it lies motionless waiting for its prey. Other sea-floor fish, such as the black-backed flounder, change colour to match the sea floor.

FLOUNDER THAT MIMICS THE SAND *A black-backed flounder – which has changed its colour and pattern to blend with the sea floor – lies awaiting its small invertebrate prey*

FEELING FOR FOOD *Fins that have become sensory organs enable the squirrel hake to locate its food on the floor of the ocean. The fish uses its mouth as a shovel to dig its prey from the mud*

BATFISH ON THE PROWL *An awkward swimmer, a batfish of the Galapagos Islands waddles across the sea-bottom on its pectoral and ventral fins. It entices small fish with a retractable lure and 'fishing pole' that is a modification of its dorsal fin*

ANGEL SHARK *The flattest shark in the seas is the bottom-dwelling angel shark – named by pious medieval observers who mistook the broad pectoral and pelvic fins for angel-like wings. They prey on flatfish and crustaceans*

THE FOUR SEASONS OF THE SEA

Seasons in the bountiful coastal seas are as well defined as those on land.

In spring, the life-giving sun begins to rise higher over the ocean and its rays penetrate deeper into the water. The cold water is rich in nutrients stirred up by winter storms, and the plant plankton which begins the food chain blooms abundantly.

This spring blossoming of microscopic plants is followed by a population explosion among the animal plankton. Countless baby fish, crabs, starfish and other tiny creatures temporarily join the animal plankton to gorge in the sea's rich pasture.

By summer, the sun's rays are stronger still. But the bloom of plant life has removed most of the fertilising minerals from the water. Now the plant plankton wanes. Nutrients from the bottom are prevented from rising by a layer of warm surface water that sits like a lid on the cold, richer water below. With their plant pastures reduced, the animal plankton begin to die.

In the autumn, surface waters cool and nutrients can rise again, resulting in a brief revival of the plant and animal plankton.

The sea continues to cool as winter approaches. Upwelling nutrients permeate the coastal sea from top to bottom, but plants cannot grow under the weak winter sun. Winter life is as bleak in the sea as on land. But with the coming of spring another cycle of life will begin.

The endless spring

On some parts of the continental shelves, nutrient-rich waters rise to the surface throughout the

SEASONS IN THE COASTAL SEAS

Spring is the season of greatest growth in the sea. The sun rises higher each day, and the nutrients (shown as red dots) that were brought up to the sunlit surface layers of coastal waters by winter storms are plentiful. Planktonic plants (yellow dots) flourish, and the tiny animals (black) that eat these plants are able to multiply rapidly. In summer, sunlight is ample, but the warm upper waters do not mix with the cooler deeper layers which contain the nutrients. Plant and animal populations, therefore, diminish drastically. In the autumn, as the upper waters gradually cool, the circulation of nutrients from bottom to surface begins again. Plants and animals thrive again, but their revival is limited by the shortening days and diminishing sunlight. Winter storms stir the coastal seas deeply, bringing huge amounts of nutrients to the surface to lie unused until the strengthening sunlight of spring causes the rebirth of life in the sea

year. These conditions of endless spring produce unusually rich fishing grounds which provide almost half the world's supply of fish.

Upwelling occurs when offshore winds and strong currents carry water away from the land, or when two currents diverge. Mineral-laden waters from the depths rise to fill the gap. Such fertile areas of upwelling exist off the west coasts of Africa and South America and off California and southern Arabia.

In the coastal waters of Peru, westward-blowing trade winds and the cold Humboldt Current from the south combine to produce an upwelling that gives rise to the world's largest fishery. More tuna are caught here than anywhere else. The largest black marlin ever recorded – a 1560-lb. giant – was taken within sight of the Andes Mountains.

It is the silvery anchovies, however, that make the Peruvian coastal waters unique. Millions of tons of these small fish are netted every year. Large shimmering patches of foam mark the places where immense schools of anchovies graze on the microscopic plant plankton, especially diatoms.

El Niño, current of death

The trade winds from the south-east, which help push the waters of the Humboldt Current past Peru, usually slacken during the southern summer. About every seventh year, these winds diminish so much that the Humboldt Current is displaced. Warm equatorial waters with a poor supply of nutrients travel southwards along the Peruvian coast, and little upwelling occurs. The warm current is known as El Niño, 'The Child', because it arrives during the Christmas season.

El Niño brings no joy. It chokes off supplies of nutrients, and planktonic plants die. Some anchovies swim to cooler, richer waters, but many also die. Numerous seabirds starve. Guano production decreases, and there is a shortage of fertiliser.

El Niño is a dramatic example – and warning – of what happens when the first link in the food chain is weakened. In 1965, cormorants nesting on the shores and islands off Peru numbered 20 million. El Niño came that year. Four years later the cormorant population was only about 6 million, and the guano harvest had fallen from 170,000 tons to a scant 35,000. Fishermen blame El Niño for the decline. But guano harvesters blame the fishermen; so many anchovies are caught, they say, that the birds are starving. Anchovy catches jumped drastically in the 1950's and 1960's, and the drop in the number of birds may indeed be due to excessive fishing.

Conservationists fear that if the birds die off, the anchovies may disappear as well. In the truly interdependent communities along these shores, fish rely on birds as much as birds on fish. Guano, rich in nitrogen and minerals, fertilises the water when El Niño chokes off nutrients from the deep. If the supply of guano is drastically depleted, the fishing grounds may not recover the next time El Niño returns.

OYSTER FARMING *A dredging barge scrapes oysters from the bottom of the sea. Fishermen spread oyster shells on the sea-bed so that two-week-old sprats can attach to them and gain some* *protection from such predators as crabs and oyster drills. The oysters are normally left to grow for at least four years before the fishermen raise them from the depths*

FARMING THE COASTAL SEAS

The plight of the Peruvian fishing industry demonstrates that man cannot continue to take more and more out of coastal seas without eventually endangering their productivity. Improved fishing techniques alone will not guarantee larger future catches. Only by intensively cultivating coastal waters can man greatly increase the annual seafood harvest. Instead of searching miles of water and dragging nets and hooks below the surface, he can rear fish and shellfish in coastal waters just as he rears cattle and crops on land.

Aquaculture – the application of farming techniques to raising animals and plants in fresh, brackish or salt water – embraces everything from merely transplanting animals from one unattended 'pasture' to another to raising them in enclosures where they are fed automatically by machine, given hormones to increase the number of their offspring, and bred selectively.

Two American biologists, John H. Ryther and John E. Bardach, surveyed aquaculture's world potential for the US National Council on Marine Resources and Engineering Development. They concluded that more high-grade animal protein can be produced in an acre of coastal water than on an acre of fertile land. An advanced pig farm produces 25 tons of live pigs per acre yearly, while an oyster farm produces 50 tons of oysters (shells excluded) and a trout farm about 40 tons of trout. Mussels raised on rafts in Spain annually yield as much as 120 tons of nourishing, tasty meat per acre.

Ryther and Bardach foresee the construction of enormous sea farms near large coastal cities. They estimate that one such installation could produce a yearly supply of mussel meat equal to three times the total current fish catch of the world. Sewage might be used to fertilise such a sea farm. Many large cities now have to dispose of millions of gallons of raw sewage every day. Ryther and Bardach propose treating this effluent to release inorganic nutrients, then diluting these with seawater. This solution could fertilise a 10 ft deep aquafarm, 12 miles long and 3 miles wide. Such a farm would produce approximately 1000 tons of diatoms and other one-celled plants a day. These in turn would provide food for shrimps, oysters or mussels, which could be harvested at the rate of 100 tons a day.

The Japanese lead the world in aquaculture. In Japan, shrimp farming is a highly profitable business. Adult shrimps spawn in laboratories, and

GREEN TURTLE FARMS *In the Caribbean, sea farmers hatch green turtles in tanks. They are later placed in fenced coastal waters where they fatten on turtle grass before being killed*

TROUT HATCHERY *Newly hatched steelhead trout, yolk sacs still attached, cluster in a coastal hatchery. Selective breeding of trout and salmon produces meatier and hardier species*

their offspring live in controlled environments until they reach market size. The very young are fed diatoms and dinoflagellates. As they grow, they receive brine shrimps, then a diet of ground-up scrap fish, unmarketable shrimps, mussels and clams.

Japanese sea farms also produce fish, such as yellowtails, puffers, filefish, porgies and sea eels. Young yellowtails gathered from the sea mature in nylon net bags attached to bamboo frames floating in bays. The fish are fed anchovies, sand eels, horse mackerel and manufactured food.

Japanese yellowtail farms produce as much as 126 tons per acre per year, or 62 lb. of fish per square yard. This figure sounds incredible, but market-sized, 2 lb. fish are removed and juveniles added continuously. Ryther and Bardach regard the success of yellowtail farming as proof that millions of fish can be raised on sea farms in coastal waters.

In many countries, including Britain and Canada, trout and salmon are hatchery-reared from eggs to the migration stage. Turned loose to fend for themselves in the open ocean, they reach market size without expense or labour. Eventually they return to the hatchery and are selectively bred to start a new cycle of life.

Dr Lauren Donaldson, a scientist at the University of Washington, has raised salmon that grow faster and larger, produce bigger eggs, and are more resistant to disease and temperature changes than other salmon. Some of his salmon have a survival rate in the open sea as high as 30 times the natural one.

Aquaculture and improved fishing techniques can probably further increase the annual seafood harvests, but there are limits to how much man can take from the sea. John Ryther believes that the annual harvest cannot much exceed 100 million tons. Several other scientists are more optimistic and believe the harvest could reach 550 million tons – or about ten times the amount of seafood caught annually in the late 1960's – if fishermen would pursue marine species which are not being exploited at the present time.

The sea can no longer be considered a copious horn of plenty with enough food to solve any world shortage. Mackerel populations have declined in the North Atlantic; sardines have vanished from areas where they were once abundant; several species of whales, including the blue whale, the largest creature on earth, are nearly extinct. It will take greater intelligence than man has yet shown, and that most difficult effort, international co-operation, to increase seafood yields without irreparably damaging the ocean's fragile food chains. But this intelligent effort must be made; and the need is urgent, for the bountiful coastal seas will have to be skilfully cultivated and harvested if they are to play their part in relieving the ever-increasing world shortage of food.

The underwater coral cities

Little by little, year by year, a coral reef grows as millions of tiny animals deposit their limestone skeletons. This living labyrinth is massive – but astonishingly vulnerable

Nowhere on earth does life exhibit a greater diversity of colour and shape than in the coral reefs. They are the cities of the underwater world – crowded, bustling communities populated by creatures belonging to nearly every major group in the animal kingdom.

Reef dwellers have evolved a spectacular variety of adaptations and associations while competing – and sometimes co-operating – in the struggle to find food, to protect themselves from predators, and to reproduce their kind. Female crabs dwell in cloistered coral chambers, safe from predators but imprisoned for life. Fanworms, undersea relatives of the familiar earthworm, filter food from the water with gorgeous, feather-like tentacles that withdraw into leathery tubes the instant a shadow passes overhead. Fish with teeth fused into strong beaks pulverise coral rock as they graze on algae encrusting the reef. Giant clams, weighing up to a quarter of a ton, take nourishment from microscopic plants growing inside their own bodies.

Everywhere in the coral-reef world there are dazzling colours: in the myriads of scuttling, creeping, burrowing worms, molluscs and crustaceans; in the flower-like anemones and the five-armed starfish; in the darting, boldly patterned fish; in the multiform corals themselves. Some animals are made conspicuous by their brilliant coloration, which may scare away potential predators or help attract a mate. Other creatures are so perfectly camouflaged that they are indistinguishable from their surroundings.

The reef itself is the work of living things. Coral animals change minerals dissolved in sea-water into the limestone of the reef. Other animals, such as clams and tubeworms, contribute the hard parts of their skeletons to the reef. Lime-secreting plants, the coralline algae, repair the windward edge of the reef by cementing loose pieces of coral rock into near-solid ramparts. Powdered remains of one-celled organisms settle in and fill the tiniest openings.

Formed over many centuries, coral reefs expand an inch or two each year, and in volume dwarf man's mightiest structures. The largest of all, the Great Barrier Reef, extends for nearly 1300 miles along the north-eastern coast of Australia. It could supply enough limestone to build some eight million replicas of Egypt's largest pyramid.

Unlike a static construction of man, a coral reef is a dynamic creation. Some organisms are constantly building it up, while others are tearing it down. The result is a labyrinth of ledges, crannies, grottoes, crevices and tunnels.

THE REEF-BUILDERS *Large cup coral polyps extend their food-gathering tentacles. With cups more than ⅓ in. across, they are among the biggest of the world's reef-building corals. Over thousands of years the limestone skeletons of countless tiny polyps accumulate to form mighty coral reefs and islets in the waters of the tropics*

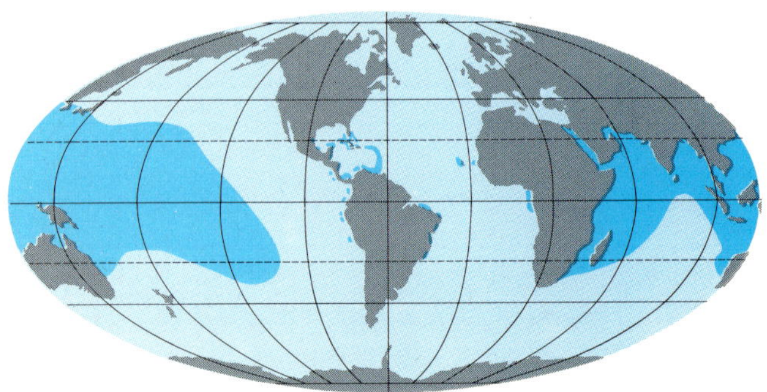

AUSTRALIAN OCEAN BREAKWATER *Heron Island, shown in the centre of the picture at low tide, is one of many hundreds of coral islands that fringe the seaward side of the 1300-mile Great Barrier Reef. The 50-acre island is used as a bird observatory*

CORAL REEF AREAS *Reef-building corals flourish only in warm, clear and relatively shallow water. The shaded areas on the map indicate the world's principal reef zones. There are three types of coral growths: fringing reefs which extend directly from the shore; barrier reefs, which are parallel to the shore but are separated from it by deep water; and atolls, which are circular reefs enclosing quiet lagoons*

How coral grows and dies

The surfaces of dried, bleached coral skeleton are riddled with tiny, cup-shaped pits. These pits were the homes of the individual coral animals, or polyps. Relatives of sea anemones and jellyfish, these stony corals are the chief architects of reefs.

A coral colony may consist of thousands of polyps, all arising from a single individual. The fertilised eggs of reef-building corals grow into free-swimming, pear-shaped larvae, the largest about $\frac{1}{25}$ in. long. A larva settles on to a solid surface, broad mouth upwards, and cements its narrow base in place, often on the remains of its ancestors. It secretes a limestone cup around itself, and spends the rest of its life in this external skeleton. The soft part of the animal within the skeleton seldom reaches more than $\frac{1}{3}$ in. in size. Surrounding the mouth are delicate food-gathering tentacles that collect plankton from the water.

The polyp sends out little branch-like buds, and these grow into other polyps that remain attached to the parent. As time goes on, each new polyp buds in turn, and a closely knit community of corals develops. In colour, it may be brown, lavender, yellow, green, violet or pink.

The colony grows upwards and outwards, gradually assuming a shape characteristic of its species. Names given to a few of the hundreds of species of stony corals suggest their diversity of shape: staghorn, elkhorn, brain, finger, flower, star and cactus.

A coral polyp's soft body tissues are housed in the protective limestone cup. When it is not feeding or is disturbed, the animal draws its tentacles into the cup.

Coral animals do not dwell alone in their stony apartments. Microscopic plants called dinoflagellates also live within their body tissues. The exact nature of the relationship between the polyps and their plant tenants is unknown, but it appears to be beneficial to both.

Coral animals offer their plant guests a sheltered environment rich in carbon dioxide. This gas, given off as a waste product by coral, is one of the raw materials necessary for photosynthesis – the process by which plants use the energy of sunlight to convert carbon dioxide and chemicals in the water into food. In using the carbon dioxide, the plants act as a built-in disposal system, and they may also help out in the secretion of limestone. Whatever is involved in this plant-animal partnership, it is certain that stony corals form reefs only when they harbour dinoflagellates.

A single coral colony does not constitute a coral reef. A reef is a complex of many colonies and many different kinds of coral, most growing on the limestone laid down by past generations. The Great Barrier Reef contains more than 350 species and millions of active colonies.

The principal coral reefs of the world exist within a 3600 mile wide band extending about 30 degrees north and south of the Equator, or roughly from the latitude of northern Florida to that of southern Brazil. Reef-building corals grow only in warm, clear water less than 150 ft deep, where there is enough light for their plant partners to carry on photosynthesis. Reefs do not grow along the west coasts of continents where cold currents chill the sea. Their growth is also inhibited by river silt, which blocks the light.

THE CORAL ANIMAL

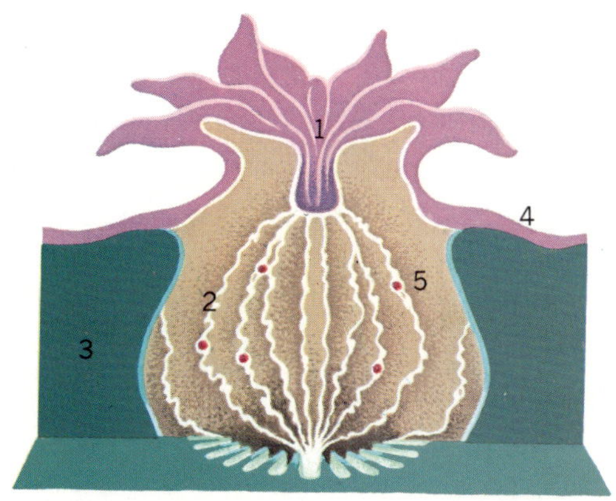

Tentacles around the mouth (1) of a coral polyp capture food, which is digested by filaments (2) lining the central cavity. The soft polyp secretes limestone partitions and a stony cup (3) to protect itself. A thin tissue layer (4) links it to neighbouring corals. Reproductive organs (5) produce both eggs and sperm. The eggs when fertilised hatch into larvae, swim about briefly then settle on a hard surface, usually dead coral. The new polyp then produces its own new colony by budding or fission

BUDDING *New coral polyps form from the cells of the live tissue which connects older polyps. The new polyps immediately start to secrete their own cups*

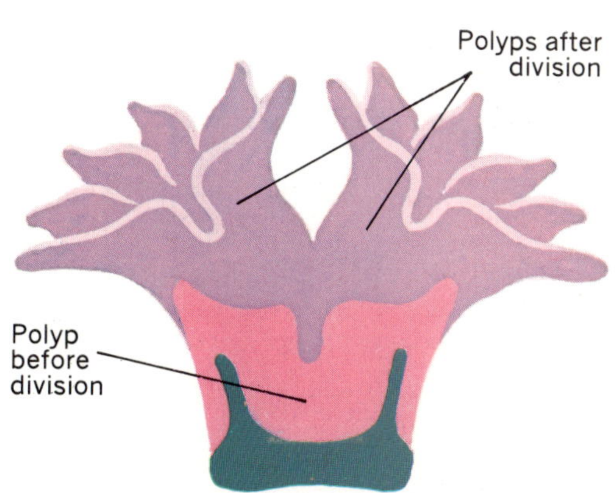

FISSION *A mature polyp lengthens, and the number of its tentacles and fleshy partitions increases. The polyp then splits in two from the mouth down*

THE COLOURFUL WORLD OF CORAL

Brilliant display of underwater life

The limy skeletons of stony coral colonies make up the bulk of most coral reefs. The carnivorous, tentacled polyps of a living colony are linked to one another by a mantle of flesh that covers the skeleton, and each feeds all the others. As the polyps divide, or as new polyps bud from the old, the colony grows upwards and outwards by as much as 3 in. each year. Soft corals, most of them stiffened by crystals or by horny, flexible skeletons, contribute little to the reef's structure. Unlike reef-building stony corals, soft corals can live far from the Equator in the waters of temperate and polar seas – for example, off Norway and Devon.

MUSHROOM CORAL *A giant single polyp housed in a stony skeleton up to 7 in. long, the mushroom coral is one of the oddest and most beautiful of corals*

CORAL POLYPS *The 1 in. long tubular polyps of cup coral emerge at night to feed. They form small colonies in shaded reefs*

FINGER CORAL *Some species of finger coral extend for acres in shallow water where sunlight is intense and currents are strong. Polyps are extended and feeding on some of the fingers in the pictures; others are withdrawn and inactive*

SEA FAN *Thousands of tiny polyps fringe a sea fan's branches at night. They contract during the day into the skeleton*

VARIETY OF CORAL SHAPES

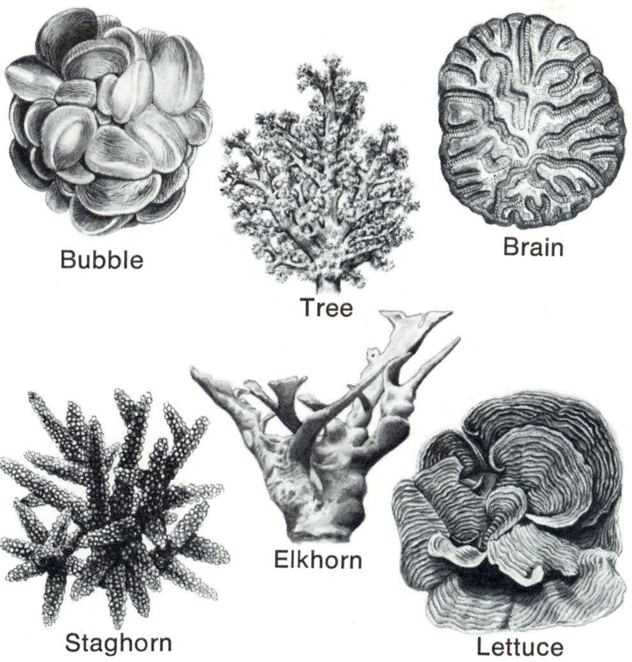

THE MAJOR ZONES OF A CORAL GROWTH

GORGONIAN CORALS *The seaward face of this Pacific atoll (explored by divers) is fringed with gorgonian sea fan corals. Gorgonians have a flexible horny skeleton which makes them able to withstand the pounding waves on seaward slopes*

There are three kinds of coral reefs, all of which are in relatively shallow water. Fringing reefs extend directly out from the shore; barrier reefs are parallel to the shore but are separated from the land by deep water; and atolls are circular reefs which have grown in the open sea on the sides and summits of hidden, ancient volcanoes.

A small atoll has several major zones both above and below the water surface, and each zone has its own association of plants and animals. At the same time, every atoll and reef in the world is different from all others.

Seaward slopes

The outer seaward slope of an atoll is often precipitous, plunging steeply towards the supporting mountain far below. Tree-like branching soft corals, called gorgonians, grow far down the slope, but they are not involved in building the reef. A small base anchors each gorgonian to the limestone slope; the rest of the coral extends far out into the dim water in a pattern of thin, branching whips that sway in the gentle currents. As live reef-forming corals approach their limit 150 ft down, their resident algal populations are greatly diminished.

The seaward slope is where the great predators lurk. Tuna, sharks and barracudas constantly patrol the shadowed rise, and huge sea basses peer out from dark caverns in the coral wall. Closer to the surface, tang or surgeon-fish, triggerfish and unicorn fish swim freely in the clear blue water; within the embraces of coral formations, smaller fish abound. Wrasses, blennies and gobies dart over the slope in the brighter water.

The slope of an atoll is cut in many places by numerous deep, narrow canyons, leaving spurs of coral extending outwards between them. The grooves, which go all the way up into the shallow tide-pool zone, shelter a wide variety of life. Rock-hugging coralline algae may be abundant enough to colour the slopes and ridges pink and violet. The channel walls abound with snails, cowries, chitons, embedded clams, and a multitude of active rock crabs, hermit crabs and shrimps. Large purple slate urchins, each with its own species of purple shrimp living among its heavy spines, conspicuously occupy indented ledges.

The ridges between the deep grooves are thickly capped with coralline algae building upon a coral foundation, the whole mass being cemented together by the algae and by the fused skeletons of the one-celled coral creatures.

The tide pools

Where an atoll reef rises close to the surface and is exposed at low tide, the pools formed in depressions harbour large concentrations of small plants and animals. Each time the tide floods in, the depressions become part of the sea, well flushed by wave action. As the water level falls again, they are isolated and become a little more difficult to live in; for instance, a pool less than about 12 ft across is likely to become too hot to support much life. High water temperature is not in itself very dangerous to heat-tolerant reef life, but warm water contains less oxygen than cool water. If a pool cannot produce enough of this vital gas through the activity of its algae, the animal population diminishes. This condition does not happen often, however, and a great many healthy pool animals are usually present. Since many tide pools trap wind-driven sand, the bottom can support a host of burrowing worms and crustaceans and, often, mottled sea cucumbers camouflaged against the sand.

The walls of a tide pool may be honeycombed by black or brown sea urchins that for generation after generation have excavated deep holes in the limestone. Sometimes they are so deeply ensconced in old holes that their outermost spines barely extend past the mouth of the cavity; no matter how violent a storm wave, they remain secure unless a whole section of coral is torn loose. Vacated holes may be occupied anew by other urchins, snails, cowries, blennies and other small animals seeking relief from the waves.

Chunks of coral lying loose on the bottom of a pool hide long-limbed brittle stars, burrowing snails, worms and many other creatures. Under a rock are hordes of fast and slow animals, as well as those that are fixed in place and cannot move at all. Sea cucumbers, resembling large sausage-shaped rocks, slowly creep across the pool floor.

Small sea anemones may decorate the sides and crevices of a tide pool, their delicate tentacles contrasting with the hard, enamelled encrustations of purplish coralline algae. Feathery hydroids stand out like tiny bushes.

A tide pool, like a reef, has zones, each preferred by its various inhabitants. Coralline algae and small sea anemones are close to the surface, where crabs, coloured exactly like the coral, are also found. Red-eyed rock crabs run over the upper levels of the walls of the pool, pausing on the lips of crevices, then scuttling rapidly and completely out of view.

Below this zone, round, short-spined sea urchins cover themselves with algae and bits of shell debris, successfully disguising their forms. Towards the bottom, along the pitted walls, black long-spined sea urchins wave their venomous needles, while peacock worms encrust lower sections of the walls, and small clumps of living coral protrude where there is room.

On the side of the tide pool, small knobs stud the limestone, each covered by coralline algae except for the wavy line running through the centre, which indicates the knob is an almost perfectly disguised rock clam. Where bushy, non-stony seaweeds hang down into the quiet pool, sea slugs may be found in quantity, the leaf-like respiratory projections along their backs wonderfully mimicking the fronds of seaweed.

Above the upper tide pools, the shore is wetted only briefly by the highest tides or perhaps only by spray. Here the life forms diminish in variety and abundance. The shore is darkened by films of blue-green algae that give the pale limestone a slippery coating. But some snails can live here, and black rock crabs scamper about, picking up small prey or scavenging for organic remains.

REEF SEA SLUG *Covered with bright, ruffled breathing organs and lacking a shell, the reef slug appears highly vulnerable. But the unpleasant taste of its flesh and its conspicuous colours are deterrents to would-be predators*

The calm lagoons, coral cradles of abundant life

On the lagoon side of an atoll reef, the shore slopes down through several successive zones to the floor of the enclosed basin. The distribution of plants and animals in each zone differs somewhat from that on the seaward side, primarily because there is no strong wave action. The lagoon waterline, which rises and falls with the tides, harbours swarms of rock crabs, hermit crabs and snails. The slow-moving snails escape the greater heat of this sheltered shore by burying themselves deep beneath the coral rubble. Where the limestone has become finely pulverised coral sand, it offers a favourable habitat for burrowing bivalve molluscs, worms, shrimps and other creatures.

Tide pools in the lagoon reef may be even more heavily populated than those on the sea side, but usually with a smaller and different variety of animals because of the quieter, warmer water. Representatives of the world's only marine insects, water striders, which range far out to sea, are often found here, skating across the still water in search of living and dead animal food trapped in the surface film, exactly as their freshwater relatives do on ponds.

Where sand stretches out into shallow portions of the lagoon, conditions permit the establishment of turtle grass, one of the marine seed plants found not only in the Pacific but in extensive beds off the Florida Keys. The sand in which it grows, as well as the blades of the grass itself, provide a habitat for a community of animals and microscopic attaching plants.

On the sloping face of the lagoon reef, corals become abundant, often assuming more elaborate forms in the absence of heavy wave action. As on the seaward side of the atoll, no one coral species dominates. Many branching corals in weird and beautiful shapes grow on the sloping floor – some slender, tapering and often intertwined, others heavy and massive. Lower down are the solid, rounded corals that occupy a great deal of space but do not rise high into the water. Some corals look like thin sheets; others curve outward as shelves. The variety seems endless. Lying loose in depressions are the mushroom-like notched discs of giant solitary polyps that produce ridge skeletons several inches in diameter.

Amid the corals are all the various members of

a reef community: giant clams, black-spined sea urchins, beautiful but dangerous cone shells with hidden, venomous, harpoon-like teeth, worms that form the staple food of the cones, handsome cowries, a miscellany of sedentary clams and, of course, crabs, shrimps, tube worms and a multitude of less conspicuous creatures. A pair of small *Trapezia* crabs living deep within each coral head remain invisible unless the head is cracked open.

A lagoon may be up to 300 ft deep, the water a clear luminous blue-green except when wind stirs the fine sand. Usually studding the lagoon floor are many pillar-like knolls or patch reefs, which rise nearly to the surface. These are constructed mostly of corals; but close to the surface, where sunlight is intense, coralline algae make a contribution. Eniwetok, one of the Marshall Islands in the Pacific, has over 2000 such knolls; even a much smaller atoll may contain hundreds.

A knoll is a great towering city of marine life, isolated from others rising mistily through the water 50 or 100 ft away. The walls, often almost vertical, support an enormous variety of living things. The corals are of many kinds, including massive colonies of *Porites* extending outward and intermixing with long arms of fragile staghorn. Each knoll has its own population of brightly coloured reef fish that hover near by, ready to dive into the protective crevices of the coral formations or to defend their own small territories on the face of the wall. At times, the large shapes of open-sea predators appear in the lagoon: a shark, manta ray, eagle ray or barracuda.

With so much growth and deposition in a lagoon, the enclosed body of water in many atolls becomes increasingly shallow, even to the extent of disappearing altogether. Jarvis and Baker islands, in the mid-Pacific, are examples of a 'pancake' atoll – a perfectly flat, dry, saucer-like island, with a slightly raised rim of sand and rubble heaped up by the action of the ocean waves against the coral. There are also, of course, many intermediate atolls with greatly reduced lagoons that have lost all connection with the sea. In such lagoons, the water may be excessively salty because of evaporation, or nearly fresh as a result of recent rainfall; in either instance, typical lagoon life is absent or severely reduced.

LIFE ON THE REEF *Currents moving across the seaward reef in the south-west Pacific carry ample food for a varied community fixed on the coral stalks. Here yellow sea squirts and green and pink sponges encrust the craggy surface of the reef*

SEA SQUIRTS *Pale blue sea squirts flourish in the warm waters of the Great Barrier Reef. Growing among them are light green sponges. The squirts and the primitive sponges filter food particles from the water*

TWO WORLDS ON A CORAL REEF
Great fish of the ocean alongside tiny insects

A wall of coral divides two different worlds as the tide retreats from an atoll or a shallow reef. On the deep seaward side sharks, barracudas and tuna prey on fish straying from their shelter among the coral. Here tree-like soft corals called gorgonians, sea squirts and sponges cling to the reef's steep face. But the other gently sloping side encloses a lagoon, with rock pools at low tide; and in this shallower, warmer water corals are more abundant and elaborately shaped. Water striders, the only marine insects, skim the surface, and turtle grass hides hordes of microscopic animals. Sea cucumbers thrive on the floor of the pools, while sea anemones, black or brown sea urchins and sea slugs, disguised like seaweed, cling to the sides. Near the surface, camouflaged hermit and rock crabs dart in and out of crevices. In the deeper pools and in the lagoon are giant clams, cone shells, shrimps and tube worms.

PACIFIC TIDE POOL *Only hardy species are able to survive the turbulent environment of a tide pool. Here, finger and staghorn corals carpet the pool floor; the gaping dark blue mantle of a clam is just visible at the bottom of the picture*

HIDING FROM PREDATORS *Colourful communities of encrusting animals and plants live beneath seemingly dead coral fragments in tide pools. On this upturned fragment are a yellow sea squirt, a red starfish, algae and bryozoans*

PLANKTON EATERS OF THE REEF

Many reef animals, including the corals, move very slowly or are entirely immobile during their adult lives. Instead of chasing after prey, they sift through the sand or catch food from the currents.

The tentacles of coral polyps, like those of jelly-fish, are studded with batteries of poison-injecting nematocysts. Any small creature unlucky enough to brush against a tentacle is quickly paralysed, and passed through the mouth into the polyp's digestive cavity.

Similar to corals in both body plan and feeding habits are the sea anemones, close relatives of the corals. Anemones are found in all the world's oceans from the tropics to the poles, and from the tidal zone to the abyssal depths. But these flower-like animals reach their fullest development on coral reefs. Cold-water anemones usually measure only a few inches across, while some reef species exceed 3 ft in diameter.

An anemone looks like a single coral polyp; its hollow, cylindrical body is topped by one or more rings of stinging tentacles surrounding a central mouth. Depending on the size and stage of maturity of the anemone, its prey may be microscopic zooplankton or small fish and crabs from the surrounding ocean.

A sea anemone usually remains fixed to a hard surface by a muscular, mucus-secreting plate, but if a location ceases to suit its needs, the animal can move around. Some anemones creep snail fashion, at a top speed of about 4 in. an hour, by rippling contractions of their base plates.

More mobile than anemones, sausage-shaped sea cucumbers crawl along the floor of the reef on tube feet resembling those of their starfish relatives. Some cucumbers plough through bottom sand feeding on debris; others use their tentacles to net plankton. Any small organism touched by a cucumber's sticky tentacles is trapped as surely as a fly on flypaper. From time to time each tentacle is pushed into the mouth and the accumulation of victims is consumed.

Several types of sea cucumbers are equipped with an extraordinary defence mechanism. When one is disturbed, it contracts its body violently, and expels a tangled mass of internal organs from its rear opening. This act of self-evisceration does not permanently harm the sea cucumber; it

SEA CUCUMBER UNDER ATTACK *A bronze-spot sea cucumber that has approached the nest of a damselfish on the sandy ocean floor near the Great Barrier Reef is repeatedly buffeted by the fish, which is intent on defending its eggs. The sea cucumber, in fright, ejects a mass of white threads, like strands of spaghetti. They become sticky in salt water and are distasteful to fish. Although the 6-in. damselfish is in no danger from them, they are strong enough to entangle and immobilise smaller fish and crustaceans*

possesses remarkable powers of regeneration and replaces the lost organs within a few weeks.

One of the reef's most beautiful plankton gatherers is the delicate feather star. Forming a lacy net with its branching, cilia-lined arms, it wafts tiny prey into a central mouth. A feather star anchors itself to a coral branch or other support with a cluster of clinging appendages, or cirri. Although it tends to remain stationary for long periods, it can creep along on its cirri or swim by slowly flapping its arms.

The feather star's ancestors carpeted great stretches of ocean floor 300 million years ago, and from them and other ancient relatives all modern echinoderms, or spiny-skinned animals, are descended. The echinoderms include starfish, sea urchins and sea cucumbers, and they are characterised by tube feet and symmetrical bodies divided into five parts or multiples of five.

The little female gall crab, one of the most curious plankton eaters of the reef, alters the coral's growth so that a limestone chamber, or gall, forms around her. Currents set in motion by her gills bring oxygen and plankton through pores in the gall. At breeding time the pores also admit a tiny male gall crab – which otherwise leads a free life on the reef's surface. The pair mate and their offspring drift out through the pores.

FEATHER STAR *The white feather star, perched on a cluster of red sea whips, fans out 20 arms to catch what food the current carries. Each arm has a groove lined with whips that pass microscopic food particles to its central mouth*

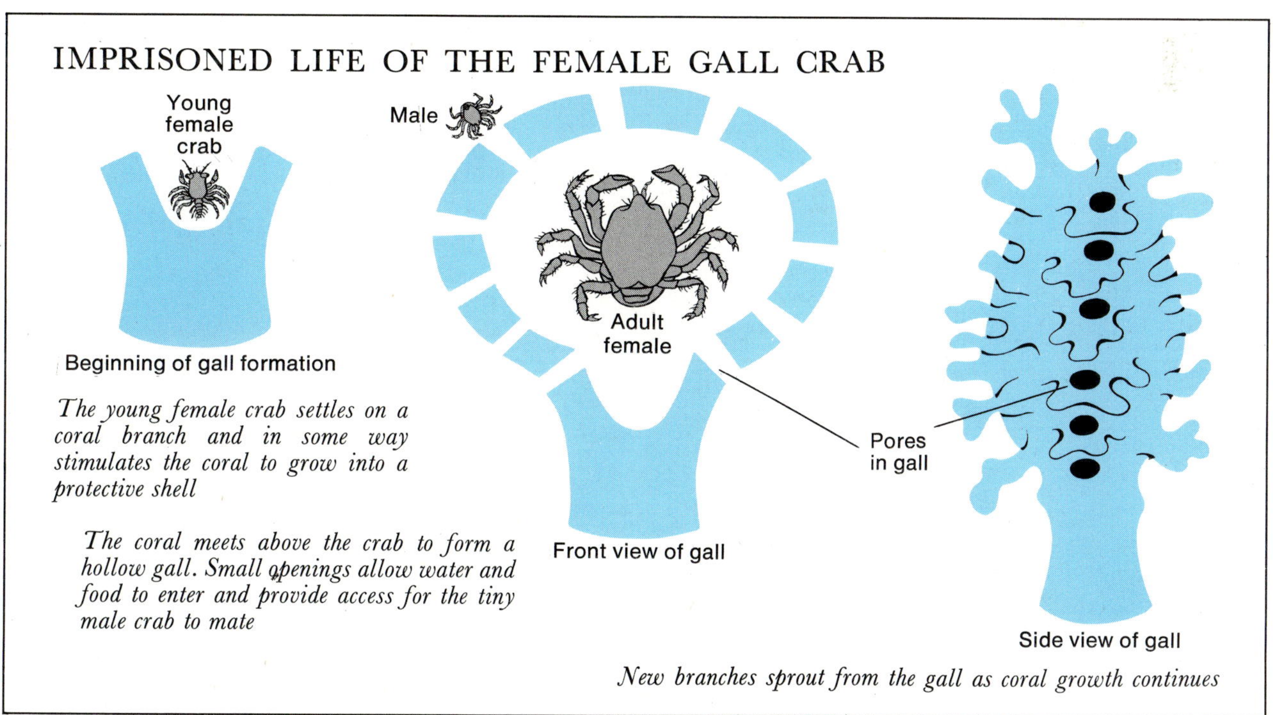

IMPRISONED LIFE OF THE FEMALE GALL CRAB

The young female crab settles on a coral branch and in some way stimulates the coral to grow into a protective shell

The coral meets above the crab to form a hollow gall. Small openings allow water and food to enter and provide access for the tiny male crab to mate

New branches sprout from the gall as coral growth continues

GAS-FILLED KILLER *A distant relative of the octopus, the nautilus rises and descends in the ocean by adjusting the amount of gas in its more than 30 chambers. It hunts food on the ocean floor and in shallow reefs, where it ensnares prey with its small tentacles*

SEA SLUG *A brightly coloured sea slug glides through the reef waters. This species, one of the more mobile, has the gill tufts (right) and sensory tentacles (left) that are typical of the sea slug family. Its body undulates as it moves*

INVERTEBRATE PREDATORS OF THE REEF

Not all reef invertebrates passively filter plankton from the water; many are skilled and active hunters. The skeleton shrimp, a gaunt, transparent animal less than an inch long, prowls the coral in search of prey. Its two pairs of hooked, grasping appendages are poised like the forelegs of a praying mantis, ready to snatch at any small creature that ventures too close.

The 'gunman' of the reefs is the 2-in. pistol shrimp. One of its claws is greatly enlarged and modified so that when the 'thumb' is snapped sharply against the 'palm,' a loud bang results. The shrimp, lying in ambush in a coral crevice, has only its sensitive antennae exposed. At the approach of a small fish, the shrimp rushes forward, aims its pistol at the victim, and pulls the trigger. The shock wave momentarily stuns the fish. Before it can recover, the victim is captured.

Reef snails and slugs are far more colourful and active than their placid dry-land and freshwater relatives. Cowries, much prized by collectors for their exquisitely patterned shells, feed on a wide variety of animals including coral polyps; the striped helmet snail devours spiny sea urchins; the trumpet snail preys upon clams and starfish.

The most dangerous of the predatory marine snails are the cone shells. Instead of the file-like feeding apparatus of most snails, a cone shell has an extending feeler, known as a proboscis, armed with minute hollow darts that inject venom. When a fish or worm comes within striking range, the cone shell swiftly pushes its proboscis against

THE 'SHARP-SHOOTER'

The pistol shrimp preys on fish smaller than itself; it concusses its victim before trying to eat it

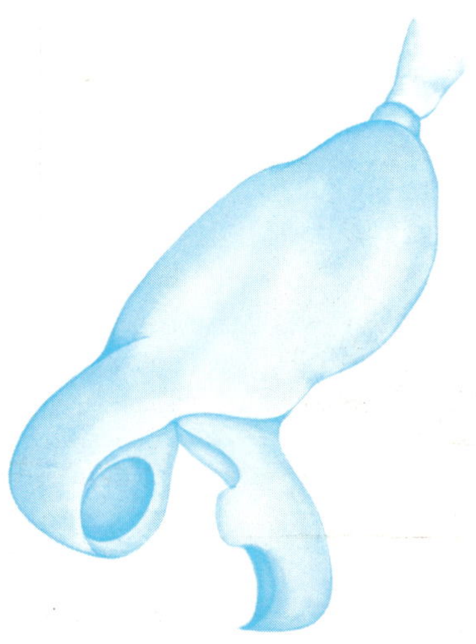

Claws at the ready, a pistol shrimp waits for prey beneath the tentacles of a sea anemone. One of the two fingers on its large right claw is equipped with a peg that fits into a socket on the other finger. When its powerful muscles snap the fingers together, the shock waves produced by the sudden explosion of water from the socket can stun a small fish

the victim and releases a dart. Once paralysed, the prey is slowly consumed by the proboscis.

Should an unwary shell collector wading on the Great Barrier Reef pick up a cone shell, he would be likely to feel a sharp, jabbing pain in his palm, followed by numbness, tingling, and spreading paralysis. Coma could ensue, and possibly death from heart failure within four or five hours. Cone shells are the only snails fatal to man.

Nudibranchs, or sea slugs, are among the most beautiful of the coral-reef's invertebrates. *Nudibranch* means 'naked gill' and refers to the feathery respiratory organs projecting from the creatures' backs. Although nudibranchs are not closely related to land slugs, both are essentially snails that have lost their shells.

Nudibranchs range in length from an inch to a foot and sport the most garish colour combinations imaginable: striped, spotted, or solid patterns of deep blue, azure, electric pink, crimson, bright orange or yellow, often with gills and head tentacles of contrasting colours. Usually a nudibranch glides along on a flat, muscular foot, but it can swim short distances.

It eats animals no other predators seem to want: sponges filled with needle-like spicules and anemones armed with stinging buds called nematocysts. As a nudibranch feeds on an anemone, the nematocysts that it swallows are somehow 'defused' and fail to discharge. But this neat trick of adaptation does not end here. As the still-intact nematocysts pass through the slug's intestine, they penetrate the intestine wall, migrate through the body, and finally come to rest in the skin of the slug's back. There they protect the slug against predators – just as they previously protected their original owner.

Eight-armed predator

The reef's largest and most formidable invertebrate predator is the octopus. Across the widest span of its eight rubbery arms, an octopus may measure from 2 in. to more than 30 ft, but the species found in coral reefs seldom exceeds 2 ft. The octopus possesses eyes as highly developed as our own, and the most sophisticated brain of all the invertebrates.

The octopus is a master of disguise. With dazzling rapidity, it changes its colour from chocolate brown to milky white to brilliant green or bright red as it glides across the shifting background of the reef. In shallow water, the waves of colour moving across its body duplicate the stippled pattern of sunlight in the ripples overhead. Fright, anger and excitement elicit vivid blushes.

The sinister reputation of the octopus is really undeserved. A timid and retiring creature, it responds to the approach of a swimmer by retreating. Sufficiently provoked, however, the octopus delivers a nasty, venomous bite from a parrot-like beak, but such occurrences are rare.

The octopus feeds on fish and crustaceans, especially crabs. Its sinuous, sucker-covered arms, as well as its high intelligence, make it a superb hunter. With cat-like stealth, it slowly moves near its prey. Then, in a flash, it pounces. After injecting the victim with a numbing venom it closes its powerful beak, and the shellfish is crushed.

The octopus hunts at night and retires to its den – either a natural crevice in the reef or a scooped-out hollow beneath a rock – during the day. Outside the entrance to the den is a pile of empty crab shells; the entrance itself is usually no wider than an octopus's arm, since the animal has an extraordinary ability to squeeze through small openings.

The octopus need fear only one enemy – the savage moray eel. Although the moray's usual diet consists of crabs, shrimps and fish, it also preys upon the octopus, if only because both sometimes seek out the same hole to hide in during the day. But the octopus is not entirely defenceless. It may confound its attacker with a cloud of ink and then scurry to safety.

THE OCTOPUS

ARMS THAT GROW FROM THE HEAD *The octopus is a mollusc, although it lacks the shell of its distant ancestors. (Some species, however, still have a vestigial internal shell.) The creature's flabby, sack-like body tapers gently to a 'neck' by which it is joined to the head. On the underside of the head, unseen from above, the octopus has a parrot-like beak. Its eight muscular arms, or tentacles, connected at the base by a thick web, emerge from its head. Octopuses normally crawl along the sea bottom on their arms, but they can also swim by forcing jets of water from the siphons in their tentacles or by flapping the webs of their tentacles*

LARGEST INVERTEBRATE *The octopus is the world's largest and most intelligent form of invertebrate life. Coiled in its burrow after a meal, it appears ferocious, with its staring eyes and sucker-covered arms: it is in fact timid and unaggressive*

REEF BUILDERS AND DESTROYERS

Many creatures live and die on the coral reef without changing it in any apparent way. Feather stars cling to it, crabs scuttle across it, octopuses lurk in its crevices, and all leave few marks. But there are creatures that profoundly affect the reef's structure. Some strengthen it by cementing loose pieces of coral together, some contribute shells and hard parts when they die. Others scrape away at the reef's surface, bore holes in it, or eat the living coral polyps, stinging tentacles and all.

Sponges are simple animals found on all coral reefs; they can be either reef builders or reef destroyers. Little more than pore-riddled masses of gelatinous material, sponges are sometimes interlaced with particles of limestone or silica. They constantly pump a current of water through their bodies by means of 'whip-bearing cells' lining their interiors. The flow of water brings in oxygen and food (bacteria, plankton and bits of decaying organic matter) and carries away waste products.

Even a small sponge passes a surprisingly large amount of water through its body in the course of a day. A sponge of the genus *Leuconia*, about 4 in. tall and $\frac{1}{2}$ in. in diameter, pumps more than 5 gallons every 24 hours. A large reef sponge may, according to recent researches, pump several hundred gallons in the same period.

GIANT CLAM *A 2 ft wide giant clam displays its mottled dark blue mantle as it filters plankton from the water. When the shadow of a predator passes over the light-sensitive mantle, the clam's fluted shell closes with great force*

Sponges thrive in the clear, silt-free waters of the reefs, and often grow 18 in. high and 18 in. across. Their presence is usually constructive, for many sponges hold together fragments of coral, much as plants on a hillside hold the soil in place. However, one family of sponges, the boring sponges, are reef destroyers. They tunnel their way into the coral, hollowing out galleries that weaken the limestone. This tunnelling process is not fully understood, but entails the removal of tiny limestone chips – a remarkable accomplishment for so lowly an animal.

There are also boring species in the reef's clam population. Some clams rasp away the limestone with the edges of their serrated shells; others dissolve it with acid secreted from their bodies. When they die, they leave their shells embedded in the reef, partially compensating for the damage they did to it while they were alive.

The giant clam of the Indo-Pacific coral reefs is a supplier of building material. The biggest clam in the world, it produces fluted shells up to 4 ft wide, weighing as much as 500 lb. Apart from its sheer bulk, the giant clam's most distinctive feature is the 'farming' it does to supplement its diet of plankton. Swarms of one-celled plants live in its body tissues, just as they do in coral polyps. These plants absorb some of the clam's wastes and provide it with oxygen and food substances through photosynthesis. In addition, surplus plants are periodically 'harvested' as food by the immense two-shelled clam.

Two families of fanworms, relatives of the common earthworm, also help to build the reef by constructing protective tubes for their soft bodies. Serpulid worms use limestone taken from sea water; sabellid worms secrete mucus to cement together grains of sand.

Both kinds of worms possess gorgeous crowns of delicate feeding and respiratory tentacles. These tentacles have light-sensitive organs so that the mere shadow of a passing fish causes the whole array to close up, like an umbrella turned inside out, and swiftly disappear into the tube. Sometimes the fanworms are not fast enough, however, for their crowns have been found by fishermen in the stomachs of reef fish.

Worms can also weaken the coral structure in a variety of ways. There are species that feed on the reef's cementing coat of coralline algae, or on the coral itself, or on the dinoflagellates living in the coral polyps. Other species, like the 16-in. palolo worm, bore into the coral. This worm passes the day deep within the reef and emerges at night to forage for small invertebrates. The palolo worm burrows through the reefs of the South Pacific; a related species does similar damage to the reefs in the Caribbean.

HOW THE WORMS OF THE REEF MULTIPLY AFTER DIVIDING

The palolo worm, which burrows in coral reefs, spawns at the same time every year: its behaviour seems to be synchronised with the phases of the moon. As the spawning season nears, a ring-like constriction develops about midway down the worm's body and the hind section fills with eggs or sperm. On the spawning night, the worms back out of their burrows in the reef and the hind sections break free and float to the surface. There the eggs and sperm pour out and mix. When the eggs are fertilised, they hatch into free-swimming larvae within hours and drop to the bottom where they take up the burrowing habit of the parent worms. At no time during this reproductive process does the adult worm ever completely leave the safety of its burrow in the coral reef

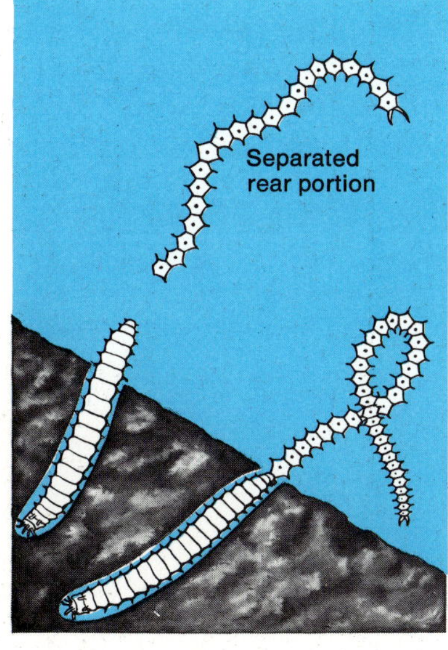

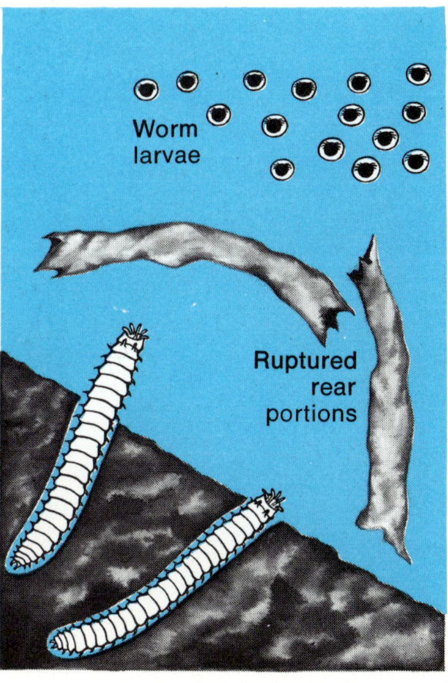

ENEMY OF THE STARFISH *The triton snail, a species that is much sought by shell collectors, feeds on the crown-of-thorns starfish in the Great Barrier Reef. The removal of the triton by divers may account for the alarming population explosion among the starfish – which has resulted in the deterioration, and in some places the complete destruction, of the Great Barrier Reef*

STARFISH – THREAT TO THE REEFS
The crown-of-thorns army leaves a bleak trail of devastation

Damage caused by boring sponges, clams and worms does not pose a great threat to the world's reefs. Any healthy coral community maintains a dynamic balance between destructive and constructive forces. In the 1960's, however, a much more dangerous reef wrecker, a starfish, appeared in alarming numbers on Pacific reefs.

Named after the red, needle-sharp spines that cover its 16-armed body, the crown-of-thorns eats coral polyps. Clinging to a coral mass with dozens of tube feet, the starfish extrudes its stomach through its mouth and floods the living coral tissue with strong digestive juices. When the crown-of-thorns moves on, it leaves behind a patch of bare white coral skeleton. When the living coral is killed, the reef is open to erosion and eventual destruction by the ever-pounding ocean waves.

Before 1960, the crown-of-thorns was something of a rarity, but by 1965 it became apparent that the species was undergoing a population explosion. Where divers had previously seen only one of the prickly stars in an hour, a hundred could be counted in ten minutes. By 1969, the crown-of-thorns had laid bare 140 square miles of Australian reefs. It was also appearing in large numbers on many Pacific reefs where it had not been seen before.

Marine biologists have been unable to isolate the cause of this population increase. Some point to human tampering with the reef's natural balance: pesticide poisoning of planktonic creatures that eat the eggs and larvae of the crown-of-thorns,

THE DESTROYER *The crown-of-thorns starfish, which gets its name from the bristling spines with which it discourages predators, eats the living coral heads of a reef*

or the indiscriminate collecting of the giant triton, a large snail that preys on the adult starfish. Others suggest that the population increase might occur regularly at intervals of hundreds of years.

To stop the destruction, groups of divers have collected the thorny marauders or killed them with injections of formaldehyde. Some scientists advocate breeding large armies of tritons or other predators to turn back the onslaught. Many other biologists, however, fear that a measure such as this might further upset the balance of reef life.

Until more is known about the life history of the crown-of-thorns starfish and the ecology of coral reefs, man's attempts to check this menace will probably not be successful. But reefs have survived natural disasters in the past, and it is probable that nature itself will correct the imbalance.

SHARP-TOOTHED PREDATORY FISH

Adapt or perish is the first rule of life. In the severely competitive setting of the coral community, fish have evolved a fantastic array of adaptations to ensure survival.

The 4-ft moray eels have muscular, snake-like bodies with no side fins. Adapted for life in the reef's holes and crevices, moray eels are found nowhere else. Using their sharp, grasping teeth and keen sense of smell, they probe for octopuses, fish, crabs and shrimps.

Morays do not make unprovoked attacks on swimmers, but a diver who reaches into a moray den may be bitten. Contrary to popular belief, moray bites are not poisonous, but they easily become infected.

When morays retire for the day, the barracudas come out to hunt. In many ways, a barracuda's tactics are the opposite of an eel's. It patrols sunlit, open waters, relying on its sharp eyesight and great speed. When a school of fish comes within striking range, the barracuda passes through it in a single rush, slashing and tearing with knife-like teeth. Then the barracuda returns at leisure to feed on the fragments. Barracudas reach 6 ft in length and have jaws capable of severing a human arm or leg in a single snap.

OCEAN KILLER *A formidable predator, the barracuda, which can be 5–6 ft in length, has been known to attack humans. To feed, it rips through a shoal of fish, slashing and tearing with its teeth. It then returns to eat the pieces*

DISGUISES TO OUTWIT THE ENEMY

The moray eels, barracudas and other large predators, superb hunters as they are, can yet be outwitted. Over millions of years small reef fish have evolved ingenious forms of defence.

Armour-plating

As a defence against the tearing, stabbing teeth of predators, some reef fish have evolved suits of protective armour, often at the expense of mobility. The trunkfish lives within a bony box almost like the shell of a turtle. Propelled only by the paddling of its small fins, it is a slow and awkward swimmer. The closely related filefish is plated with small, prickly scales that give its skin the texture of sandpaper.

A rough, leathery skin also protects the triggerfish, which carries a unique defensive weapon: a three-spined dorsal fin that can be locked into an upright position. At the approach of danger, the fish darts headfirst into a coral crevice and wedges itself in tightly by erecting this fin. No amount of tugging can dislodge it.

Several kinds of surgeon-fish carry a pair of blades on each side of their body near the tail. Each is a razor-sharp spine. When threatened, a surgeon-fish gives a warning sweep of its tail. If the intruder persists, the surgeon-fish waves its tail violently from side to side to kill or scare it away. Once the danger has passed, it returns to placidly grazing on algae.

Unpalatability

Some fish avoid being eaten simply by being unpalatable. The soapfish responds to attack by exuding a froth of mucus that may be poisonous. Certain species of parrotfish spend the night in loose-fitting 'cocoons' of stringy mucus secreted by their skin. A number of reef fish are poisonous to man, so it seems likely that some are poisonous to other fish as well.

Trickery

Many reef-dwelling fish rely on tricks of behaviour to elude predators. Brightly coloured, 4-in. damselfish never venture far from tangles of coral branches, and they quickly seek shelter at the first sign of trouble. When the pearly razorfish, which feeds on small molluscs of the reef floor, senses danger, it simply dives headfirst into the sand and buries itself. The yellowhead jawfish excavates a 12-in. burrow in the sand and lines it with pebbles. Hovering tail-down just outside this home, it feeds on plankton. Should a large creature approach, the jawfish backs into the safety of its fortress.

Camouflage

Colour plays a vital role in the lives of fish that live in the coral reef's shallow, well-lit waters. Camouflage – colour that conceals – serves as a device both for attack and for defence, hiding predator as well as prey.

A barracuda hovering motionless near the surface is made less conspicuous by a type of camouflage known as obliterative shading. Seen from above, the barracuda's grey-green, slightly dappled back matches the tone and texture of the water; from below, its silvery belly blends equally well with the brightly lit surface water. Virtually all fish that spend much time in open water have this dark-above, light-below shading.

Brightly coloured reef fish, on the other hand, would seem to advertise their presence. But strong colours, which may stand out glaringly in open water, can break up a fish's outline and make it almost invisible against the dazzling, multi-coloured background of a coral reef. Butter hamlets sport liveries of blue, pink, yellow and orange, yet do not appear conspicuous against a multi-coloured backdrop of corals.

Another type of camouflage common among reef fish is disruptive shading: contrasting bars, stripes and blotches that obscure prominent body features. Butterfly fish, damselfish, gobies and blennies hide large, staring eyes with black patches or bands, and wear 'false eyes' on their bodies or tails that are simply dark spots ringed with contrasting colours. When a predator lunges for what appears to be the head, it often comes up with just a bit of the tail or a mouthful of water as the intended victim speeds away.

AN 'EYE' IN THE TAIL *The conspicuous black spot on the tail of the golden long-nosed butterfly fish is a form of camouflage, to lure would-be predators away from the fish's head. Attackers usually get only a mouthful of water or a piece of tail as the fish darts away. Its real eyes are near the front and are themselves camouflaged by the first coloured band*

HOW FISH CAN CHANGE COLOUR FOR CAMOUFLAGE

Fish change colour when pigment-bearing cells, called chromatophores, expand or contract. The section of fish shown here has black and red chromatophores, plus cells with a white or iridescent appearance. In fig. 1, both red and black cells have contracted; fig. 2 shows red contracted and black expanded; fig. 3 illustrates the reverse; and in fig. 4 both colours are expanded.

These changes are triggered by reflexes as the fish passes near rocks or plants. Because both colours can be expanded or contracted in many combinations, the fish can achieve a wide variety of camouflage colours to suit each environment it passes through. It can produce both solid colours and mottled patterns to blend with variegated backgrounds

Quick-change artists

Some fish camouflage themselves by changing colour. Cells in their outer skin, called chromatophores, contain black, brown or grey pigment. Other cells have yellow, red or orange substances called carotenoids (the same substances that give carrots their characteristic colour). The skin also contains crystals of guanin, a reflective substance that catches the light and emits a silvery sheen.

The pigments in the chromatophores are controlled by a finely branched network of nerves and can be expanded or contracted to expose more or less colour. As the pigments are spread out or drawn in, hues and markings flow across the fish's body. Tropical groupers can flash through a whole repertoire of spotted, striped and blotched patterns in an instant. The Nassau grouper is particularly adept at changing its colour scheme to match its surroundings.

Several reef fish wear bright colours by day and take on duller colours at night – possibly to make themselves less conspicuous to predators. The Spanish grunt's silver and yellow become dull and blotched; the porkfish's large yellow dorsal fin turns black to match the darkened water.

RANGE OF COLOUR *The Nassau grouper fish can vary any of its eight colours. The most lurid display of bands and stripes (above) may signify alarm; more typical is a uniform austere colouring (right)*

DEADLY 'STONE' *The 2 ft long stonefish assumes the appearance of a weed-covered rock as it lies half buried in the sand, awaiting some unwary prey. The venom in its spines is the most poisonous found in any fish*

Masters of disguise

Shape as well as colour can contribute to a fish's camouflage. The incredibly elongated trumpetfish and pipefish spend much of their time poised head-down among stands of soft corals, slowly swaying in unison with the waving stalks of coral until some suitable prey is within reach.

The stonefish of the Indo-Pacific reefs is a master of disguise. Lying half buried in the sand, it looks exactly like a pitted, irregularly shaped rock, complete with encrustations of algae. When a smaller fish approaches, however, the 'rock' suddenly develops a capacious mouth, and the startled victim is abruptly swallowed. Human beings, deceived by the stonefish's camouflage, may fare little better: concealed along the fish's back are spines, whose venom can cause excruciating pain and sometimes death.

When threatened by a predator, the spiny puffer pumps water into itself until it looks like a thorn-covered ball. Faced with the impossibility of swallowing something like this, the attacker retreats. When taken out of water by fishermen, the puffer gives a similar performance, using air to inflate itself and so bristle its spines.

RUNNING FOR SHELTER *Barely 1 in. long and nearly defenceless, cardinal fish try to evade their predators by hiding in the branches of a colony of staghorn coral. Their normal daytime red colouring fades to silver so that they merge into the light grey of the staghorn. Vertical black bands round their bodies help to break up their outlines*

Camouflage is a life-or-death matter for many members of the coral-reef community, enabling them to elude their predators or dupe their prey. Both deceptive shapes and disguising shading can be effective; misleading behaviour may also help. The sabre-tooth blenny, for example, can imitate the dance of the harmless cleaner-wrasse to deceive its prey. Some fish have a permanent coloration that masks them against the coral reef. Groupers change colours, adjusting to their backgrounds. Certain crabs masquerade by attaching bits of shells or seaweed to themselves. Gobies and blennies have false eyespots to fool predators; stripes hide their real eyes. Long, thin trumpet fish sway in the current, posing as coral stalks as they await the appearance of crabs or other prey. Vertical black bands round their bodies make the tiny cardinal fish almost invisible as they swim among the branches of staghorn corals.

DANGER IN THE SAND *To take its prey unawares, the peacock flounder lightens its dark rings and overall body colour to match the sandy bottom of a Caribbean coral reef*

HIDDEN BY COLOUR *An irregular shape (above) and ability to change colour enable the scorpionfish to assume the appearance of algae-encrusted rocks on a South Pacific reef (left). Two spine-fringed fins can be seen near the centre of the picture; one eye is visible near the pale pink sponge on the left. The fish lies on the bottom and snaps up smaller fish and crustaceans as they pass. Venomous spines can inflict painful wounds*

EFFECTIVE DISGUISE *Barely visible against the bottom, a lizard fish gulps down a victim, tail first*

MATCHING THE SURROUNDINGS *Striped shrimpfish hover head-down among the spines of a giant sea urchin*

LIONFISH *Like other members of the scorpionfish family, the 8-in. lionfish is armed with sharp and highly venomous dorsal spines. It defies predators by flaunting aggressive candy-striped patterns and strikingly bold colours. Lionfish cruise along the bottom of tropical coral reefs, using their non-stinging pectoral fins to sweep smaller fish into their mouths*

COLOUR TO REPEL OR COMMUNICATE

Scientists are cautious about assigning reasons for the brilliant hues and patterns of reef dwellers. Presumably, each colour scheme has survival value, or the highly selective processes of evolution would not have produced it. However, the precise nature of this survival value is not always easy to determine.

No one seeing a stonefish can doubt the effectiveness of its extraordinary camouflage. But who can judge the extent to which predators are 'fooled' by a trumpet fish's imitation of a stalk of soft coral? And a more fundamental question can be posed: if concealing coloration carries such a high degree of survival value, why are there so many coral-reef fish with colour schemes that, far from concealing them, actually seem to advertise their presence?

Fish with prominent defensive weapons often use conspicuous colour as a warning to predators to keep their distance. On the lionfish's back is a row of brightly striped venomous spines. An animal that comes too close is likely to be severely stung and will be much less likely to come as close in the future. Similarly, the colour of a surgeon-fish's scalpels sometimes contrasts sharply with the rest of its body.

Bright colours frequently announce unpalatable flesh. Triggerfish – whose flesh is poisonous or at least highly distasteful to other species – are among the most boldly coloured reef fish.

Fish learn to avoid poisonous spines and flesh the hard way; they do not inherit such behaviour. Silversides are a favourite food of grey snappers on Caribbean reefs. In a laboratory study, silversides were coloured red and injected with an unpleasantly flavoured chemical. At first the grey

CONSPICUOUS BUT WELL ARMED *Brightly striped Pacific cat-fish may be conspicuous, but they are well equipped for defence, with highly venomous spines to protect them from predators*

snappers preyed on the red fish, but soon began avoiding them altogether. The snappers quickly learnt to associate red silversides with bad taste, and remembered this association.

Conspicuous coloration sometimes serves as a sort of safe-conduct pass. The 4-in. cleaner-wrasse eats parasites and dead tissue that it finds on other fish, including predators that normally snap up small fish. The wrasse's vivid blue-and-black stripes and its dancing movements signal to other fish that it is a friend rather than a meal. Even voracious predators such as moray eels submit to the picking and scraping of cleaner-wrasses.

Fish wishing to be cleaned may communicate by changing colour. The Atlantic surgeon-fish, approaching the station of a bluehead wrasse, changes from purple-brown to olive, which evidently tells the wrasse that it has come to be cleaned and will not use its sharp blades to attack and destroy the wrasse.

Some fish mimic the conspicuous coloration of other species. The sabre-toothed blenny can

WEAPONS IN THE TAIL *The surgeon-fish's conspicuous black-and-white blades contrast sharply with its sombre, mottled-grey body. Under threat it lashes its bladed tail to rip its enemy*

imitate not only the colour but also the invitational dance of one kind of cleaner-wrasse. If an unsuspecting fish approaches for a grooming, it gets instead of the expected cleaning a series of painful bites that chop away pieces of skin and fin. One quite edible filefish is carefully avoided by predators because its bright colours duplicate those of the poisonous four-saddle puffer.

Defence and mating

Many reef fish stake out territories and vigorously defend these feeding or breeding areas against members of the same species. Such fish are frequently marked with distinctive colour patterns that, in effect, serve as 'no trespassing' signs. While territorial defence may cause occasional bloodshed between individuals of the same species, it benefits the species as a whole by keeping its members spaced out so that they need not conflict with one another in the struggle to find their food.

Many fish assume special courtship colours to attract mates. As is common among birds, the male is often more colourful than the female. At mating time, the male redtail parrotfish has a gorgeous green and blue body and orange-red fins; the female is a uniform pale red with a touch of bright yellow on her tail fin. Male gobies, which attract females to their lairs by elaborate courtship dances, are brilliantly coloured during the breeding season.

In territorial fish, bright patterns that ordinarily keep individuals well separated may undergo changes at mating time. Some species lose their aggression-provoking colours during courtship and regain them afterwards. Others permanently abandon their gaudy livery and territorial behaviour on reaching sexual maturity. The blue angelfish and the beau gregory, however, do not change their colours at mating time; males and females are paired off permanently and jointly defend territories.

COLOURS THAT WARN 'KEEP OFF' *Vivid rainbow colours make the harlequin tusk-wrasse immediately recognisable by other members of its species and may warn against infringing its territory. The tusk-wrasse feeds on shrimps and small fish*

PINK-TAILED TRIGGERFISH *The gaily coloured tail of this triggerfish may be a warning to predators that its flesh is distasteful. Like other species of triggerfish it can lock its dorsal fin erect to wedge itself into a crevice while it awaits prey*

WHITE-SPOTTED TRIGGERFISH *The bold marking on the white-spotted triggerfish may help members of the species to identify each other. Flamboyant patterns may also serve to suggest to predators that the fish is prickly and difficult to eat*

RITUAL INHIBITS ATTACK *A squirrelfish allows a black-and-yellow cleaner-wrasse to scour its body. Normally the nocturnal squirrelfish feeds on smaller fish, but the prancing ritual of the approaching wrasse apparently stops any attack*

TINY CREATURES GROOM THE KILLERS

An amazing underwater partnership

Gatherings of various species of fish at certain locations on a reef once puzzled scientists. Now it is known that the fish are waiting to be cleaned. Some small fish and shrimps specialise in nibbling parasites and dead tissue from larger fish. The territories of these cleaners become established as 'cleaning stations', to which clients return at regular intervals. Many client fish normally feed on smaller fish and crustaceans; but bold colouring and a ritual dance identify cleaners to prospective clients and safeguard them from attack. Clients have their own ritual postures and motions – some even change colour – to signal their readiness to be cleaned or to end the process. Both sides benefit from the relationship. Cleaners are guaranteed an ample food supply, and clients are kept healthy. One small fish – the blenny – closely imitates the colour and movement of the cleaning fish but when potential clients approach it takes a bit of flesh or fin and darts away. Observers have found that adult fish are rarely deceived, but young inhabitants of the reef are often caught unawares

CLEANING A WHOLE SCHOOL *A cleaner-wrasse ministers to a school of goatfish. One goatfish, top right, indicates that it is ready by extending its fins and barbels. The cleaner-wrasse's tail is just below the hanging barbels*

READY FOR CLEANING *A coral trout opens its gill covers for a cleaner-wrasse. The fearless wrasses also enter their clients' mouths to search for parasites and dead tissue. When the trout wants to move off, it warns the wrasse by ritual movements*

MENACING BUT HARMLESS *A giant moray eel parts its jaws to welcome the attention of a cleaner-shrimp. The spider-like shrimp, waiting just to the right of the moray's head, removes small parasites and pieces of loose skin*

SAFE AMONG THE ANEMONES *A brightly patterned clownfish shelters from predators among the tentacles of sea anemones. The relationship is mutually beneficial, for clownfish keep sea anemones free of debris and nip off diseased tentacles*

SHRIMP AND 'LODGERS' *A snapping shrimp emerges from the burrow it shares with gobies. The shrimp makes the burrow and keeps it clear of debris; its partners do not appear to make any contribution towards the arrangement*

STRANGE REEF PARTNERSHIPS

One of the most fascinating aspects of the coral community is the vast number of partnerships that exist – in some instances between animals, in others between animals and plants. Scientists use the term *symbiosis* (meaning 'living together') to describe the associations, and use other, more specific terms for the various types of symbiotic relationships between different species.

Commensalism is a partnership that may benefit one or both organisms but is not essential to the welfare of either. *Mutualism* is a relationship in which two organisms are so truly interdependent that neither can live successfully without the other. Finally, *parasitism* is an association in which one partner benefits at the expense of the other organism.

In a great many cases, scientists simply do not know enough about the details of a partnership to give it an appropriate label.

Is the relationship between the coral polyps and the one-celled plants that live in their tissues an instance of mutualism or of commensalism?

In nature they are always found together, which would point to mutualism. In the laboratory, however, coral polyps can be made to release the plants; the corals will continue to grow, and the plants take up life as free-swimming dinoflagellates. This suggests commensalism. Such distinctions, of course, are largely academic, and the fact remains that in the reef world many creatures live together in strange and unusual ways.

Frequently, one creature takes up residence in the living body of another. The pearlfish, for example, spends its entire adult life in the lower digestive tract of a sea cucumber, only emerging from the anal opening at night to feed. Remarkably, the fish's comings and goings do not provoke the sea cucumber to eject its internal organs, even though this reaction is easily triggered by what seem to be lesser irritations. Some observers believe that the pearlfish gets board as well as lodging, supplementing its diet by eating the cucumber's easily regenerated internal organs. Some pearlfish use clams, oysters, starfish and

THE HOME THAT IS ALIVE *A slender pearlfish, with its spine and internal organs showing dark through a transparent body, wriggles tail first into its host, a sea cucumber. The cucumber derives no benefit from the arrangement; but the pearlfish not only has a lifetime's shelter from predators, it occasionally also feeds on the host cucumber's internal organs – which are naturally and quickly regenerated*

CREATURES OF A CORAL REEF

All reef dwellers depend not only on the coral, but on each other. This assemblage includes species from Caribbean and Indo-Pacific reefs, crowded together much more closely than they would be on a real reef. As innumerable coral polyps slowly build up the reef, aided by discarded shells and coralline algae, predatory clams and starfish rasp away at the limestone and destroy the polyps. The delicate, deadly tentacles of corals and sea anemones sway in the current. Starfish, snails and worms crawl over the reef; more mobile molluscs and fish flit in and out of the narrow crevices

1. Sea whip
2. Flamingo tongue
3. Basket star
4. Sea cucumber
5. Brain coral
6. Fanworm
7. Sergeant major
8. Encrusting sponge
9. Sea slug
10. Sea anemone
11. Cowry
12. Tube sponge
13. Giant clam
14. Coral shrimp
15. Starfish
16. Boxer crab
17. Sea fan
18. Damselfish
19. Staghorn coral
20. Parrotfish

other animals as homes. Another fish, the 2-in. conchfish, lives inside the queen conch snail.

Sometimes two different kinds of animals share the same burrow. In Indo-Pacific reefs, small gobies station themselves at the entrance to the sandy burrows of snapping shrimps. When an enemy appears, the fish dart inside. The shrimp may benefit by the warning; but the gobies also eat young snapping shrimps.

A number of creatures use the shells of dead animals for protection. Lacking the hard outer covering of their relatives, hermit crabs move into the shells of dead snails. As the crabs grow, they periodically 'trade in' their old shells for larger ones they have found.

Guarded by anemones

Some hermit crabs gain added protection by placing sea anemones on their acquired shells. The anemones profit from the association by feeding on scraps from the crabs' meals. When the crabs move to larger shells, they carefully transfer their anemones to the new homes. Anemones also protect boxer crabs, which brandish the stinging creatures in their claws.

Many other reef animals live under the protection of other organisms. Clownfish shelter among the venomous tentacles of sea anemones. These fish secrete mucus that inhibits the discharge of the anemones' stinging nematocysts, thus enabling them to live in direct contact with the tentacles. Possibly the anemones benefit from the presence of their tenants, which have been observed cleaning debris from the anemones and nipping off diseased tentacles. There is no doubt about the advantage for the fish. Conspicuously marked with bands of red-brown and pale blue, clownfish are quickly snapped up by predators when removed from the anemone.

Long, needle-like spines of sea urchins often serve as a protective thicket for a dozen or more thin shrimpfish. The fish hover in a vertical, head-down position, with their longitudinal stripes looking very much like urchin spines. One species of cardinal fish also seeks the sanctuary of spines when threatened, and repays the protector by cleaning its body.

These examples constitute only a minute sampling of hundreds of such associations known to exist in coral reefs. Indeed, the coral-reef community can be regarded as a huge, complex symbiotic system in which thousands of plants and animals live together, bound by an intricate web of interdependencies.

The open ocean

From outer space, the open ocean dominates the astronaut's view of Planet Earth. Little explored by man and inhospitable to human life, it covers an area almost three times greater than all the continents and islands

Beyond the shallow coastal seas lies the pathless immensity of the open ocean: over 300 million cubic miles of salt water covering nearly three-quarters of the earth's surface. Averaging more than 2 miles deep, the open ocean provides several hundred times the living space of all the continents and islands. On land, most life is concentrated in a relatively thin film between the tree-tops and the first dozen feet or so of the soil. In the ocean, on the other hand, life can be found from the sunlit surface waters to the bottom of the deepest known trench, almost 7 miles down.

The open ocean is mankind's last unexplored frontier. Its dark, near-freezing depths, where pressures mount to tons per square inch, are as inhospitable to human life as the dry 'seas' of the moon. Even the relatively accessible surface of the open ocean has its forbidding and mysterious regions, and some of them are still shunned by man.

Perhaps the most famous of these is the Sargasso Sea, lying in the mid-Atlantic between Florida and the Azores. Christopher Columbus was probably the first European explorer to sail through this mid-ocean sea and describe its floating weed. His crew became alarmed when they sighted the plants, believing that their ship was near shore and might run aground.

Over the centuries, exaggerated tales of the Sargasso Sea were accepted as fact. Early writers pictured the sea as a vast, impenetrable tangle of seaweed. According to popular accounts, any ship venturing there would be trapped for ever in an eerie graveyard of rotting ships and unspeakable monsters.

In 1855, Matthew Fontaine Maury, America's first oceanographer, described the Sargasso Sea: 'Covering an area equal in extent to the Mississippi Valley, it is so thickly matted over with gulf weeds . . . that the speed of vessels passing through it is often much retarded.' In 1897 a British magazine warned: 'It seems doubtful whether a sailing vessel would be able to cut her way into the thick network of weeds even with a strong wind behind her.' Even now lone sailors crossing the Atlantic in small craft plot their courses to avoid the ill-omened Sargasso.

The legend persists, although for many years ships have regularly navigated the Sargasso Sea. They sight plenty of weed – but rarely in patches larger than a door mat. Scientists estimate that there are 7 million tons of the weed floating in the mid-Atlantic, but the plants are spread across nearly 2 million square miles of the ocean's surface. Far from trapping a ship, the Sargasso weed would not even slow a dinghy.

HAMMERHEAD SHARK *A voracious predator that feeds on bony fish, rays and other sharks. It grows up to 15 ft long and has been known to attack human beings. The lobes of the animal's flattened head, with a nostril and eye at each tip, may act as steering planes to increase its agility and balance when searching for food*

MOTTLED PROTECTION *Camouflaged by its mottled colora-tion, a young sargassum triggerfish shelters among the tangled weeds. Its prickly dorsal spine snaps erect to deter predators*

SLENDER AND BRANCH-LIKE *The toothless pipefish resemble the branches of the weeds surrounding them. Unable to open their jaws, the fish suck in their tiny prey*

CREATURES OF THE SARGASSO SEA

A mid-Atlantic refuge and breeding ground

The Sargasso Sea, a vast expanse of brown sargassum seaweed floating in the mid-Atlantic, provides a refuge and breeding ground for about 16 species of animals. Dolphinfish and jacks migrate there to breed and eels from rivers in America and Europe make the long journey to lay their eggs in the depths. Many of the smaller fish – for example the sargassum pipefish, sea slugs and crabs – have a natural camouflage resembling the colours of the weed; but they are nevertheless the prey eaten by the larger porcupine fish, dolphin-fish and sargassum fish. The sargassum weed derives its name from the Portuguese word for a grape – a reference perhaps to the small air bladders that keep the plant afloat.

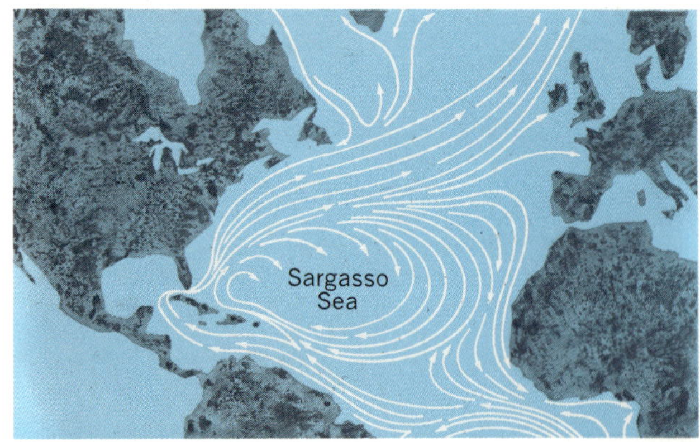

TRAPPED BY THE CURRENTS *The Sargasso Sea is an oval expanse of warm, seaweed-strewn water trapped by the major currents that flow clockwise around the Atlantic*

FISH THAT CAN CRAWL *The grotesque sargassum fish, whose bumpy skin mimics the fronds of the weeds, jiggles a fleshy lure over its mouth to attract prey. The fish crawls over the weeds on specially adapted, arm-like pectoral fins*

HIDING FROM PREDATORS *A $\frac{1}{2}$-in. sargassum crab clings to weeds beside beaded strings of sea slug eggs. Partly hidden behind the branch just to the right of the eggs is the well-camouflaged body of the sea slug that laid them*

SURVIVAL IN THE OPEN OCEAN

The sparsity of floating weed in the Sargasso Sea emphasises the true nature of the open ocean. Most of it is a biological desert incapable of supporting the dense populations of coastal seas.

The sunlit surface waters of the open ocean are generally poor in minerals and contain relatively low concentrations of the nutrients essential for plant growth. The deeper waters, where there is too little light for plant growth, are rich in nutrients, but only in a few areas do they become mixed with the surface layers.

Thus plant production is much lower in the open ocean than in coastal waters fertilised by minerals washed down from land or brought up from the deep by upwelling currents. Plant production in the mid-South Pacific, for example, is only one-sixth of that recorded in the sea off parts of Saudi Arabia and western Africa.

Despite its low fertility, the open ocean provides a more stable and less demanding environment than, for example, the seashore. Sudden or severe fluctuations in temperature or salinity do not occur. And even violent storms have little effect a few yards beneath the surface.

But inhabitants of the clear, well-lit surface waters face a threat to survival that does not confront most seashore creatures: a total lack of hiding places. The animals of the open ocean cannot dart into a rock cranny or burrow into a sandy bottom. There are no sheltering kelp meadows, no coral citadels. Yet the open-ocean inhabitants are hardly defenceless. Several of the larger drifters, such as jellyfish and their relatives, bear nematocysts – stinging capsules that ward off potential enemies. Other forms rely on superior speed to elude predators; flying fish and squids even take

COMB JELLY *The transparent comb jelly uses its eight rows of hair-like strands – called cilia – to paddle through the water. The jellies, close relations of jellyfish and sea anemones, gather in swarms and eat fish fry and plankton*

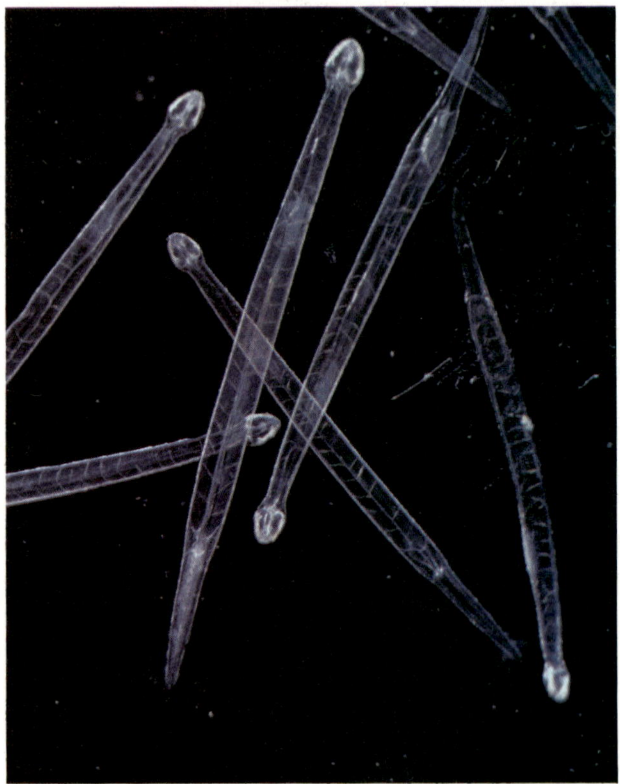

SEE-THROUGH WORMS *Transparent arrowworms are not easily seen by the small crustaceans and fish on which they feed. Many species of arrowworm sink to deeper water by day and rise to feed near the surface during the night*

FLOATING HOME *A shrimp-like crustacean with huge claws, the amphipod Phronima lives inside a transparent barrel which it makes by hollowing out the body of a salp. Phronima spends its whole life and breeds its young within the barrel*

to the air. Some fish find safety by banding together in large schools.

Nevertheless, the odds against survival in the open ocean are tremendous. Many species produce prodigious numbers of eggs to compensate for the high mortality rate. A female ocean sunfish may discharge as many as 30 million eggs in a single breeding season; if just two of the hatchlings reproduce, the maintenance of the population is assured.

Invisible creatures

With no hiding places, camouflage plays an important role in open-ocean survival. The larger creatures – from sea slugs and snails to the biggest sharks and whales – have blue or green backs and silver or white undersides. Many planktonic creatures – copepods, comb jellies and larval fish – achieve an astonishing degree of invisibility by being nearly transparent. Arrowworms are clear and colourless and live in surface waters throughout the world. One species grows to 4 in. in length.

An arrowworm often hangs motionless and unseen in the water, waiting for a copepod or young fish to come within striking distance. It detects its prey through the hairs on its surface which sense any nearby vibrations in the water. Once a meal is detected, the creature shoots forwards with startling speed and grabs the victim with the hard, pincer-like bristles around its mouth.

Transparent as arrowworms and almost as abundant, comb jellies are also carnivorous. They propel their gelatinous bodies by rhythmically beating hair-like cilia arranged in eight rows. By day the comb jellies are almost invisible, except for a glassy iridescence when the cilia reflect the sunlight; at night they shine with flashes of blue-green luminescence. Most comb jellies catch their food with long, trailing tentacles bearing 'lasso' cells, whose sticky coating traps prey.

Like comb jellies and arrowworms, the delicate salps are nearly as insubstantial and transparent as the water they inhabit. Relatives of the bottom-dwelling sea squirts, salps move freely through surface waters, feeding on plankton. The most common varieties are shaped like kegs open at both ends. Muscular 'barrel hoops' rhythmically expand and contract, pumping water in the front opening and out the back, and propel the animals through the sea.

Individual salps vary in length from $\frac{1}{2}$ in. to about 6 in., depending on the species. But colonies of salp-like animals, known as pyrosomes, have been reported to reach a length of 8 ft. Common in all oceans, pyrosomes seem at first sight to be huge solitary salps. But each pyrosome is really composed of hundreds or thousands of tiny individuals joined together to form a hollow cylinder. The individuals face the exterior surface of the cylinder and suck in water and food. The water is discharged into the centre of the cylinder and flows through the end, moving the colony along with a sort of gentle jet propulsion.

THE STRUGGLE TO STAY AFLOAT

All plants and animals living near the surface of the open ocean must somehow avoid sinking into the abyss. To carry on photosynthesis, using the sun's energy to manufacture food, plants must stay within a few hundred feet of the surface. Many animals, in turn, must remain near the drifting single-celled plants that form the first link in the ocean's food chain.

A few seashore animals manage to find a place of attachment even in the open ocean. Barnacles fasten to driftwood, ships' hulls, large fish, turtles and whales. Whales carry a wide variety of external parasites including barnacles, algae, remoras, copepods and some 18 different kinds of 'whale lice'.

Floating clumps of weed in the Sargasso Sea provide a habitat for a unique community of animals. Shrimps, crabs, tubeworms, sea squirts, nudibranchs, barnacles and sargassum fish live in the weed.

These hitchhikers are exceptions, however. Most open-ocean organisms, with nothing to cling to, must depend on built-in adaptations to stay afloat. Many planktonic plants and animals possess needle-like spines, fine hairs, or wing-shaped structures that add little weight but increase resistance to sinking. Just being small is an advantage. The smaller an organism, the greater the ratio of its friction-producing surface area to its body volume, and the more slowly it sinks.

But adaptations that increase friction can only slow down the process of sinking; they cannot prevent it. Even the tiniest plants and animals must somehow counteract gravity in order to stay in the sunlit surface waters. Diatoms contain oil globules that reduce their weight. Microscopic radiolarians buoy themselves up with bubble-like chambers full of water and carbon dioxide. Dino-flagellates 'tread water' by lashing their whip-like flagella. Copepods kick their legs and antennae back and forth as fast as 600 times a second.

Floating jellies

A creature perfectly adapted to a drifting life should weigh only as much as the water it displaces, so that it neither rises nor sinks. Jellyfish, with tissues that are 95 per cent water, come close to this ideal.

A jellyfish's umbrella-shaped bell is a thin, double-walled sac filled primarily with a gelatinous substance. In the centre of the bell is a digestive cavity, and hanging from it like a tattered curtain is the creature's funnel-like mouth. Simple sense organs responding to light or gravity help to keep the animal in an upright position. Rhythmic

THOUSANDS OF MOUTHS *A floating jellyfish draws in minute planktonic prey through thousands of mouths on its hanging lobes. Canals from the lobe openings lead to its stomach. The fish's fringed dome-like bell may be 12 in. wide*

contractions of the bell give the jellyfish a slight upward push to keep it from sinking.

Trailing beneath the bell, tentacles studded with nematocysts paralyse small animals that come into contact with them. Tiny adhesive pads hold the victims until the nematocysts have delivered their poisonous darts. The paralysed prey is then hoisted into the jellyfish's mouth and digested.

One of the most widely distributed jellyfish, the moon jelly, has a transparent bell, 6–18 in. in diameter, which is made conspicuous by hair-like tentacles fringing the edge, and four oval reproductive organs on the underside. The moon jelly preys on copepods and other tiny animal plankton, but the sting of its short tentacles is harmless to humans. In addition to tentacles, it has sticky bands of mucus on the surface of its bell. The mucus, together with any food that has become stuck to it, is constantly moved towards the edge of the bell by the beating of the cilia. There, accumulating blobs of food are licked off by tongue-like extensions of the central mouth.

Considerably more dangerous, to man as well as to other animals, is the lion's mane jellyfish. In the Arctic Ocean, one very large species attains a bell diameter of 8 ft and has tentacles that extend downward for 100 ft or more – making it the largest planktonic animal in any ocean. The lion's mane jellyfish inhabiting temperate latitudes of both the Atlantic and Pacific reaches bell diameters of from 1–3 ft and may trail tangles of tentacles 75 ft long. Its sting can kill a foot-long fish and raise angry red welts on human swimmers.

For sheer virulence, however, no jellyfish can equal the sea wasps of the tropical Pacific and Indian Oceans. These creatures have dome-shaped bells that may grow to a height of 10 in. or more, but are usually considerably smaller. From the bottom of the bell hang clusters of relatively short tentacles, the sting of which can kill a man. An American marine toxicologist, Bruce Halstead, has called one species of sea wasp the most deadly organism·alive.

Brilliantly luminescent and beautifully tinted with delicate shades of rose, lavender and ivory, *Pelagia noctiluca* is adapted to life in the open ocean. After budding from the body of an adult, most young jellyfish settle on the ocean floor, gradually grow into saucer-shaped larvae, and then float to the surface. *Pelagia noctiluca*, however, discharges its larvae directly into the open sea, where they mature without going through a floor-dwelling stage.

MOON JELLYFISH *The four white circular reproductive organs of the moon jellyfish are visible as it swims by contracting its shallow bill. Larvae settle on seaweed or rocks to develop into fixed polyps, then into mobile adults*

STINGING KILLER *A tropical sea wasp is one of the most lethal creatures of the sea. It is a jellyfish with stinging tentacles that kill almost on contact: human swimmers have been victims. Sea wasps feed mainly on small fish*

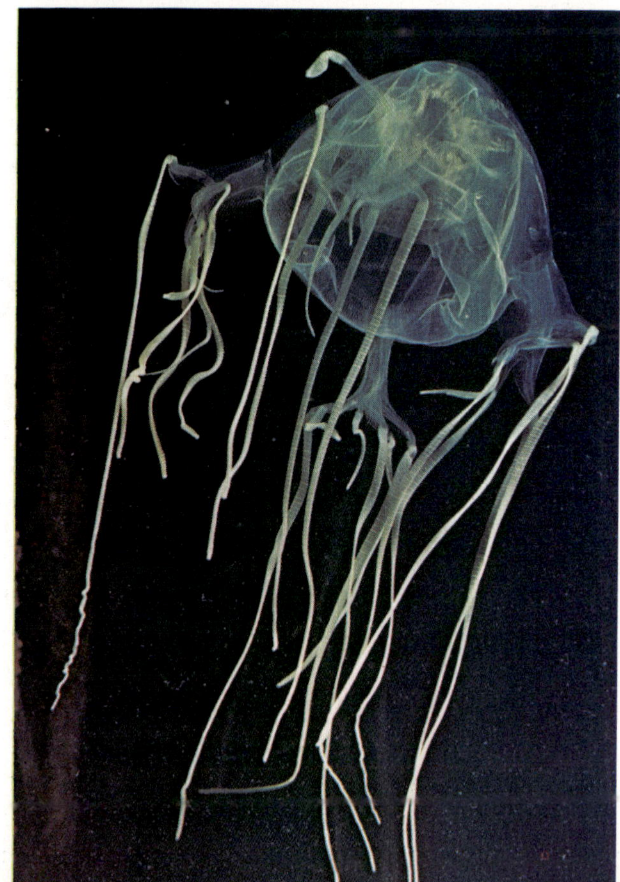

Clustering for survival

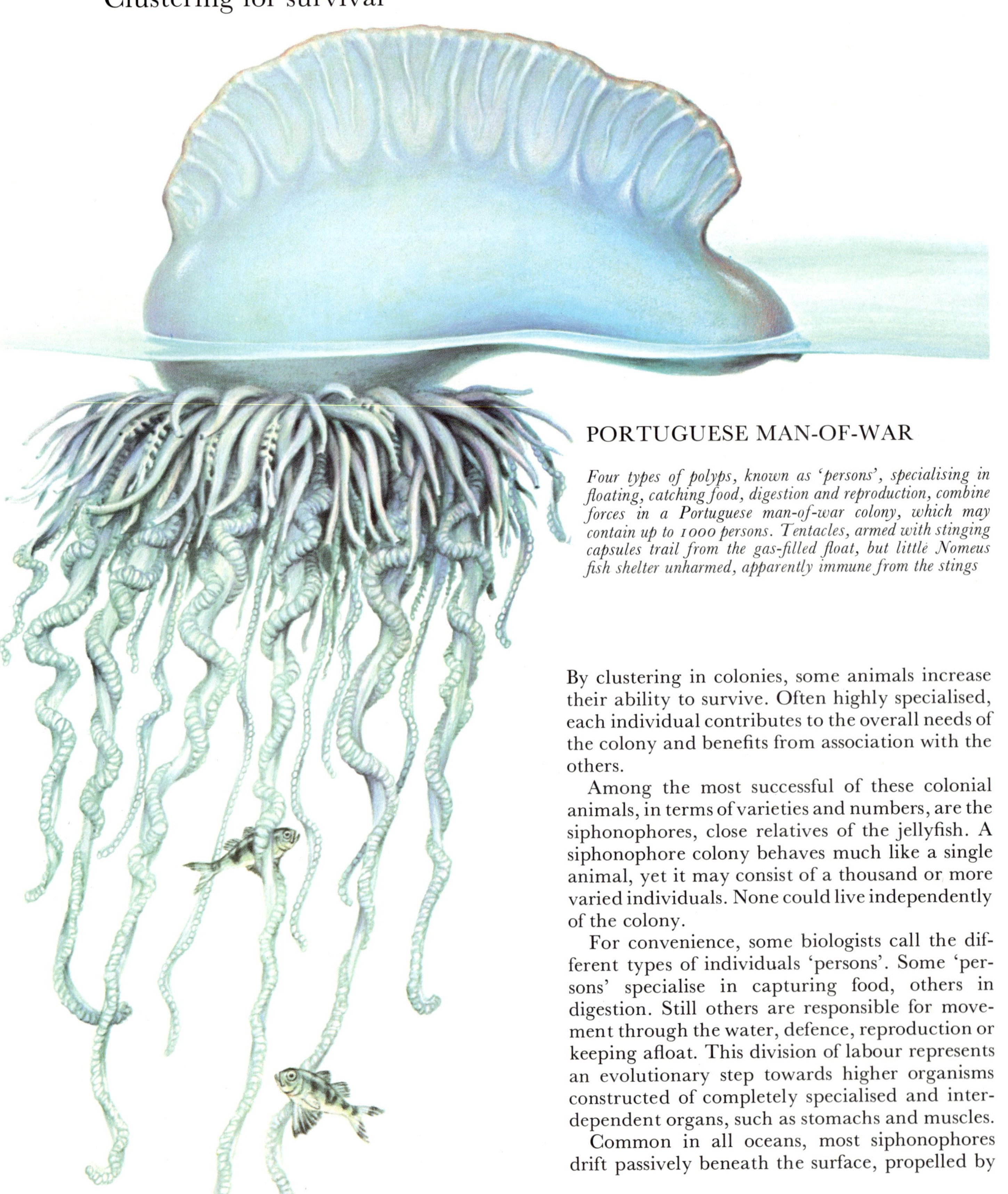

PORTUGUESE MAN-OF-WAR

Four types of polyps, known as 'persons', specialising in floating, catching food, digestion and reproduction, combine forces in a Portuguese man-of-war colony, which may contain up to 1000 persons. Tentacles, armed with stinging capsules trail from the gas-filled float, but little Nomeus fish shelter unharmed, apparently immune from the stings

By clustering in colonies, some animals increase their ability to survive. Often highly specialised, each individual contributes to the overall needs of the colony and benefits from association with the others.

Among the most successful of these colonial animals, in terms of varieties and numbers, are the siphonophores, close relatives of the jellyfish. A siphonophore colony behaves much like a single animal, yet it may consist of a thousand or more varied individuals. None could live independently of the colony.

For convenience, some biologists call the different types of individuals 'persons'. Some 'persons' specialise in capturing food, others in digestion. Still others are responsible for movement through the water, defence, reproduction or keeping afloat. This division of labour represents an evolutionary step towards higher organisms constructed of completely specialised and interdependent organs, such as stomachs and muscles.

Common in all oceans, most siphonophores drift passively beneath the surface, propelled by

BY-THE-WIND SAILOR

This relative of the Portuguese man-of-war is known as a by-the-wind sailor. Like the man-of-war, it is not a single animal but a colony of individual polyps, known as 'persons'. Stinging persons, which fringe the central disc, numb its prey. The large mouth, below the disc, is surrounded by reproductive persons. The air-filled buoyancy chambers are shown in the cutaway view (below)

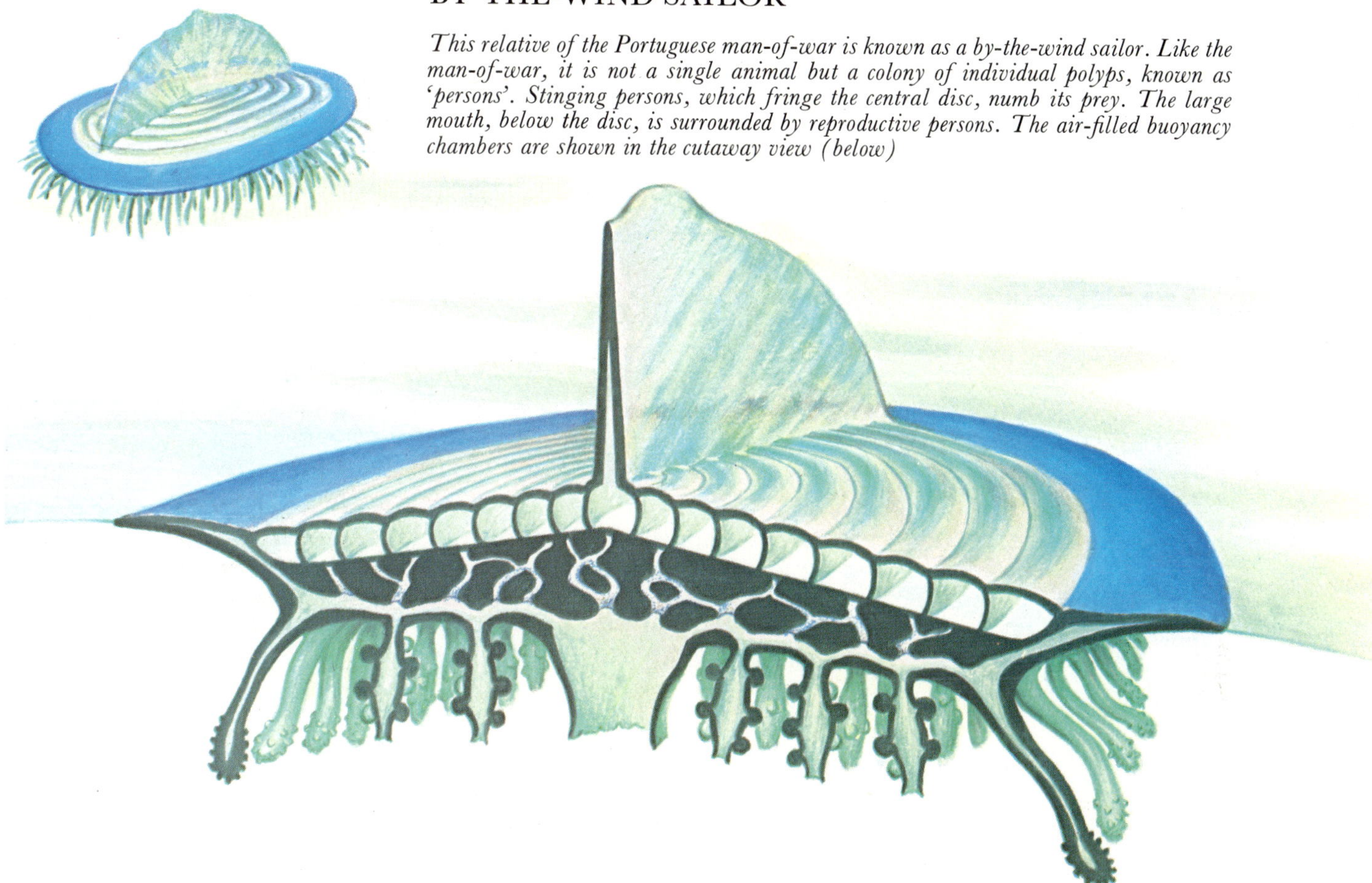

several gently pulsating swimming bells. But one of them, the Portuguese man-of-war, has a gas-filled float that reaches above the surface and catches the wind. Sometimes men-of-war gather in huge flotillas, stretching for miles.

The Portuguese man-of-war's balloon-like float, tinted pink, purple and blue, and crested with a narrow, frilled sail, may measure nearly a foot in length. A gas-secreting gland at its base inflates the float. Every few minutes, the man-of-war momentarily 'capsizes' itself to moisten the delicate float membrane. Beneath the float, stinging tentacles trail through the water for 30 ft or more. The venom they eject, chemically similar to that of the cobra, can kill fish the size of a fully grown mackerel. As the tentacles contract, the prey is hoisted up to the many mouths of the feeding 'persons', where it is digested and distributed to the entire colony.

For human beings, any contact with the tentacles – even dried, disintegrated ones mixed with beach sand – causes a painful, nettle-like rash. A swimmer encountering a living man-of-war can receive disabling or fatal injuries. Man-of-war stings have been blamed for a number of human deaths, but some may not have been caused by the venom itself. A swimmer entangled in the tentacles could react with panic and drown.

The sailor's companion

Curiously, one animal is a constant companion of this deadly sailor. *Nomeus*, the man-of-war fish, spends most of its life swimming with apparent impunity among the toxic tentacles, and comes inshore only when the man-of-war drifts in. The fish is deep blue with vertical black stripes and it blends in well with the siphonophore's tentacles.

Just how a finger-length man-of-war fish manages to stay alive in its lethal stronghold is not clearly understood. It nibbles the man-of-war's tentacles, and possibly this diet partially immunises it to the venom. In laboratory experiments the fish has survived venom injections ten times as potent as those that kill fish of similar size. On the other hand, men-of-war have been caught in the act of devouring their little companions.

MOLLUSCS OF THE OPEN OCEAN

The molluscs – soft-bodied animals characterised by a protective shell and a large fleshy foot – have spread out in several directions from the shallow seas of their ancestors. Some species have migrated to dry land or fresh water; a few have taken up a free-floating life in the open ocean.

The pteropods – modified snails – have developed wing-like extensions of the foot and have drastically lightened, or altogether eliminated, the shell. Popularly called sea butterflies, these dainty creatures flutter along through the water. Cilia on their 'wing' surfaces drive diatoms and other small organisms toward their mouths.

Abundant in the open ocean, pteropods often form dense shoals that attract hungry herring, cod and haddock. In parts of the South Atlantic they are present in such enormous numbers that the frail, cone-shaped shells of dead pteropods cover large areas of the ocean bottom.

Unlike its pteropod relatives, the purple sea snail has retained a relatively large shell, but compensates for this burden by secreting a raft of mucus-covered bubbles to which it clings upside-down. This upside-down life has led to a reversal of the typical open-water colour pattern: what would normally be the snail's upper surface is white, and what would be its lower surface is deep purple-blue. Found in all warm seas, the purple sea snail eats planktonic animals.

Swimming slugs

The shell-less nudibranchs, or sea slugs, have contributed the swimming slug, *Glaucus*, to the open-ocean community. Its relatives creep over the bottom in tide pools and coral reefs, but *Glaucus* swims upside-down at the ocean's surface. Since it spends much time upside-down, the slug has evolved an inverted colour scheme similar to that

IMMUNE TO POISON *Floating upside-down at the surface of the ocean, the swimming slug Glaucus feeds on a siphonophore Porpita. The deep blue slug has become immune to the poisonous stinging polyps of the siphonophores and incorporates them in its own tissues as a defence against predators. After feeding, it may leave its eggs on the derelict float*

of the purple sea snail. *Glaucus* also feeds on by-the-wind sailors. After devouring their tentacles and feeding polyps, *Glaucus* may leave behind a clutch of its own eggs on the derelict float.

The squid is the most highly evolved of the sea molluscs and of all the ocean invertebrates. Like its close relative the octopus, the squid has eyes which look almost like human eyes, a powerful parrot-like beak and high intelligence. The heavy shell of its snail ancestors has been reduced to a thin, leaf-shaped internal 'shell', adding little weight but providing rigidity. The cumbersome ancestral foot has evolved into a streamlined 'head' and ten sinuous, sucker-laden arms.

By forcibly expelling water through a narrow siphon beneath its head, the squid jet-propels itself faster over short distances than the fastest fish. Usually the squid darts backwards. But it can reverse direction by shifting the position of its siphon, as it does when it moves back and forth in a school of fish. For more leisurely movement, or for hovering, the squid slowly flaps two triangular fins near the rear of its body.

Male and female squids meet to mate

THE SWIFT SQUIDS

The sharp-eyed squids are among the most agile animals in the oceans. Using jet propulsion, they squirt water from a siphon to speed through the sea, sometimes with such force that they rocket out of the water into the air. They can change direction by moving their siphons. Some squids emit inky or phosphorescent liquids to confuse an enemy; others rely on speed. After mating, females deposit their membrane-enclosed eggs on the ocean floor

SEA SNAIL *A purple sea snail clings upside-down to a frothy raft of mucus bubbles. When attacking, it squirts out a purple fluid that seems to anaesthetise its planktonic prey*

Young squids, almost ready to hatch, show their eyes, black ink sacs and pear-shaped yellow yolks

THE PERFECTION OF A FISH

The structure and streamlining of a fish are the culmination of 400 million years of constant adjustment, selection and refinement. What has emerged is a marvel of dynamic beauty and supple grace. Man is slow and awkward on land compared to a fish in water.

More than 95 per cent of the 25,000 species of fish have hard, bony skeletons. Among these bony fish are the most versatile feeders and over long distances they are the fastest swimmers in the sea. Sharks and rays – fish with skeletons of cartilage – are sometimes big and fast, but they do not exhibit the variety of forms, habits and adaptations found among their bony cousins.

To swim, a fish sends waves of muscular contractions down its body from head to tail, pushing its curving sides backwards against the water. A sweep of the tail ends each wave and gives an added push.

The fins of a bony fish are used primarily for orientation and balance. The dorsal fin on the back and the anal fin on the belly are keels that keep the fish from wobbling. The caudal, or tail, fin adds graceful smoothness to the swimming motion; a fish with an injured tail fin 'staggers' erratically. The pectoral fins just behind the gill openings make it possible for fish to slow down suddenly and turn sharply. The pelvic fins, set below and sometimes behind the pectorals, are also for braking.

Some fish use their fins for other purposes. Ocean sunfish, with their ponderous, rigid bodies, cannot swim conventionally, and propel themselves forwards by sculling with their dorsal and anal fins. Remoras attach themselves to larger fish with a dorsal fin modified as a suction cup on top of the head. In angler fish the dorsal fin has become a living fishing rod, often complete with a luminous bait.

Fins make possible the spectacular aerial excursions of flying fish. Among 'two-winged' species, the greatly enlarged pectoral fins can stretch out

STREAMLINED AND COMPACT DESIGN

Most bony fish are streamlined to move swiftly under water. The perch is typical, with its reproductive, digestive and excretory organs tightly packed into a muscular body. Oxygen is extracted from the sea-water as it flows over the gills; the gas-filled swim bladder controls the buoyancy of the fish and regulates internal pressures. A lateral line system senses water flow so that the dorsal and ventral fins balance the fish as the tail fin steers it

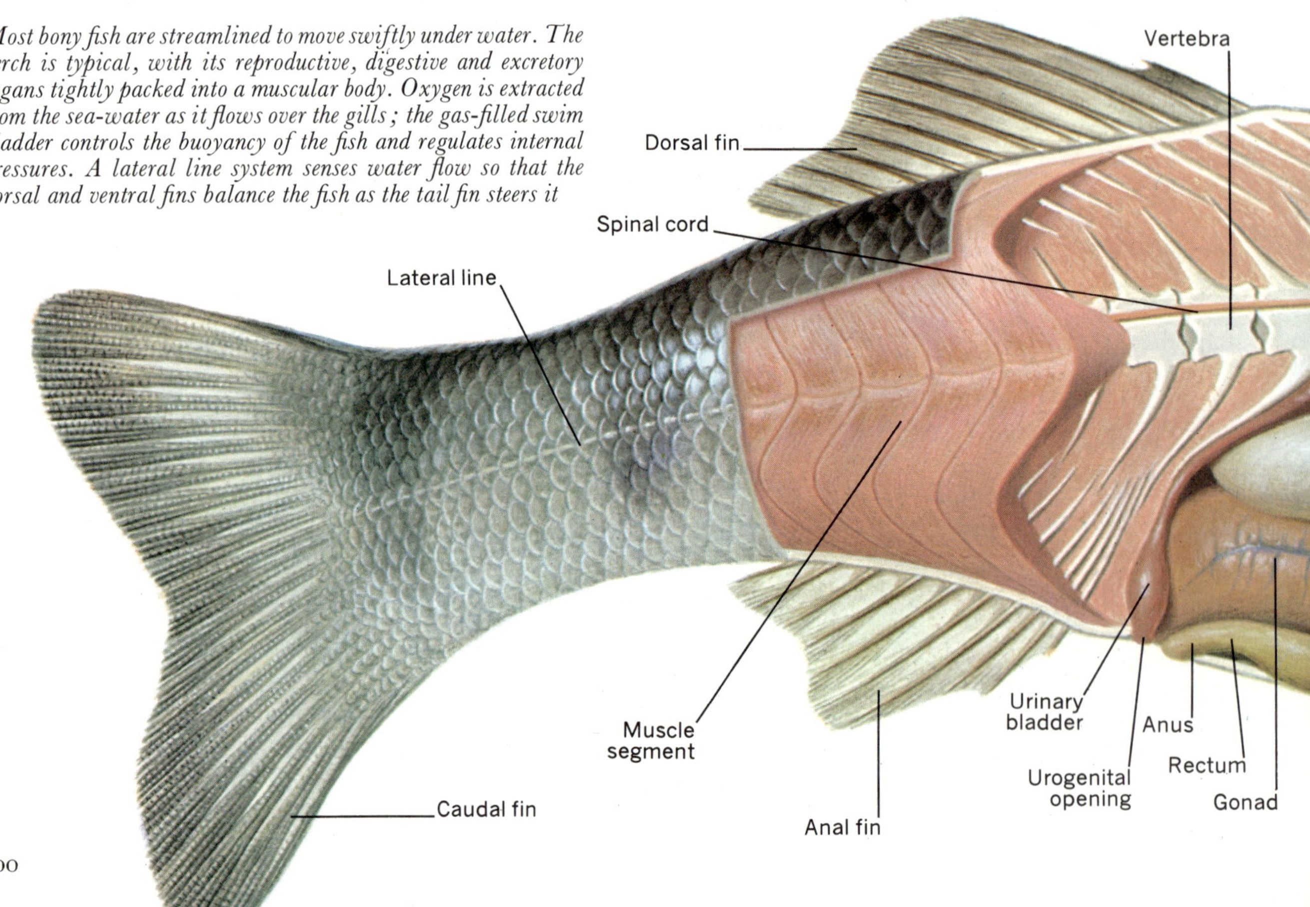

rigidly to form gliding planes. In the 'four-winged' species, the pelvic fins are similarly modified. In both types, the lower part of the tail fin vibrates rapidly to get the fish airborne. Once in the air, a flying fish can soar for 150 ft or more, at speeds of up to 40 m.p.h.

Most bony fish that feed in the open ocean have mouths at the fronts of their bodies, usually with lower jaws that jut out further than the uppers, so that they can grab a meal from any position. Some fish, such as the John Dory, have jaws which can shoot forwards as the mouth opens. When the prey is seized, the jaws quickly retract.

The aggressive swordfish

The upper jaws of marlin and sailfish are drawn out into rounded bills for clubbing other fish. Swordfish have broad, flat bills. Flailing left and right, they swim into schools of mackerel, menhaden, herring, small fish and squid. Then these toothless marauders swallow their beaten victims whole. They will attack almost anything, often without apparent purpose. Numerous broken-off swords have been found in the planking of wooden ships: they can penetrate through two thicknesses of sturdy oak timbers. In 1967 a 9-ft swordfish attacked the United States research submarine *Alvin* at a depth of about 1900 ft off the Georgia coast. The entire length of the 38-in. sword penetrated the outer fibreglass hull of the vessel, which was forced to surface. The crew extracted the trapped fish and ate it for dinner.

Swordfish also apparently attack whales. Broken-off swords have been found in the sides and backs of blue whales and fin whales captured in the Antarctic Ocean, around the Aleutian Islands and off Newfoundland.

Fish which prey on other fish 6 in. or more in length possess sharp, slender teeth for grabbing and holding their meal. Barracudas have both long 'canines' for seizing victims and smaller, dagger-like teeth that slice up victims as cleanly as a fishmonger's knife. Lancet fish, which may reach 6 ft in length, have curved teeth the size of a man's finger. Stout teeth, $\frac{1}{8}-\frac{1}{4}$ in. long, arm the jaws of bluefish. Schools of bluefish often attack schools of herring or other small fish, leaving behind nothing but tiny fragments of bone and flesh. These killers have no known natural enemies which regularly feed on them.

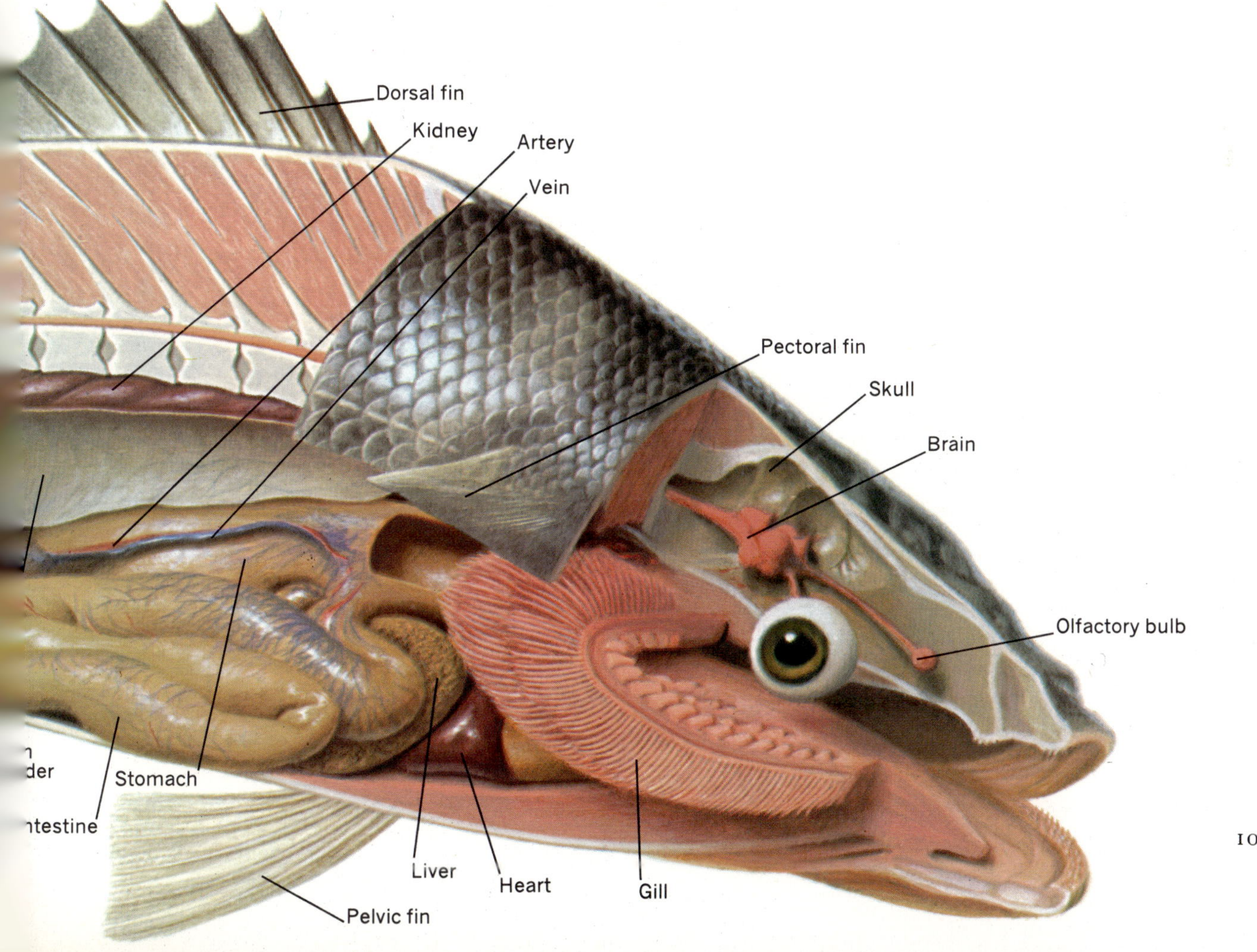

THE TUNA FAMILY

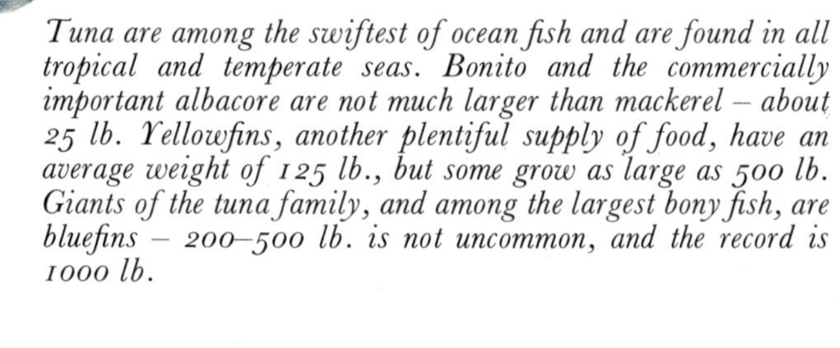

Tuna are among the swiftest of ocean fish and are found in all tropical and temperate seas. Bonito and the commercially important albacore are not much larger than mackerel – about 25 lb. Yellowfins, another plentiful supply of food, have an average weight of 125 lb., but some grow as large as 500 lb. Giants of the tuna family, and among the largest bony fish, are bluefins – 200–500 lb. is not uncommon, and the record is 1000 lb.

A MARLIN TAKES OFF *In an attempt to break loose from a fisherman's hook, a marlin bursts from the water. These fish, armed with lethal spikes, can swim at about 40 mph*

The swiftest fish

No bony fish range more widely through the open ocean than the tunas. Being heavier than the water they displace, tunas and their smaller relatives – mackerel, bonitos and albacore – must swim constantly to remain afloat. If they stop, they drift tail-first towards the bottom. To get enough oxygen-rich water flowing across their gills, they must swim fast. Their deeply forked, bone-hard, sickle-shaped tails drive them through the sea with vigorous side-to-side strokes.

No fish can swim faster than the fish of the tuna group. Everything about their appearance indicates speed: the beautifully streamlined and satin-smooth body; the gill covers tightly fitted against the sides; the fins lying retracted in special grooves where they offer minimum resistance. Yellowfin tuna have been known to swim at 45 m.p.h., and the wahoo (a tuna relative) at 48 m.p.h. – probably faster than any other fish. At

PERCH AND SUNFISH *A pair of perch clean parasites off the skin of an ocean sunfish, which signals them to begin eating by tilting itself backwards. A fully grown sunfish may reach a length of 10 ft and weigh more than a ton. A slow swimmer, lacking pelvic fins, the sunfish floats sluggishly at the surface. Tough gristle beneath the skin protects it from predators*

the sight, smell or sound of food, tunas accelerate quickly and reach top speed in less than a second. Their appetites are boundless and their diet versatile: herring, mackerel, anchovies, sardines, flying fish and squid are all gulped down as a tuna speeds through the water.

Tunas maintain a body temperature of 25–30°C (77–86°F) in sea temperatures from 10–30°C (50–86°F). This adaptation speeds up metabolism and so contributes to the tunas' speed and strength – their rates of digestion are high, their nerve impulses travel faster through their bodies, and their massive swimming muscles contract and relax about three times as rapidly as those of other fish.

The slow sunfish

Contrasting sharply with the swift tuna, the ocean sunfish passes much of its adult life lying on its side at the surface, apparently basking in the sun. Lacking a true tail, the languid sunfish swims so slowly that it might almost be considered a planktonic drifter despite its enormous size. Several specimens exceeding 10 ft in length and over a ton in weight have been caught.

Feeding primarily on jellyfish and other drifters, the ocean sunfish seems to have few predators. A 2–3 in. layer of tough gristle just beneath the skin protects the fish from many enemies, but parasites take their toll. Virtually every specimen examined has been riddled with a variety of parasites. Remoras even live inside the gill chambers of the ocean sunfish.

Recent investigations suggest that this creature may not deserve its reputation for laziness. The stomach contents of dissected sunfish sometimes include animals found only in deep waters, indicating activity far below the surface. Some scientists who have studied the fish contend that individuals at the surface are diseased, and that far from enjoying the sun, they are actually dying.

THE SHARKS AND RAYS

Despite 2000 years of accumulated lore and observation, sharks are among the least-understood of marine animals. Like snakes and octopuses, they arouse irrational emotions in humans. But dispassionately viewed, sharks must be ranked among the most beautifully designed, superbly functional animals in the sea.

Several characteristics distinguish sharks from other fish. They have no true bones; their skeletons are made of cartilage, a flexible, gristly substance. They have five to seven gill openings on each side, instead of just the one gill opening characteristic of bony fish. Their tough skin is covered not with scales, but with tiny tooth-like projections called denticles. Many sharks give birth to live young, whereas most bony fish lay eggs into the sea.

Because they lack true bones, sharks have left few fossils, but in the 1800's the remains of a shark that lived some 350 million years ago were discovered near Cleveland, Ohio. It is now clear that sharks are one of the oldest forms of life; they have changed little in the last 65 million years. Their simple and efficient body plan and several special adaptations account for their durability.

Reproduction is one of these adaptations. Male and female sharks mate directly – unlike most bony fish, which discharge sperm and eggs into the water. Direct mating is an effective mode of reproduction in the open sea, for the eggs are almost certain to be fertilised. In most sharks, the fertilised eggs hatch in the female's body, and the young remain for a time in the mother's reproductive tract, where they are nourished by a milky fluid.

A few species of sharks have a process of reproduction similar to that of mammals. These sharks have a placental connection with the young in the reproductive tract, and an 'umbilical cord' that supplies the embryos with nourishment from the mother's blood. The blue shark, the hammerhead shark and the smooth dogfish reproduce in this way. Only a few species, such as the whale shark, the cat shark and possibly the Greenland shark, lay eggs in the water. A leathery capsule sometimes known as a 'mermaid's purse' envelops their eggs.

Compared with bony fish, which may produce thousands of offspring each breeding season, sharks bear few young. A blue shark, for example, gives birth to about 30 young, which are born about 2 ft long with a full set of teeth, ready to fend for themselves. Sharks will eat anything, including

HUNTERS AND KILLERS OF THE OCEAN *Sleek, streamlined sharks, their mouths set with razor-edged teeth, are the ocean's most efficient hunters and killers: they eat anything, including their own young. The great white shark – which may grow up to 40 ft long – is certainly a man-eater, and the smaller species – the blue shark and the white-tipped shark – may also attack human beings. The rare frilled shark and the thresher shark are not dangerous to man*

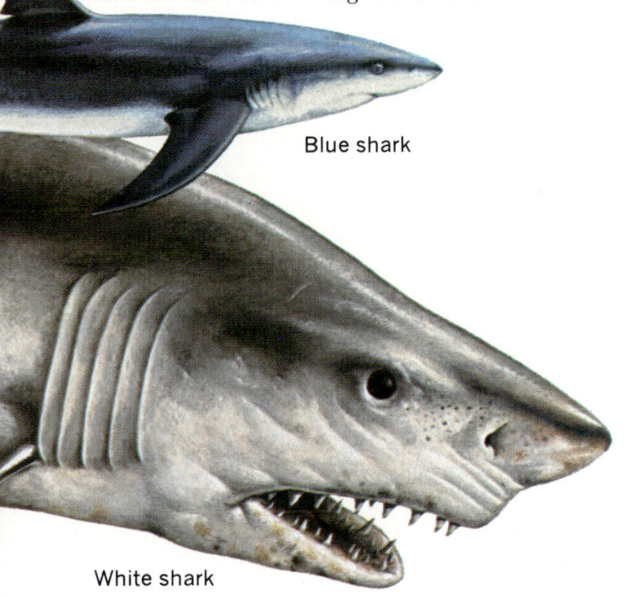

their own young, but during the season when the young are born the males leave the schools and females cease to feed, subsisting on oil in their livers.

Tell-tale teeth

Shark teeth vary so much in shape and size that a species can be identified by one or two teeth left embedded in a wooden boat, or by tooth marks left on a victim. The teeth are in rows, one behind the other. Sometimes the teeth in the first two rows are set at angles to each other, like the teeth of a crosscut saw. When a tooth is lost, another one moves up to take its place from the five or more rows of replacement teeth. New rows of teeth are always developing. In some species an entire row moves up to replace a worn row.

Most sharks prefer to feed on fish, but they will eat almost anything, including squids, seals, birds, sea turtles, smaller sharks, dead or injured whales, garbage and, on occasion, humans. The stomachs of captured sharks have contained the head and forelegs of a crocodile; a whole reindeer, minus the antlers; coal; unopened tins of food; grass; broken clocks; empty beer cans; and parts of cows, pigs, sheep and dogs.

AGGRESSIVE SHARK

A white shark, jaws filled with seven rows of sharp, serrated teeth, swoops on an underwater photographer. The man-eating fish, weighing more than 3 tons, can swallow creatures half its size and often devours them intact. When the shark loses one of its front teeth, which stand 2 in. out of the gum, another from the row behind moves forward and a new tooth is grown at the back. The killer is found in all parts of the world and feeds mainly on seals and sea lions

QUIET GIANT *A whale shark cruises near the surface of the sea, its chin festooned with clinging sucking fish, called remoras. Whale sharks, which are found in tropical seas and occasionally in more temperate zones, may grow up to 45 ft long. Despite their size, they do not attack humans; they live mainly on small planktonic creatures in the water*

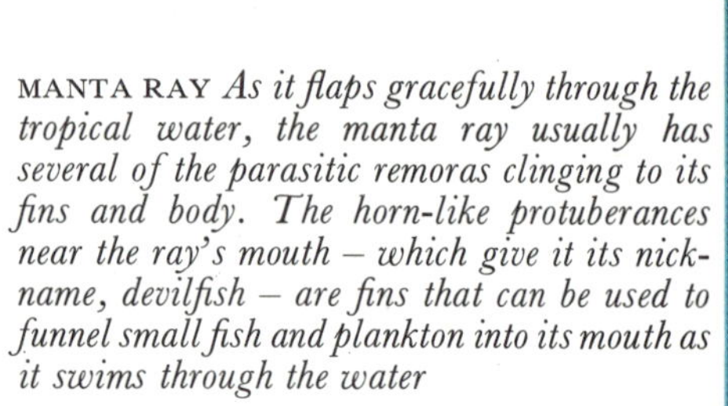

MANTA RAY *As it flaps gracefully through the tropical water, the manta ray usually has several of the parasitic remoras clinging to its fins and body. The horn-like protuberances near the ray's mouth — which give it its nickname, devilfish — are fins that can be used to funnel small fish and plankton into its mouth as it swims through the water*

Largest but least dangerous of the oceans' giants

Incongruously, the largest sharks – whale sharks and basking sharks – are plankton feeders. Whale sharks reach lengths of up to 45 ft and weigh 15 tons; basking sharks are only slightly smaller. Both species cruise slowly at the surface, filtering food from the sea with hairy mats, or gill rakers, which lie inside their long gill openings. Neither of these sharks has the formidable teeth of the flesh eaters, but they have attacked boats when provoked.

The fearsome devilfish

Flatten a shark from top to bottom, broaden its pectoral fins until they extend like wings from head to tail, move the gill openings beneath the body, and the result is a ray. This branch of the shark group includes skates, stingrays, sawfish and electric rays. Most rays live on the bottom in shallow water, but several kinds, including the largest rays of all, live in the open ocean. These are the manta rays, or devilfish.

Fully grown manta rays may have 'wingspans' of 18 ft or more and weigh as much as $1\frac{1}{2}$ tons. Mariners once believed that mantas seized ships by their anchor chains and towed them to an unknown and fearful doom. It was also believed that mantas wrapped unlucky swimmers in their 'wings' and then devoured them. Such tales are quite groundless, but a giant manta ray that has been harpooned or otherwise provoked can reduce a fishing boat to splinters.

If left to their own affairs, however, the devilfish are among the most placid creatures in the sea. Like the largest sharks – and, indeed, like the largest whales – the manta rays feed only on small fish and plankton, which they channel into their mouths with two fleshy extensions of their pectoral fins.

Life in the depths

*No part of the sea is more alien to man than the abyssal deep.
This chill, dark, forbidding world is sparsely inhabited by some
of the strangest creatures to be found anywhere on our planet*

All night long the dripping cable wound back on to the huge reel aboard the Royal Danish Research Vessel *Galathea*. For many hours, the ship had been hove to over one of the deepest ocean trenches in the world, a long chasm in the floor of the Pacific running parallel to the east coast of the Philippine Islands. Scientists aboard awaited the reappearance of the sturdy trawl that had been secured to miles of heavy wire and lowered into the abyss 32,565 ft below. Would the trawl bring back evidence that living creatures could survive in eternal darkness and immense pressure?

At last the ship's searchlights picked up the outline of the trawl's bulky triangular bag. Anton Bruun, leader of the *Galathea* expedition, later described the tension and excitement of that moment: 'During the work of taking in the trawl ... we prepared for the disappointment of seeing a bag without any bottom animals in it.... "There's clay on the frame!" somebody cried. "It's been on the bottom!" ... Everybody on board who could leave his job gathered round. ... We hardly noticed the red prawns ... or black fish; we all knew these to be pelagic animals, caught on the way up. ... But there, on a rather large stone, were some small whitish growths – sea anemones! Even if no more animals had been found, this would still be the outstanding haul of the expedition! It was proof that higher animals can live deeper than 10,000 metres (more than 30,000 ft). Is it surprising that all were overjoyed? And that pleasure became excitement when out of the greyish clay with gravel and stones we picked altogether 25 sea anemones, about 75 sea cucumbers, five bivalves, one amphipod, and one bristle worm?'

Such commonplace creatures have seldom created more of a stir. Their recovery from the bottom of the Philippine Trench on July 22, 1951, pushed the known limit of life another 8000 ft into the ocean depths.

Eight and a half years later, animals were found at even greater depths. The Swiss scientist, Jacques Piccard, and US Navy Lieutenant Don Walsh took the bathyscaphe *Trieste* to the bottom of the Challenger Deep, 200 miles east of the Mariana Islands, in the Pacific. At 35,800 ft, their searchlight picked out a flatfish on the bottom. Apparently undisturbed by the *Trieste*'s arrival, the foot-long fish rested quietly for a minute or so and then swam off slowly into the darkness. The explorers also saw a little red shrimp. Like the fish, it resembled shallow-water relatives. Here was proof that animals can live in the greatest depths of the world's oceans.

LUMINOUS FISH *Most of the creatures that inhabit the twilight zone of the ocean are luminous, partly to attract prey and partly to help them see. This star-eater has a 'spotlight' below each eye and rows of light organs along its jaws and sides. A luminous barbel on its chin may serve to attract prey*

"

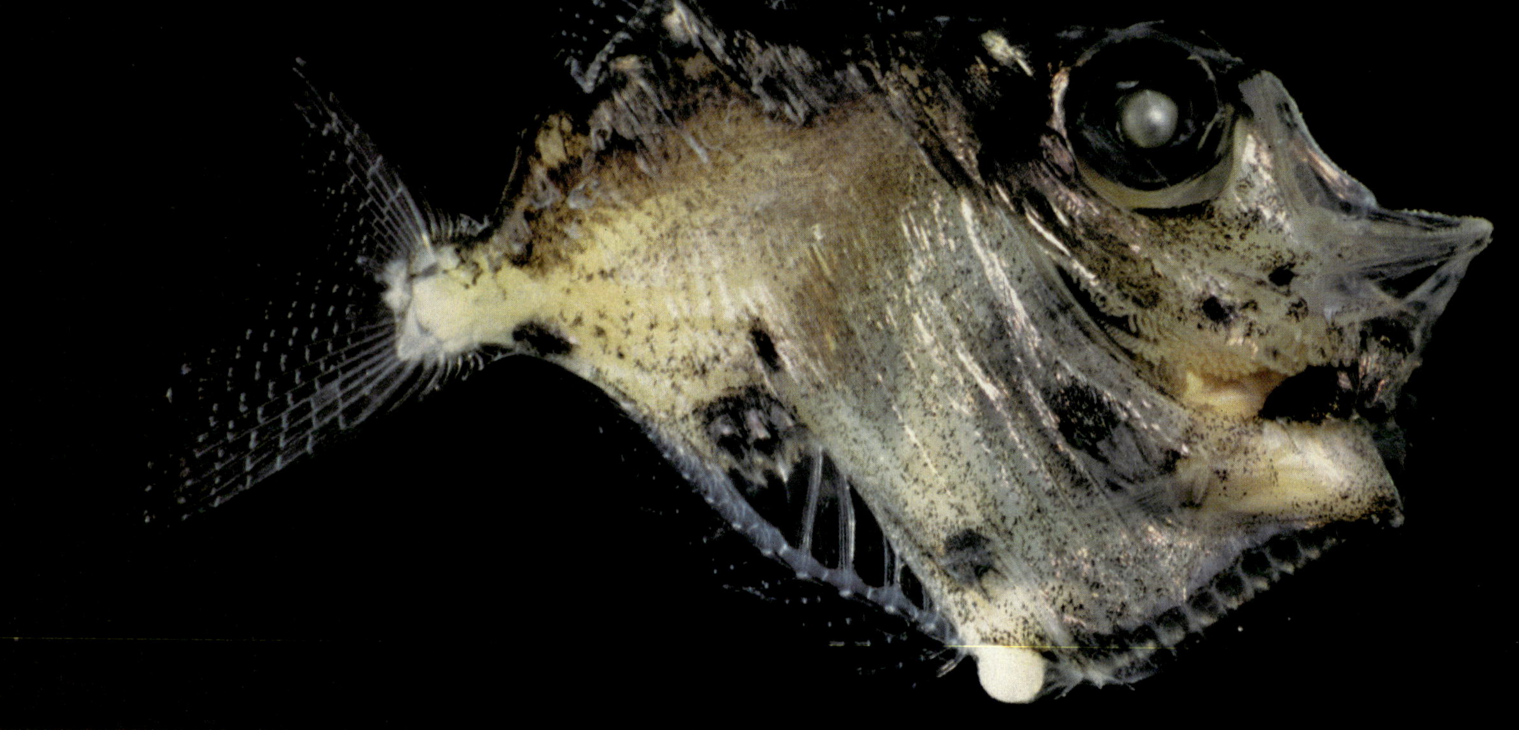

SURVIVAL IN PERPETUAL NIGHT

Early 19th-century scientists would have been astonished by the findings of the *Galathea* and the *Trieste*. As late as 1843, the distinguished British naturalist Edward Forbes concluded that no animal could survive below about 1800 ft. His theory was quite plausible, for conditions in the abyss appear inhospitable to life.

Sunlight fades away rapidly with increasing depth, so that animals living in deep water must find food and seek mates in darkness. More important, even in the clearest water there is not enough light for photosynthesis below 400 ft. The sun's warmth, too, is quickly absorbed by surface waters, and temperatures in the abyss remain uniformly cold throughout the year.

In addition to darkness and cold, there is tremendous pressure. For every 33 ft a creature descends, the pressure exerted on each square inch of its body increases by almost 15 lb. At 3300 ft therefore, the pressure is 1500 lb. per square inch – a hundred times greater than at the surface. At 35,000 ft it is almost $7\frac{1}{2}$ tons per square inch.

As forbidding as these conditions may seem, they are at least stable. During the course of many millions of years, life has spread slowly throughout the oceans, from the sunlit surface waters to the deepest trenches.

Curiously, the enormous pressure – which early scientists believed would crush any animal to pulp – required few special adaptations. Certain sea urchins, starfish, worms and other organisms live both in very shallow water and at depths of 10,000 ft or more. These creatures are little affected by pressure because their tissues are permeated with fluids at the same pressure as the surrounding water. Body fluids push outwards just as hard as sea-water pushes inwards, and the two forces offset each other.

Nor has the chilling cold restricted the spread of life into the depths. Nearly all deep-sea animals are cold-blooded and have body temperatures close to the low temperature of the water. As a result, they probably grow more slowly and reproduce later and less frequently than warm-water species. Because of this reduced tempo of life, the animals require less food, which is scarce in the depths.

The lack of light has had far more evolutionary impact than the high pressures or low temperatures. As depth increases and light diminishes, the inhabitants of the sea change dramatically.

Countershading

In well-lit waters near the surface, fish bear light-coloured bellies and green or blue backs. From about 500 ft down to about 1500 ft – in the twilight zone of the sea – there is still enough light for countershading to help in camouflaging an animal. At these depths, finger-sized lantern fish, carrying batteries of luminescent organs, are greyish, light brown or silvery; many kinds have silvery-white undersides and pink-tinged sides. Hatchet fish, so named because of their shape, frequently have silvery or beautifully iridescent bodies with brownish backs.

But not all the creatures of the twilight zone are countershaded. Living there, and in the darker zone below, are animals with a variety of colours. Black stomiatoid fish gleam with an iridescent sheen. Fish called gulpers and swallowers, which prowl below 6000 ft, are velvety black or deep brown. Whalefish, resembling their namesake in shape but only a few inches long, bear jaws and fins of bright orange or red. Scarlet shrimps snap up orange arrowworms and scarlet copepods. Squids, so quick to change colour near the surface, wear a livery of deep red, purple or brown, sometimes studded with luminous organs.

Why are some creatures camouflaged by countershading or dark bodies while others seem to flaunt their presence with bright colours? The answer involves the way light is absorbed as it passes through water.

Sunlight is a mixture of different colours, each with a characteristic wavelength and energy level. On a bathysphere descent near Bermuda, Dr William Beebe, the American marine biologist, observed that red light waves, the least energetic, penetrated only about 20 ft. By the time the bathysphere had descended to 50 ft, a large scarlet shrimp that he carried with him appeared black. The shrimp could reflect only red light, so it appeared black below the depth penetrated by red light waves.

Even in the clear water off Bermuda, orange light disappeared at about 150 ft, yellow at 300 ft and green at 350 ft. At 800 ft, nearly all the wavelengths were absorbed, and Beebe saw only 'the deepest blue-black imaginable'.

OPOSSUM SHRIMP *Many of the deep-water shrimps are identical to their surface counterparts, but the opossum – so-called because the females have pouches like the American mammals – can emit a luminous cloud to confuse predators*

HOW COLOUR VANISHES

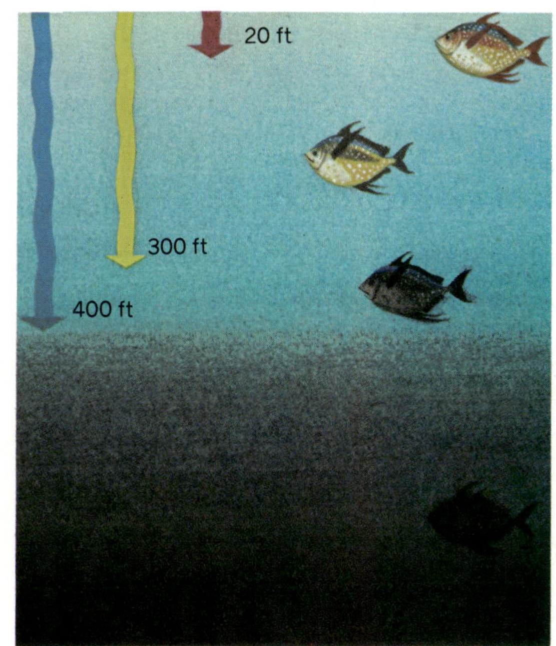

As sunlight is filtered through sea-water some colours are absorbed more quickly than others. For this reason the creatures in the sea appear a different colour at different depths. Red is the first colour to disappear as this fish – an opah – moves down from the surface; it is followed by yellow, then by blue

Creatures who carry their own lights

Although the sunlight penetrates only a few hundred feet below the surface, many deep-sea creatures produce their own light. This bioluminescence – 'living light' – is especially common in animals found from about 1000 to 8000 ft below the surface.

Some organisms, including certain crustaceans, squids and fish, have their own light-producing cells. Other deep dwellers harbour colonies of luminescent bacteria, always in the same locations on a particular host species. This arrangement aids both parties: the bacteria provide light for the host, and the host provides food and shelter for the bacteria. Some hosts switch on the bacteria by pumping oxygenated blood to them, and extinguish them by reducing the blood flow.

Bioluminescence occurs when a substance known as luciferin (from the Latin word meaning 'light-bearing') combines with oxygen in the presence of an enzyme called luciferase. During a sequence of chemical changes, energy is released in the form of light. Electric-light bulbs produce considerably more heat than light, but bioluminescence can be virtually 100 per cent efficient, generating light and practically no heat.

Most bioluminescence in marine organisms is blue-green, although red, yellow and other colours have been reported. The lights may be located almost anywhere on the body. A number of deep-sea fish carry rows of lights on their bellies and sides. Some wear lights on their heads or on fleshy growths – barbels – suspended from their chins. Squids have luminous spots on their arms and circling their eyes.

Such brightly illuminated creatures moving along in the dark would seem conspicuous targets for predators. However, bioluminescence in the twilight zone may take the place of countershading in surface waters. If the intensity of a fish's belly lights matches the intensity of sunlight filtering into the twilight zone, the fish will create neither silhouette nor shadow when seen from

RECOGNITION AND BAIT *Twilight-zone fish use their light organ sometimes as bait for predators, sometimes to recognise their own species. Lantern fish, hatchet fish and stomiatoids, for example, see their prey by their lights: angler fish use them as lures*

the water below by hungry predators on the prowl.

Bioluminescence also helps to attract potential prey, as anyone who has ever fished at night with a lamp can understand. Angler fish and stomiatoids actually 'fish' with luminous lures and these apparently serve to entice animals whose normal prey is luminescent.

At breeding time, the pattern and colour of an animal's lights could serve to identify it as male or female. Male lantern fish, for example, bear powerful lights over their tails; females carry only small lamps on the underside of their tails. Just as female fireflies flash signals to males during the mating season, deep-sea creatures may announce their readiness to mate by flashing lights.

Closely related animals sometimes have different sizes and arrangements of luminous spots. One species of lantern fish has three rows along its belly; another has only two. Such differences may help the animals to recognise others of their species as they form schools or find mates. Dr Beebe could easily differentiate between species of lantern fish by their pattern of lights. 'In absolute darkness,' he said, 'I could tell how many of each species were represented in a new catch, solely from their luminous hieroglyphics.'

Bioluminescence may help animals to see prey. Dr Beebe watched copepods caught in a sheet of light from the belly of a lantern fish. The fish suddenly twisted around and began to gulp them down. Lights near the eyes of one type of stomiatoid fish illuminate whatever it looks at. Such creatures have been seen to 'catch' schools of krill in these eyebeams and snap them up with a sudden rapid movement. Krill, too, possess lights around their eyes that may enable them to find still smaller food in the dark.

Bioluminescence can also be used defensively. In the gloom of the depths, the ink cloud with which some squids confound their enemies would be ineffective. At least one deep-sea species expels a brilliantly luminous cloud which confuses any would-be attacker long enough for the squid to dart to safety. Some deep-sea shrimps employ a similar tactic by releasing a substance that bursts into a miniature Milky Way of luminous particles.

ILLUMINATED TRAP *The long-fanged viperfish has more than 300 light organs in the roof of its mouth to tempt the shrimps and fish on which it feeds. The light organs are repeated in rows around its lower jaws and along its belly*

SEEING IN THE TWILIGHT

Fish of the twilight zone possess highly sensitive eyes – perhaps the most sensitive in all the animal kingdom. Dr N. B. Marshall, the British biologist, has examined the brains and eyes of species living as deep as 300–5000 ft. He concluded that despite the virtual absence of light, these fish 'can probably see their kind, their food and their enemies' – making the utmost use of whatever faint light penetrates the twilight zone.

Although most vertebrates have eyes which are roughly spherical, those of many deep-sea fish are tubular and contain two retinas, or light-sensitive fields – one for images of distant objects and one for images of closer objects. Fish with tubular eyes are equipped to make accurate judgments about the distances of potential prey or predators.

However, at a depth of almost 5000 ft the world of the deep sea becomes too dark even for the most sensitive eyes, and below this point the eyes of many fish are smaller and weaker. Some inhabitants of the abyss – fish and invertebrates – have tiny, weak eyes or are altogether blind. They must rely on touch, taste and smell.

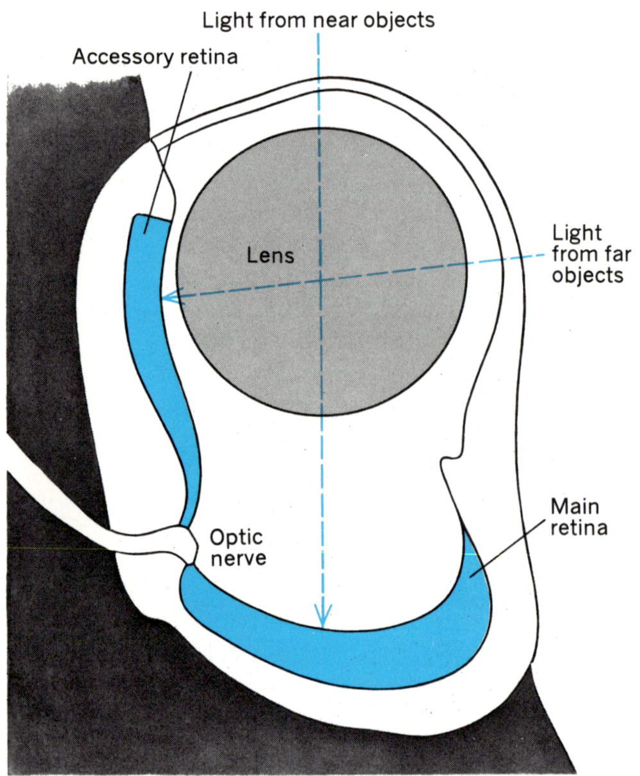

The upward-staring eyes of the Scopelarchus have two retinas: one at the side for all-round vision and a main retina at the back of each tubular eye for sharp vision over a limited distance. Unlike most fish, the Scopelarchus sees through both eyes in the same direction – giving it binocular vision and enabling it to judge the position of its prey accurately

SWALLOWER FISH *The grotesquely bloated belly of the swallower fish contains a recent victim which has been gulped whole for digestion later. The 6 in. long fish can stretch its jaws to seize creatures much larger than itself*

FINDING FOOD IN THE DEPTHS

Food is scarce in the sunless depths, for there are no green plants. Scraps, waste products and dead organisms drift down slowly from surface waters. Some of this meagre rain is consumed during its fall. But enough food eventually reaches the bottom to support a community of animals attached to, burrowing through or creeping along the ocean floor. In turn, these detritus feeders are food for larger, more active animals.

Not all deep-sea food chains are based upon this rain of particles. Countless invertebrates and small fish make daily vertical journeys, rising at night to feed nearer the surface and descending again into deeper water before daylight. Larger animals follow and prey on the small migrants. Such movements extend to great depths, so there is also a 'bucket brigade' transfer of food from rich surface waters down to the barren abyss.

Garbage from ships and decaying vegetation swept down from the land supplement the meagre deep-sea menu. Wherever the *Galathea* trawl came up with tree branches, coconut shells or other debris from rivers and swamps, the catch of deep-sea animals was exceptionally good. In general, however, a meal is hard to come by in deep water, and deep-sea animals have evolved some bizarre adaptations for getting whatever is available.

Gulpers and swallowers
Many fish of the depths are little more than tooth-filled mouths attached to expandable stomachs and long, thin tails. The gulpers, common at depths below 6000 ft, have trap-door jaws and elastic, bag-like stomachs that can expand to several times their normal size to accommodate prey even bigger than the gulpers themselves. One 6-in. gulper had a 9-in. fish coiled inside.

Gulpers grow up to 6 ft long, but their whip-like tails account for most of this length. The tips of the tails are often luminescent, and may lure animals within reach of the gulpers' mouths.

Fish with whiskers
Stomiatoids are another group of deep-sea fish that can engulf surprisingly large prey. Usually 6–7 in. long, they have fang-like curved teeth, large heads and elongated bodies that taper into small tails.

SWALLOWING LARGE PREY

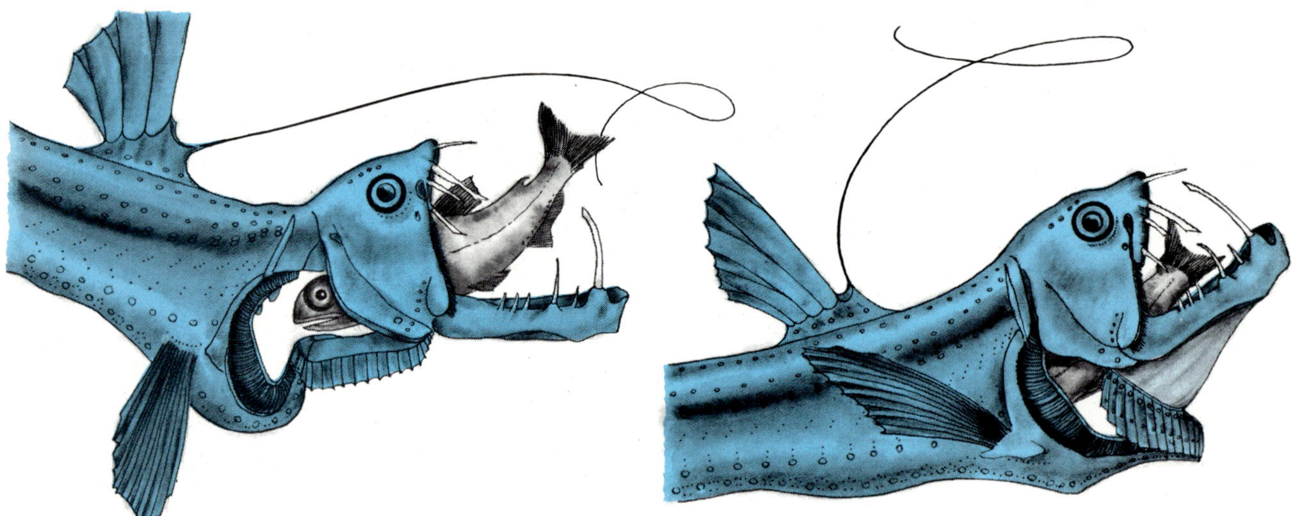

The skull of the stomiatoid fish swings up and its lower jaw protrudes at the same time to enlarge the throat opening

As muscles pull the fish's heart and gills aside, it raises its jaw. Movable teeth in its throat push the prey into its stomach

A typical stomiatoid has a barbel dangling below its chin or throat. Depending on the species, this barbel ranges from a short 'whisker' to a sinuous whip ten times the length of the body. There is a $1\frac{1}{2}$-in. species with a barbel about 15 in. long, and an $8\frac{1}{2}$-in. species with a barbel nearly a yard long. Barbels vary in form as well as in length. Some are merely single strands, others branch and rebranch like leafless twigs, and still others resemble bunches of grapes or exotic flowers. Many are luminescent.

Scientists are still debating the function of these elaborate whiskers. They may be used to lure prey, or they may be species-recognition devices, enabling stomiatoids to identify mates in their dimly lit environment.

Fish with fishing rods

There is no doubt that the bony appendages of the black, scaleless, deep-sea angler fish found at depths below 2000 ft attract prey. Most species have a luminescent lure at the tip of a 'fishing rod', which is a modified spine of the dorsal fin.

To an angler fish's prey – possibly a lantern fish or another angler fish – the lighted lure dangling on the rod might very well look like a luminous worm or shrimp. As the prey moves closer to investigate, the angler brings the lure back nearer its mouth. The lower jaw suddenly drops open, and the gill covers expand to create a cavity large enough to accommodate a meal bigger than the

fish itself. This rapid action creates a powerful suction that sweeps the victim into the angler's mouth. The jaws snap shut, impaling the prey on long, curved teeth. The prey is swallowed whole and pushed into the angler's expandable stomach by teeth in the throat.

Over millions of years, anglers have evolved into highly specialised fish. Through a series of adaptations, the first spine of the dorsal fin separated from the others, moved forwards, and lengthened into a fishing rod that lies flat in a groove on top of the head. Some species have short, stubby rods; others have long, slender ones. Muscles swing the rod forwards, back towards the mouth, or out of the way when jaws and teeth come into play.

Perhaps the most unusual angler of all was brought to the surface from a depth of 12,000 ft off the western coast of Central America. This jet-black, broad-headed, 18-in. species dredged up by the *Galathea* has achieved the ultimate in the placement of its luminescent lure. It dangles from the roof of the mouth behind curved teeth.

Possibly because of this sedentary method of fishing some species of angler fish have become weak swimmers. But they have developed an effective means of stalking prey by crawling, using limb-like fins. Although the species of angler fish caught so far in the deep sea have been small – often only 2–3 in. – large species have been found in the stomachs of sperm whales, together with

the remains of animals of the deepest waters.

Certain deep-dwelling anglers possess adaptations for reproduction as astonishing as those for feeding. In the dark, sparsely populated depths, these fish may have difficulty finding a mate. Therefore, when a male meets a female, he sinks his teeth into her and hangs on – sometimes for the rest of his life. If he belongs to the species whose males are lifelong parasites, the skin around his mouth and jaws fuses with the female's body; only a small opening for breathing remains on either side of his mouth. His eyes degenerate, as do most of his internal organs. His circulatory system becomes connected with hers, and from then on he is nourished by her blood. In effect, the male becomes an external appendage producing sperm to fertilise the female's eggs.

Although the female probably releases her eggs in the deep sea, they quickly float upwards and hatch in surface waters, where the newly hatched fish find plenty of copepods and other small plankton to eat. As they grow, females begin to develop fishing rods and males acquire gripping teeth on their snouts and chins. When the anglers change from larvae to adults, they descend to deep water and begin their strange way of life.

1000 Feet
4000
8000
12,000
16,000
20,000
1
2
3
12
13
14
15
16
28
29
30
31
32
33
34
35
36
37
38
Meltzer

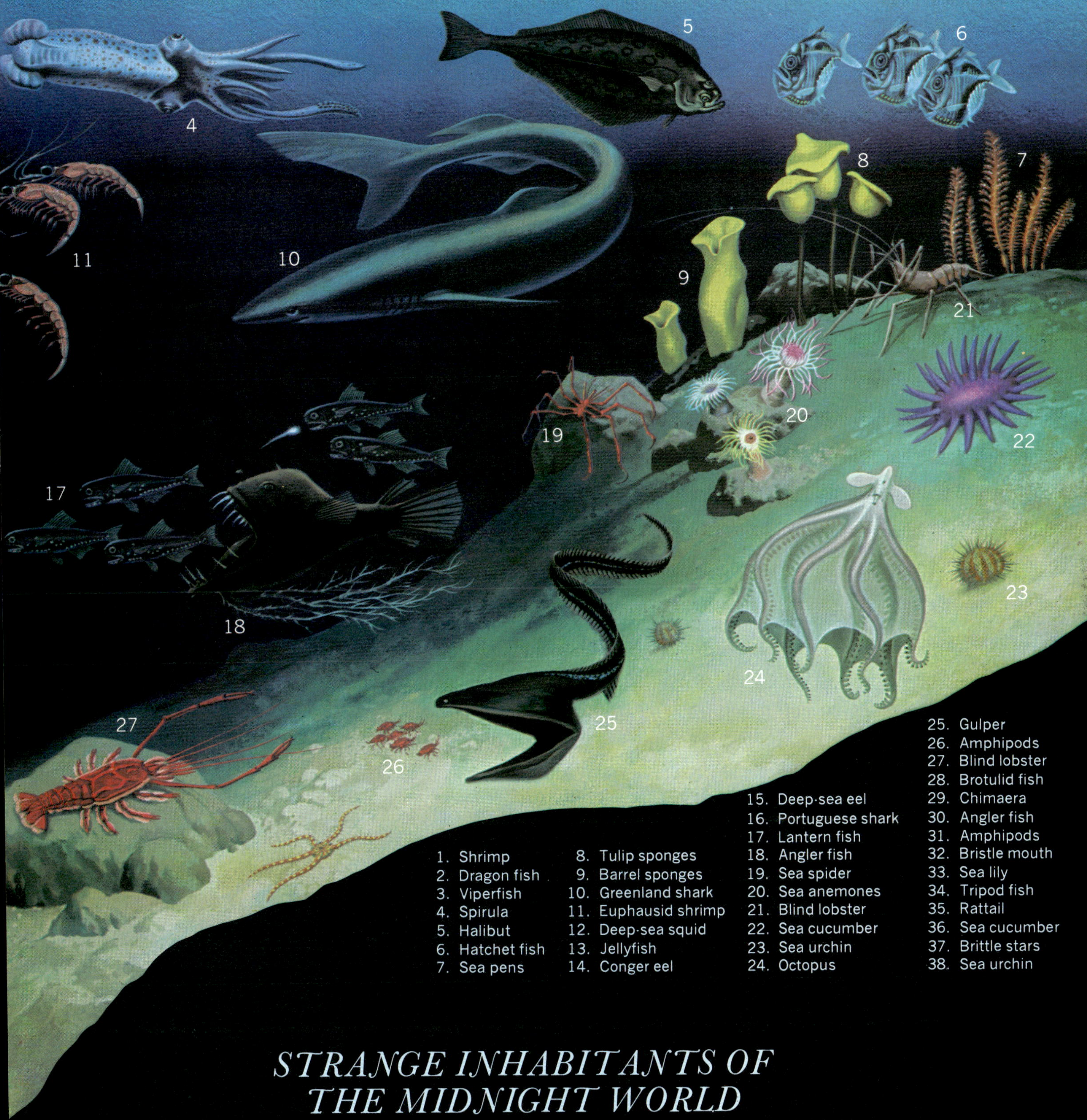

STRANGE INHABITANTS OF THE MIDNIGHT WORLD

How nature has evolved a number of solutions to common problems

Evolving in a world of darkness and sparse food, the creatures of the deepest seas have become among the most curious animals on earth. Many of the inhabitants of the twilight zone are luminescent. The darker waters below are the territory of the fantastic swallowers, gulpers and anglers. Spiny creatures, such as brittle stars and sea cucumbers, populate the deep-sea floor.

WHERE FOOD IS SCARCE *A pink rockfish and a long-legged crab search for food on the ocean bed 3000 ft down off the south Californian coast. Food is scarce at such depths, and the animals are consequently few in number*

GIANT OF THE DEEP *This sea pen flourishing at 16,000 ft is 3 ft tall. The longest shallow-water species is only about 12 in.*

HOW A SQUID PROPELS ITSELF THROUGH THE DEEP

The squid, travelling here from right to left, normally moves itself through the water by a kind of jet propulsion

When it wants to slacken speed or make some underwater manœuvre the squid brings its fins into use

FREAKS OF THE DEEP SEA

Most deep-sea fish are small, but the invertebrates are frequently giants compared to their shallow-water relatives. On the floor of the abyss live sea urchins with bodies a foot wide; most shallow-water species are only a few inches across. Hydroids, usually just a fraction of an inch high, reach the prodigious height of 8 ft in the deeps off Japan. Sea pens, ordinarily 12 in. high, also grow to 8 ft in deep water.

Shrimps, too, grow unusually large in the deep sea. Some scarlet and brilliantly luminescent species are nearly a foot long. Some have antennae twice as long as their bodies.

The giant squid

No deep-sea creature is more spectacular than the giant squid, the largest of all living invertebrates. Its eight muscular arms, sometimes as thick as a man's thigh and studded with thousands of suckers, may reach 12 ft. Extending beyond this writhing mass are two thin tentacles, sometimes more than 40 ft long. The flattened ends of these tentacles bear 100 or more suckers with serrated rims. The arms and tentacles carry prey to the squid's beak, capable of tearing the toughest flesh into small pieces.

The largest squid ever found was washed up on a New Zealand beach in 1888. It measured 57 ft, with two stretched-out tentacles accounting for 49 ft. This specimen was of the genus *Architeuthis*, which includes several species of giants.

Little is known about giant squids. They live in all the world's seas, but spend most of their time at depths of 600 ft or more. Scientists do not even know what the creatures prey on. It is known, however, that they are sometimes involved in battles with huge sperm whales – battles which must be the most titanic struggles since the days of dinosaurs. Whales have been found with scars from the tooth-rimmed suckers of squids. The whales are probably always the aggressors, for the remains of giant squids turn up in the stomachs of whales, and never the other way around.

Giant squids use the same method of swimming as their smaller, faster relatives – by shooting water out of a tube under the head in a sort of jerky jet propulsion. Giant squids have reportedly overtaken ships at 12 knots (about 14 mph), yet their anatomy suggests they are actually weak swimmers, unable to pursue and capture active prey. Flamboyant accounts of these monsters grabbing wooden ships as large as 150 tons and dragging them under are certainly not true.

Squids have highly developed nervous systems and excellent eyesight. These contribute to their fast reflexes, but they do not necessarily imply high intelligence. Probably the intelligence of the giant squid, like its ferocity, has been exaggerated.

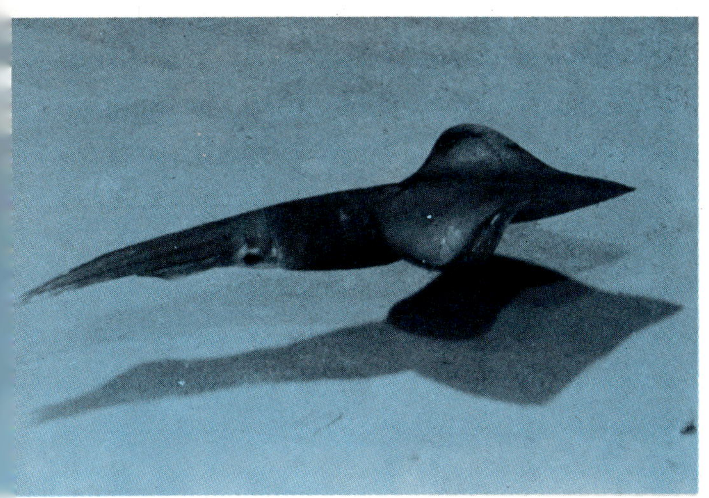

To propel itself slowly along the bed of the ocean in search of food, the squid 'flies' with its fins

Unlike many other animals in the sea, the squid can use its fins to move backwards or forwards at will

NEMO OBSERVATORY *The plastic hull of the American Nemo observatory gives unobstructed views of marine life. It can stay as deep as 1000 ft for 12 hours*

AQUANAUTS EXPLORE THE 'LAST FRONTIER'

Improved equipment allows deeper exploration

Since 1934, when the Americans, William Beebe and Otis Barton, made the first bathysphere descent, half a mile into the sea off Bermuda, man's knowledge of the depths – and of the fascinating array of life found there – has increased enormously. Modern technology led to the development of deeper-diving bathyscaphes. The *Trieste* descended 35,800 ft (nearly 7 miles) in the Marianas Trench in the Pacific off Guam Island, in 1960, a record still unequalled. Scientists now probe the depths in scaled-down submarines with thick portholes. Using these minisubs – about the size of large cars – underwater explorers are able to range much more freely over the ocean floor than they could do in bathyspheres with their restricting cables. Minisubs are equipped with mechanical arms, which can be used both to gather specimens for laboratory study or to carry out deep-water salvage operations.

DEEP-DIVING SAUCER *The SP 350, designed in 1959 by Jacques-Yves Cousteau and Jean Mollard, allows its two-man crew to stay 1050 ft down for 24 hours. It has hydraulic pincers to take samples when the pressure is too great for divers to work outside it*

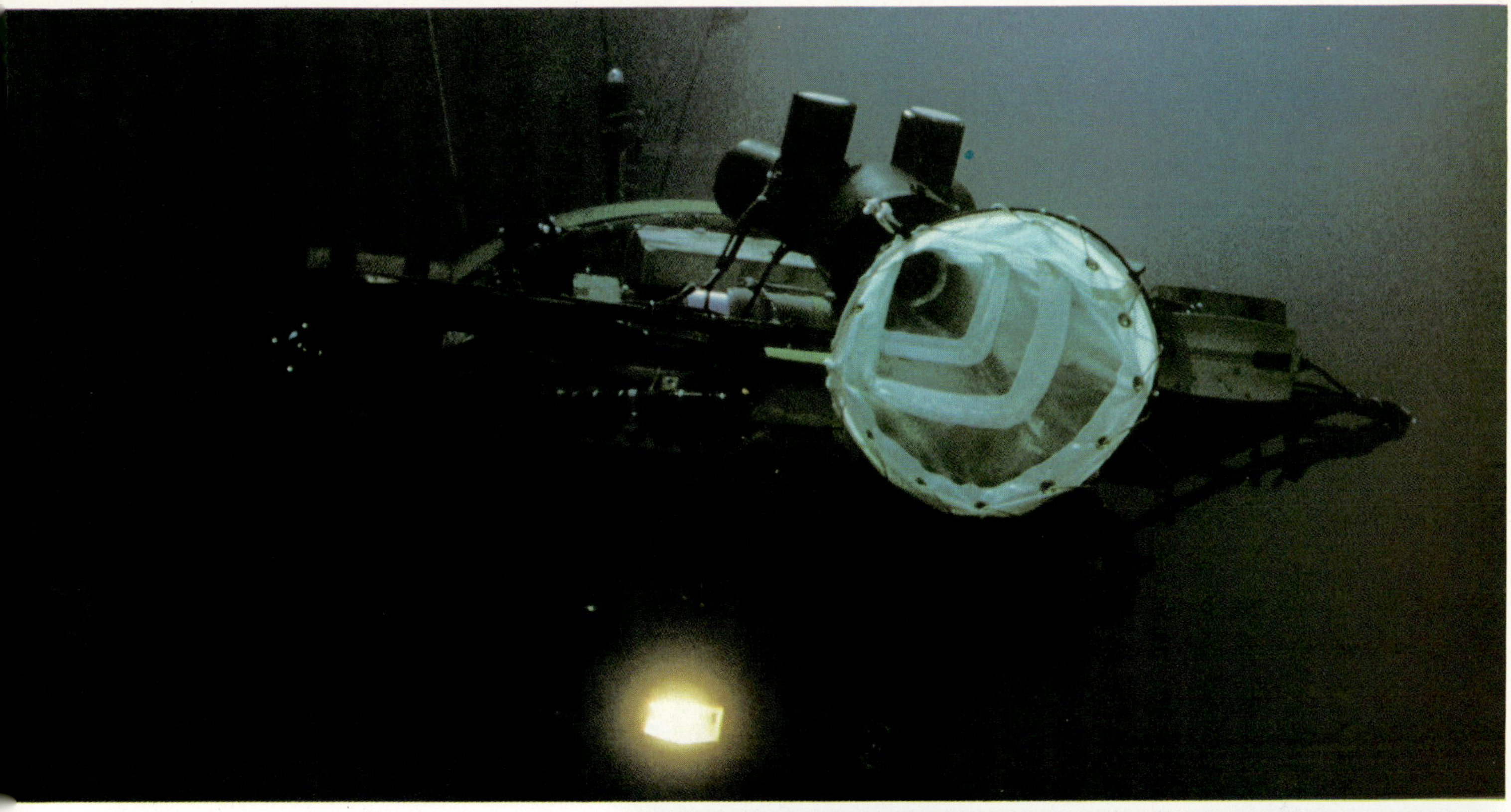

DOWN TO THE DEPTHS *The American submarine* Deepstar 4000 *(above) prepares for a spiralling descent to the floor of the Pacific. The 18-ft craft can take a crew of three (right) to depths of 4000 ft. It is packed with scientific instruments and can scoop up samples of vegetable and animal life*

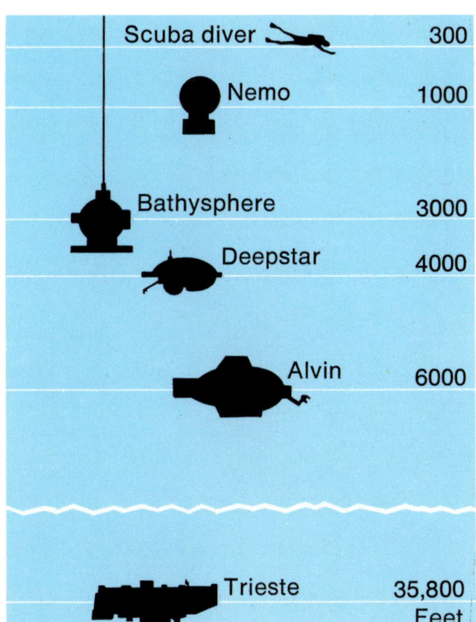

DEEPER AND DEEPER *Technological developments have extended man's exploration range. Without a submarine or bathyscaphe, a diver can dive 300 ft; the record depth was reached by the U.S. bathyscaphe* Trieste, *in 1960*

LIVING FOSSILS

Scientists have long been intrigued with the possibility that the ocean depths conceal creatures long thought extinct. Animals and plants evolve in response to changes in their surroundings. Since conditions in the abyss are believed to have been nearly stable for hundreds of millions of years, it is reasonable to suppose that some creatures survive there with the same form and habits that they had in the remote past.

The idea that the depths harbour 'living fossils' won many adherents after 1864, when Norwegian oceanographers dredged up a sea lily from 1800 ft. Sea lilies, close relatives of feather stars, are colourful, multi-armed 'blossoms' on brittle stalks. Living sea lilies of this particular very primitive sub-order had never before been seen, although their fossil remains had been recovered from rocks 120 million years old.

Shortly after the 1864 discovery, two more living fossils were dredged up from the deep sea. A large scarlet sea urchin, previously known only as a fossil in the 100-million-year-old chalk cliffs of Dover, was found in the North Atlantic in 1870. Then a curious 3-in. mollusc with short, thick arms and a jar-shaped body was netted by the *Challenger* expedition of 1872–6. Called *Spirula* because of its internal spiral shell, it swims head downwards. The shell is divided into chambers filled with gas. Belemnites, molluscs with similar internal-chambered shells, were common in the sea 100 million years ago, but they had disappeared by 50 million years ago – all except the ancestors of *Spirula*.

The vampire squid

When the ancestors of *Spirula* roamed the seas, large numbers of molluscs that were neither squid nor octopus, but a little of both, flourished. Some had arms joined by broad webs and round bodies equipped with paddle-shaped fins. All vanished from the fossil record some 100 million years ago, but in 1903 one velvety black animal was discovered in the deep sea. Its ancestors apparently survived in the cold, dark depths from 3000 to 9000 ft.

This archaic animal is evidently a poor swimmer, for its muscles are soft and pulpy. It probably drifts or moves along feebly in the eternal darkness, head down, arms hanging limply and the webs forming a sort of loose bag around the

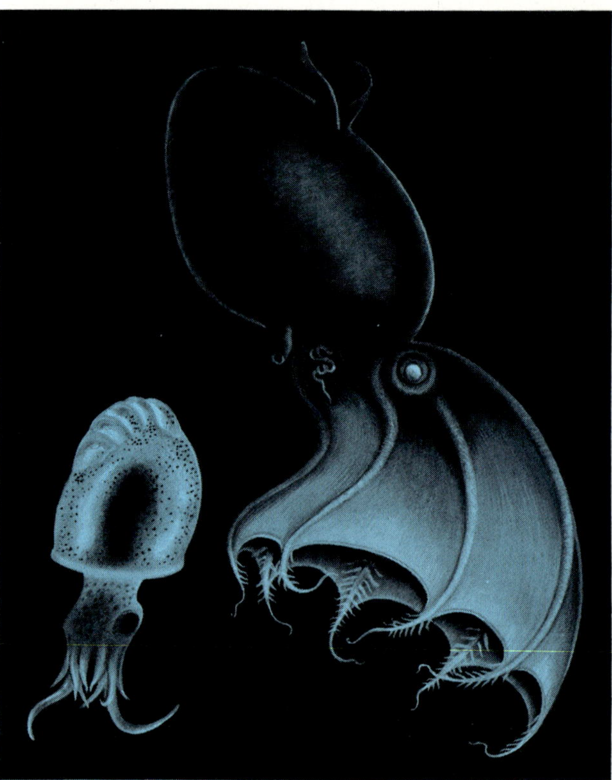

LIVING FOSSILS *The 3-in.* Spirula *(left) and 5–8½ in. vampire squid (top) are survivors from the remote past*

mouth. Its name – the vampire squid – befits its bizarre appearance.

The vampire squid was first thought to be an octopus, but then its two additional arms were discovered. Apparently used only as feelers, these long, suckerless arms can be coiled into special pockets on the web. Ten appendages are characteristic of squids, but these 'feelers' are so different from the strong, prey-grabbing tentacles of other squids that biologists have placed the vampire in a category of its own.

Vampire squids have been found in very deep waters in all the oceans of the world. Apparently they spend their entire lives in the deep; one was netted 9850 ft down in an undersea canyon north of New Zealand. They have good eyesight and are studded with luminous organs, including two particularly large ones that can be covered with flaps of skin. Their prey probably consists of small, relatively inactive animals which they detect through their feelers.

A male vampire squid fertilises a female by inserting a packet of sperm into her with one of his arms, in much the same way as the more familiar squids and octopuses. Spawning and hatching

take place 6500–8500 ft down. The young at first have only eight arms. But by the time they are $\frac{2}{3}$ in. long, the two extra arms have developed, as well as the web and light organs. The largest vampire squid ever captured was an $8\frac{1}{2}$-in. female; no males larger than 5 in. have been found.

The coelacanth

The most exciting living fossil was discovered in 1938. Fishermen working in the Indian Ocean, off South Africa, brought up a strange metallic-blue fish almost 6 ft long. It had large, thick scales covering its body and strong, meaty fins that it probably used to move along the bottom. The creature was a coelacanth, the descendant of an ancient line of fish thought to be extinct for 70 million years. Coelacanths are not strictly deep-sea dwellers. The first one was discovered at a depth of about 240 ft, but the few other specimens subsequently captured have come from depths between 600 and 1200 ft.

Another living fossil emerged from the dark, muddy clay 12,000 ft down in the Pacific off Costa Rica in 1952. Oceanographers aboard the *Galathea* made the catch: ten living specimens of a limpet-like animal believed extinct for 350 million years. Similar to the limpets of rocky shores, it has a fragile, pale yellow, conical shell $1\frac{1}{2}$ in. long and $\frac{1}{2}$ in. high. The creature probably moves along the sea bottom on its large pink-and-blue foot, possibly aided by structures also used for breathing. A pair of short, fleshy tentacles behind the mouth may gather food from the mud.

This primitive animal, named *Neopilina*, is believed to be a link between worms and molluscs. Its body is segmented, as is that of a marine bristle worm. But at the same time, its shell, body structure and rasping tongue show that it is also related to molluscs. In other words, both worms and molluscs are probably descended from a common ancestor much like *Neopilina*.

Past experience has demonstrated that living fossils, even in the changeless world of the abyss, are the exception rather than the rule. But oceanographers, now probing the depths with greater intensity than ever before, will no doubt discover more of these startling creatures. The possibility of finding them will always quicken pulses and raise hopes whenever a trawl returns from the mysterious ocean depths.

Ocean odysseys

Many sea creatures, from tiny shrimps to giant whales, respond to an instinctive urge to migrate. Their journeys may cover only a few hundred feet, or span entire oceans

Each afternoon at sunset, one of the grand biological rhythms of the marine world begins anew. With the waning light, myriads of sea creatures start to move upwards. By midnight, much of the planktonic and free-swimming life of the ocean is concentrated near the surface. As night ends and the sky begins to lighten, the travellers reverse and descend once again into the depths.

From the shallows to the abyss, members of nearly every major group of sea animals join in these vertical migrations. Legions of protozoans, copepods, arrowworms and jellyfish; swarms of snails, shrimps and krill; schools of squids, lantern fish and other deep-sea dwellers move up in the evening and down in the morning.

Feeble swimmers make surprisingly rapid and lengthy journeys. Copepods no larger than grains of rice climb from 350-ft depths to the surface at rates of 50–200 ft an hour. Minute larval barnacles move upwards at 50–75 ft an hour. Such daily round trips may take eight hours or more. In the process, plankton animals sometimes endure pressure changes of several atmospheres and

temperature fluctuations equivalent to travelling in surface waters from Iceland to the Equator and back.

What impels these creatures to make such arduous migrations? As far as scientists can determine, their behaviour is a reaction to light. Each species moves up and down following a certain intensity of light as the sun sets and rises. Even bright moonlight drives plankton animals to lower levels at night, and they stay further from the surface on sunny days than on cloudy ones.

Assuming that waning or increasing light causes plankton migrations, the question remains: what are the advantages of these movements? One answer is that tiny vegetarians such as copepods have developed a behaviour pattern that both provides them with food and protects them from predators. Under cover of darkness, they swim up to the surface and feed on the abundant plant life there; in the daytime, they sink into the shadowy depths where their enemies have more difficulty finding them. Bigger, carnivorous shrimps, copepods and worms could be migrating up and down merely to catch the vegetarians. Perhaps these predators in turn attract the more active shrimps, fish and molluscs.

This plausible explanation, however, does not fully account for the luminescent organs, which advertise the presence of many vertical migrants. Around Antarctica, for example, large numbers of luminescent krill glow with an eerie green light. Obviously darkness provides no protection for

LONG-DISTANCE VOYAGER *After laying her eggs on an island off the US coast, a female loggerhead turtle swims back to mate with a male waiting offshore. Like other turtles, these lumbering reptiles, which weigh several hundred pounds, migrate long distances every two or three years from feeding grounds to breeding areas*

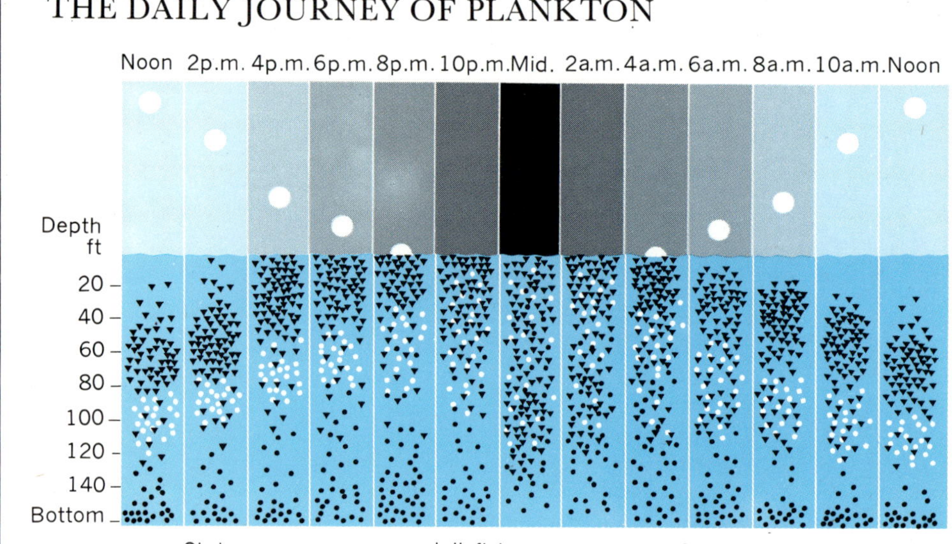

THE DAILY JOURNEY OF PLANKTON

Much of the animal plankton rises to the surface of the ocean at night to feed, then migrates a few hundred feet down during the day – particularly in summer when sunlight is strongest. Scientists believe that vertical migration may help the plankton to feed, avoid predators and to take advantage of currents to drift to new feeding grounds. The three migrants whose journeys are shown here, all migrate daily

these creatures, and apparently conspicuousness is not a serious drawback or they would long since have become extinct.

Sir Alister Hardy, the marine biologist, argues that vertical migration may give weak swimmers the added mobility needed to reach fresh feeding grounds or return to rich pastures. Surface waters usually move faster than deeper waters, and often in different directions. A copepod making a night climb of only 100 ft may find itself a mile or two away from where it was the day before, surrounded by a new food supply. On the other hand, krill in the Antarctic Ocean climb 650 ft during the evening into cold, northward-flowing waters. At dawn, they descend into a warmer current flowing southwards. This migration pattern keeps them in rich pastures of their diatom food.

Riddle of vanishing shoals

While the vertical migrators intrigue biologists, they have played havoc with the instruments and charts of navigators and mapmakers. During and after the Second World War, ships' captains using echo-sounding equipment (electronic devices that bounce sound waves off the ocean bottom to measure depth) reported shallows where the water was supposed to be deep. Over the years, nautical charts came to display hundreds of shoals designated *ED* – 'existence doubtful'. Evidently something in the mid-depths scattered the sound waves and reflected a portion

of them back before they reached the true bottom, thereby misleading navigators.

Whatever these shoals were, they seemed to be linked in some way to the vertical migrations of animals. A study made off the California coast in 1942 revealed that these deep scattering layers, or DSLs as they came to be called, rose towards the surface at sunset. Then, at the first light, they descended to depths of 700 ft or more. Further investigation showed that DSLs stretching continuously for hundreds of miles were present throughout the Atlantic, Pacific and Indian oceans.

Biologists could only assume that DSLs were caused by echo-sounding signals bouncing off the bodies of countless marine animals. But what animals? The bands on the echograms were more clearly defined than those that planktonic organisms might be expected to produce. For a time, commercial fishermen were excited by the possibility of huge, hitherto-unknown schools of fish. Large fish, however, make dark, individual marks on echograms, and the DSLs were showing up as continuous traces.

Nets and remote-control cameras lowered into the layers produced no conclusive results; the mysterious sound scatterers, apparently frightened away by the camera lights, were agile enough to avoid the nets.

Finally, researchers descended in bathyscaphes to investigate. During day trips off the California coast, Dr Eric Barham of the US Navy Electronics Laboratory found great numbers of hakes, the deep-sea fish that are related to cod, at depths between 600 and 1000 ft, and dense concentrations of siphonophores hovering motionless between 850 and 1500 ft, with their long tentacles stretched out in every direction like 'a living net'. Between 1200 and 1500 ft, Dr Barham saw enormous shoals of closely packed shrimps. About 200 ft below the shrimps he spotted great schools of luminous 3-in. lantern fish, and between 2150 and 2300 ft he observed thousands more of these small fish that rely on luminescence for their whole existence.

The locations of these animal concentrations generally correspond to depths at which deep scattering layers have been found off the coast of California – at about 1000 ft, 1400 ft and 1700 ft. Although much is still to be learnt about their composition and movements, clearly some of the DSLs at least are caused by the reflection of sound waves from shrimps and other crustaceans, from the gaseous floats of siphonophores, and from the swim bladders of fish.

FISH THAT CAN GLOW *Millions of luminous lantern fish migrate up and down in the ocean each day, following their main food, the copepod. They descend as deep as 3500 ft where they cast a glow in the dark waters. Their gas-filled swim bladders deflect sound waves and give the appearance of a false ocean bottom, called by scientists a deep scattering layer*

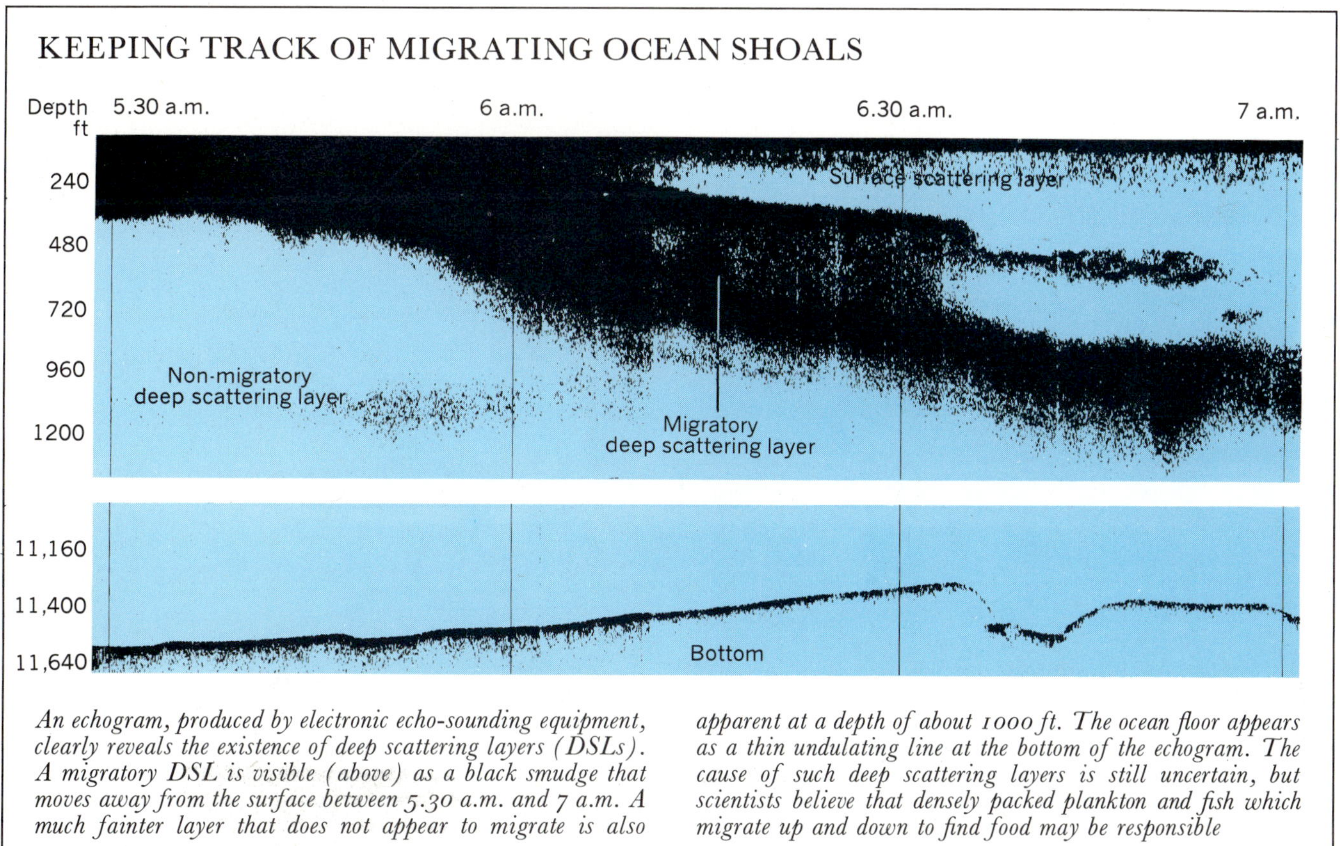

An echogram, produced by electronic echo-sounding equipment, clearly reveals the existence of deep scattering layers (DSLs). A migratory DSL is visible (above) as a black smudge that moves away from the surface between 5.30 a.m. and 7 a.m. A much fainter layer that does not appear to migrate is also apparent at a depth of about 1000 ft. The ocean floor appears as a thin undulating line at the bottom of the echogram. The cause of such deep scattering layers is still uncertain, but scientists believe that densely packed plankton and fish which migrate up and down to find food may be responsible

THE MYSTERY OF MIGRATION

Vertical migrations of planktonic organisms are caused by automatic responses to changes in light intensity. A tiny copepod, for example, does not 'choose' to swim upwards at sunset; its nervous system simply reacts to the decreased light level, initiating muscle movements that carry the creature towards the surface.

In the same way, the more extensive migrations undertaken by many marine animals are triggered by environmental changes – changes in water temperature, in day length, in the abundance of food – and by internal chemical changes, such as the secretion of hormones. Animals setting out on migratory journeys do not think forwards through time and across space to definite goals; they merely respond to an instinctive urge to set off. Along the way, they follow signs in the sea and sky that have guided generation after generation of their ancestors over hundreds or thousands of miles to the same destinations.

Like the great animal migrations on land, the long-distance journeys of marine creatures are usually seasonal movements to places where food is more abundant or where conditions are more suitable for the birth and growth of young. Large sea animals, such as whales and tunas, cross entire oceans in their wanderings; salmon travel between the open ocean and rivers and streams; eels swim from inland brooks and lakes to the depths of the Sargasso Sea. Sharks, marlin and sea

LOBSTER CAVALCADE *Spiny lobsters walk in procession along the sea floor in search of new supplies of food off the Florida coast. The queue formation, in which each clutches the tail of the one ahead or flicks it with antennae to maintain contact, is defensive; each animal's vulnerable belly is protected from any possible predators by the lobster next in the column*

turtles travel singly or in small groups, while herring, squids and mackerel migrate in immense schools, which may help to protect them.

Many coastal crustaceans – particularly crabs, lobsters and shrimps – move into deeper waters to spawn. Their offspring are planktonic for a while, then settle to the bottom, crawl to shallow water where food is plentiful, and grow to maturity. Each autumn off Bimini in the Bahamas, thousands of spiny lobsters migrate from shallow reefs to deeper offshore waters. During this annual journey, the lobsters assume a 'marching order' when they cross relatively open areas. Lining up in single file, each individual hooks one pair of its front legs around the tail of the animal in front. Males and females join together to form chains more than 50 lobsters long, which move over the sea floor as fast as a man can swim.

Many ocean animals take advantage of prevailing currents in their breeding migrations. The typical pattern is for the strong, experienced adults to travel against the current to reach the spawning grounds; the weak, unprotected young then drift with the current on the return trip to adult feeding areas.

North Sea wanderers

The plaice – a bottom-dwelling flatfish found in many parts of the North Sea – exhibits this behaviour pattern. The large population of plaice that lives in the sandy shallows off the Netherlands migrates south-westwards against the current to spawning grounds at the northern end of the English Channel. Millions assemble there each year between January and April to produce hundreds of millions of floating eggs. These hatch while drifting north-eastwards with the current back to the feeding grounds.

EYE CHANGE

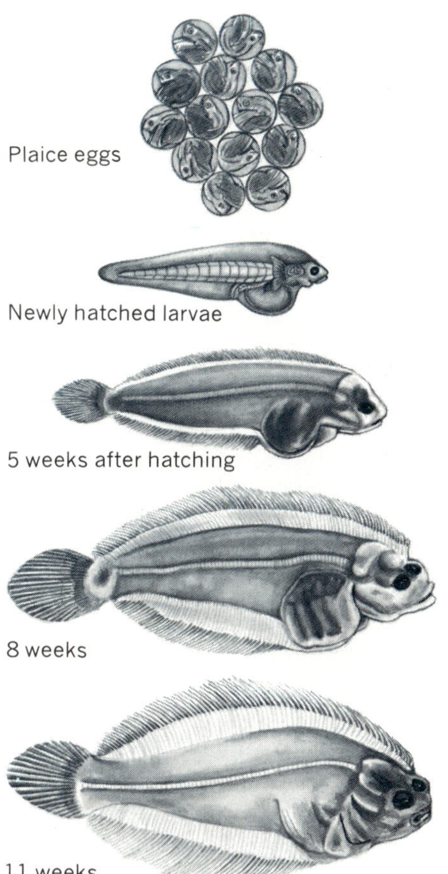

As a plaice grows, its left eye moves to the right side, which becomes the fish's back when it is about 11 weeks old

MIGRATING TO SPAWN *Adult plaice, which swim by undulating slowly against the current, spawn in the southern North Sea and some return later to feeding grounds off Holland. Their young drift northwards by themselves when they hatch*

MIGRATING SEALS AND SEA LIONS

Sea creatures that breed on land

Protected by hairy coats and thick layers of blubber, seals and sea lions range the world's oceans, thriving in polar seas or the cold currents of temperate seas. Like seabirds, they spend most of their lives at sea, but each year they return to the islands or shores where they were born, to give birth and mate. In some species, embryo development is delayed so that the single pup is born about a year later – when its mother is back on land. Fur seals and sea lions paddle with powerful fore-limbs; seals such as the harp, crabeater and leopard scull with their hind flippers. All dive under water to seize their prey such as fish, squids and crustaceans. One of the most fascinating of the creatures that inhabit the desolate coldness of Antarctica is the Weddell seal. These seals navigate by sound waves – like some members of the whale family – and they communicate with each other in a wide variety of underwater sounds. Their bodies can withstand the extremely cold Arctic water.

SEAL HAREM *A bull fur seal may have 50 mates and as many pups. These gregarious animals, pictured on St George Island off Alaska, migrate far south each winter, but return to the island in June to form breeding colonies*

CRABEATER AND LEOPARD SEALS *Despite its name the crabeater (left) lives exclusively on krill. Often seen with it in the Antarctic, the leopard seal (right) feeds on penguins. Both migrate to Australian waters in the winter*

NEVER ON LAND *Harp seals never come ashore, even to breed; their pups are born on drifting ice packs in the North Atlantic. When it is just a few weeks old, the suckling pup (left) is weaned, and its colour changes to the dark grey of the adult*

NURSING MOTHER *Off San Miguel Island, near Santa Barbara, a Californian sea lion suckles her year-old pup (foreground) as a newly born pup also tries to feed. Some sea lions rear their young before the birth of new pups*

SEA LION STAMPEDE *The sea lion has few natural enemies – mainly whales, sharks and man – but at the first sign of danger on land they stampede towards the safety of the sea. After the breeding season, the bulls leave their many mates and migrate north*

EPIC JOURNEY HOME

A number of different fish migrate from the sea into fresh water to spawn. Striped bass and shad live close to shore and travel only relatively short distances up rivers to lay their eggs. Salmon, however, make truly remarkable journeys, sometimes travelling as far as 3000 miles between their feeding grounds in oceans and their spawning sites in rivers, streams and lakes far inland.

The most spectacular spawning migrations of salmon are those made by five Pacific species: sockeye, silver or coho, king or chinook, pink and chum. These fish mature in the frigid, food-rich waters of the Gulf of Alaska. At the age of two to four years, they leave these feeding areas and return to freshwater breeding grounds ranging from northern Korea to Siberia, and from central California to Alaska.

The journey is long and arduous. Once they enter fresh water, salmon cease to feed and are sustained wholly by the fat and muscle tissues of their bodies. They often swim for weeks against swift currents and through turbulent rapids, leaping and swimming up waterfalls.

Covering up to 50 miles a day, salmon reach points as far inland as almost 2000 miles up Alaska's Yukon River. Nothing but death can stop these beautiful streamlined fish from reaching their destinations – and death awaits them at the end of their journey. Only rarely does a Pacific salmon survive long enough after spawning to return to the sea.

End of the journey

The gaunt, emaciated salmon arriving at the spawning grounds bear little resemblance to the fat, sleek fish that left the ocean a few weeks earlier. Their bodies are cut and bruised, their fins ragged and torn, their eyes and gills infested with fungus and other parasites. Weakly, the female digs a shallow trench in the bottom gravel and deposits her eggs; the male fertilises the eggs, and the female covers them with gravel.

Then the adults usually die. Their bodies litter the stream and lake bottoms and line the banks, sometimes piling up in heaps several feet high. Scavengers come from miles around to take advantage of the feast.

The fertilised eggs buried in gravel hatch in three to 12 weeks. The young stay in the nest for about three weeks longer, nourished by yolk

THE STRUGGLE TO REACH HOME *Hundreds of salmon, after spending years in the open ocean, fight against the current to return to breed in the mountain streams where they were hatched*

sacs that remain attached to their bodies after they leave the eggs. Then they work their way to the surface and take up life as free-swimming fry. After a period ranging from less than a month to two years, depending on the species, they start their journey to the sea. This migration, which begins in the spring, is apparently triggered at least in part by the lengthening day. The increasing amount of daylight stimulates the salmon's hormone production, bringing about a general increase in bodily activity.

Young salmon travel primarily in twilight and darkness, spending the bright hours hidden from predators in the sand or gravel bottom or among the rocks. The fish feed and grow along the way, undergoing physiological changes necessary for the years they will spend in salt water.

The strong currents against which their parents battled to reach the spawning grounds now favour

RIVER-BED NESTS *Returning to spawn in the very stream-bed where they themselves were hatched, female Pacific salmon scoop nests in the stream-bed and lay their eggs. The hook-* *jawed males (above) patrol the area to ward off any inter-lopers, then they fertilise the eggs by covering them with milt. Only rarely do the salmon survive to return to the open sea*

the young, guiding them swiftly and surely to the ocean. Those hatched far upstream, however, must pass through large lakes where there are no river currents. Here they probably rely on such guides as the position of the sun or the polarisation of light from the sky. Ultimately, young salmon reach the sea and proceed to feeding grounds where they will live until they, too, are spurred by instinct to return to their ancestral spawning streams.

The Atlantic salmon – actually an ocean-going species of trout – makes similar migrations up rivers and streams in continental Europe, Great Britain, the north-eastern United States, and eastern Canada. Unlike Pacific species, a few Atlantic salmon reach the sea again where they regain condition before re-entering fresh water to spawn once more. It is known that some Atlantic salmon survive long enough to return to their breeding grounds to spawn a third, or even a fourth, time.

DEATH AFTER SPAWNING *Male Pacific salmon die soon after spawning; females guard the nests for a few days before death*

How a tagged fish revealed one of nature's strangest secrets

To chart the salmon's migrations, techniques have been developed for keeping track of the movements of individuals. Young salmon captured alive either during their seaward migration or before their release from hatcheries are marked by clipping away portions of their fins. This does not harm a fish, and provides positive identification of its origin. Over many years, millions of salmon have been given 'home addresses' in this way and then released.

A marked salmon, recaptured at sea, is tagged with a numbered disc and then released. If it is captured once more – perhaps two or three years later in its spawning grounds – scientists can consult their records and determine where the salmon has been. A remarkable fact emerged from these tagging programmes. An adult salmon, migrating from the sea to fresh water, does not travel merely to the river system where it originated. It seeks out the very stream where it hatched several years earlier.

The first conclusive evidence of this extraordinary homing ability was gathered in the early 1940's. Some young Atlantic salmon were captured from the north-east Margaree River in Nova Scotia in 1938. They were marked by removal of a small fatty fin from their backs, then returned to the river. In June 1940, one marked fish was caught in the Atlantic, 550 miles from the site of its first capture. Tagged and released again, it was caught 96 days later – back in the north-east Margaree River. In 1943, two tagged Pacific salmon returned to their spawning creek in Vancouver Island, British Columbia, after years at sea.

Over the years, the recapture of many tagged salmon confirmed absolutely that they return home to their birthplace with almost unerring accuracy. They seem driven by one of the strongest instincts in nature.

Sometimes this drive to get home requires heroic exertion. A one-year-old silver salmon was released from the Prairie Creek Hatchery in Humboldt County, California. A year later, it returned from the ocean, swam up Redwood and Prairie creeks, flopped across a nearly dry stream-bed,

FIGHTING EVERY OBSTACLE *Homeward-bound salmon challenge any obstacle – here a waterfall in Oregon – in their desperation to return to spawn. After repeatedly trying to leap man-made barriers, such as dams, they often exhaust themselves and die*

made its way through a culvert under a highway, traversed a storm sewer, got through another culvert, and travelled 80 ft up a water channel. It then wriggled through a 4-in. vertical pipe with a right-angled bend, knocked a screen cover off the mouth of the pipe, leapt over a nearly impassable wire net, and plopped wearily into its old rearing tank.

Sense that leads them home

How does a salmon find its way home? As the fish travels hundreds of miles up a complex system of rivers and streams, how is it able to choose repeatedly and accurately between one branch or another at a fork?

The salmon is guided largely by its highly developed sense of smell. It literally sniffs its way home, singling out from all others the distinctive scent of the stream in which it was hatched. Since fish live in a world of limited visibility, they rely on the chemical senses – taste and smell – to a much greater extent than do most land animals. Their anatomy reflects this: the olfactory lobes of the brains of most fish are enormous compared to those of other vertebrates. In laboratory experiments, minnows have demonstrated a sense of smell 500 times as acute as that of man.

In the middle 1950's, Warren Wisby and Arthur Hasler, at the University of Wisconsin, devised an experiment that proved conclusively that the sense of smell can guide a salmon to its spawning stream. About 300 returning silver salmon were taken from two branches of Issaquah Creek in the state of Washington. The salmon in each branch were divided into two groups; the nostrils of one group were plugged with cotton wool, while those of the other group were not. All the fish were released downstream at the junction of the two branches and allowed to resume their journey. Nearly all salmon with unplugged nostrils swam into the branch where they were captured, but less than a quarter of the fish that could not smell found the branch.

Apparently salmon 'memorise' the characteristic scent of their home stream – its distinctive blend of vegetation, soil particles, minerals and animals. This 'memory' is not inherited. Rather, it is learnt at a very early age, perhaps during the first week of life. If salmon eggs are transferred to another stream or to hatchery tanks, the adult fish return to the place of hatching and not to the place where the eggs were laid.

More perplexing than how salmon find their home streams is how they find their way from far out at sea to the rivers that lead to their home streams. In the open ocean, a home-stream scent must be diluted far beyond the ability of even the salmon's keen sense of smell. There salmon must get their bearings at least partly from the position of the sun. But salmon do not rely solely on the sun for orientation; some of them travel at night.

Perhaps fish, like many birds, have developed a way of navigating by the stars. Sensitivity to changing water temperature may also play a role. Each water mass has a characteristic temperature; hence a fish could sense its drift from one mass into another, and alter its heading accordingly. Perhaps salmon use a combination of these clues to hold a course, compensate for drift, and find their way home.

NAVIGATING BY SMELL

As a salmon swims, water flows into its front nostrils past olfactory receptors, which are linked by nerves to its brain, and out of the rear nostrils. Tests made by blocking the nostrils of migrating salmon have shown that they could not find their long way home to the spawning streams without their acute sense of smell

THE LIFE JOURNEY OF AN EEL

Fish such as salmon that migrate into fresh water to spawn are called *anadromous*, or 'running up'. A smaller number of species are *catadromous*, or 'running down'. The best-known catadromous fish are the freshwater eels, which mature in streams and lakes on both sides of the Atlantic, swim into the sea, reproduce and die.

For centuries, the breeding habits of freshwater eels remained one of the great unsolved mysteries of nature. Both Pliny the Elder and Aristotle insisted that eels arose from mud by spontaneous generation. It was not until 1777 that anyone saw a female freshwater eel with fully developed sex glands, and not until 1824 was a comparably mature male found.

In 1896, the Italian naturalists Giovanni Battista Grassi and Salvatore Calandruccio uncovered the first real clue to the origin of the eels. In an aquarium they had several specimens of a small fish known as *Leptocephalus brevirostris* ('short-snouted puny head'). Except for two tiny black eyes, the fish were transparent as glass and had the flat, elongated shape of a willow leaf. Grassi and Calandruccio observed that as the fish grew they became rounder, lost their transparency, and were eventually transformed into elvers, or baby eels. *Leptocephalus* was only the larvae of the common eel. Since the leaf-like young were found only in salt water, the naturalists theorised that baby eels hatched somewhere at sea. This supposition tallied with the yearly disappearance of adult eels into the ocean. But the breeding season and breeding grounds remained mysteries.

The brilliant biological detective work of Danish oceanographer Johannes Schmidt solved part of the mystery. In 1904, while studying the breeding habits of North Atlantic cod, Schmidt discovered eel larvae between Iceland and Great Britain. Intrigued, he recruited hundreds of fishermen, who towed deep nets in many parts of the Atlantic and kept the hauls for examination.

By analysing the hauls from the nets, Schmidt traced the distribution of different-sized eel larvae throughout the Atlantic. As his data accumulated, an unmistakable pattern appeared on his charts: eel larvae taken nearer and nearer the Sargasso Sea were smaller and smaller; those caught in the depths of the Sargasso Sea itself were the very smallest. Here were the eels' breeding grounds.

North American eels hatch in an area to the west of, but overlapping, the birthplace of their European relatives. Spawning takes place in spring, at a depth of about 1500 ft. In the summer, the larvae rise to the surface into the currents of the Gulf Stream. The North American eel requires about a year to transform from a *Leptocephalus* to an elver – exactly the time required for the Gulf Stream to carry it to North American shores. The European eel, on the other hand, remains a larvae for two-and-a-half to three years – again, exactly the time required for currents to carry it to the mouths of European rivers.

The Gulf Stream sweeps both kinds of eels along together, but at some point in their migration they part company. One enters North American rivers while the other continues to Europe.

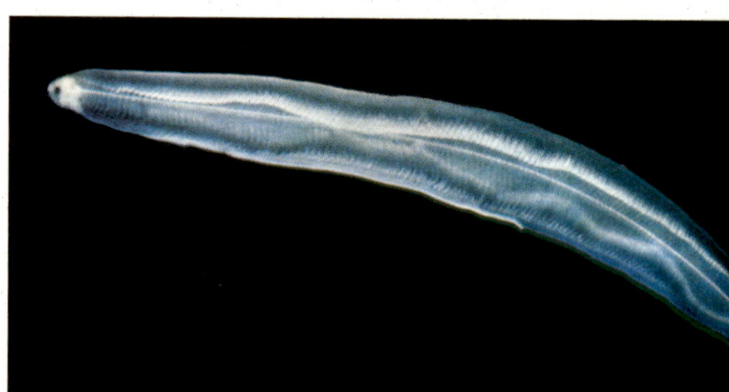

FROM SEA TO RIVER *Freshwater eels begin life in the Sargasso Sea as leaf-shaped larvae (above). By the time they reach the mouths of North American and European rivers they have become transparent 3-in. elvers (below)*

When the swarms of young eels reach their destinations on both sides of the Atlantic, males settle in the brackish water of estuaries, but the females swim far upstream to headwaters and lakes. Some even slither through dew-dampened grass to ponds with no links to the sea.

Once established in fresh water, eels remain for five to 15 years. They hide by day and hunt at night, feeding on anything from carrion to small water birds. Females grow to 4 or 5 ft, but males rarely to more than 2 ft.

Return to the Sargasso

As they mature, eels undergo changes in preparation for their return journey to the sea. Some are external changes: the bronze-green or yellow skin is replaced by glistening silver, a colour that makes the eels more difficult to see in the open ocean. The eyes grow until they nearly cover the sides of the head, presumably in preparation for the passage through the dimly lit ocean depths. Internal changes enable the eels to withstand life in salt water. Stores of body fat accumulate to sustain them on their trip to the Sargasso Sea, when they apparently do not eat.

In autumn, mature females leave streams, lakes and ponds and join males in estuaries. Together they set out for the Sargasso Sea on a trip that may take them across 4000 miles of ocean. None will return. Like Pacific salmon, eels die after spawning.

Many questions about the eels remain unanswered. How do the adults find their way to the breeding grounds? An instinctive urge to swim against the current may play a part, yet females must swim *with* the current to reach the males in estuaries. Do the eels use visual guides in the sea or sky? What triggers the urge to migrate?

The life history of freshwater eels still remains, to a large extent, a mystery.

MARATHON MIGRATIONS FROM THE SARGASSO TO COASTAL WATERS

After hatching in the Sargasso Sea, young eels are carried by ocean currents to coastal areas on both sides of the Atlantic. The dotted lines show the distribution of American eel larvae; the solid lines, European larvae. No larvae smaller than the size given on a line have been found beyond it. The outermost lines show the greatest distances from the Sargasso Sea that larvae are found. Larvae change into elvers when they reach their home shores and eventually become adult eels in fresh water

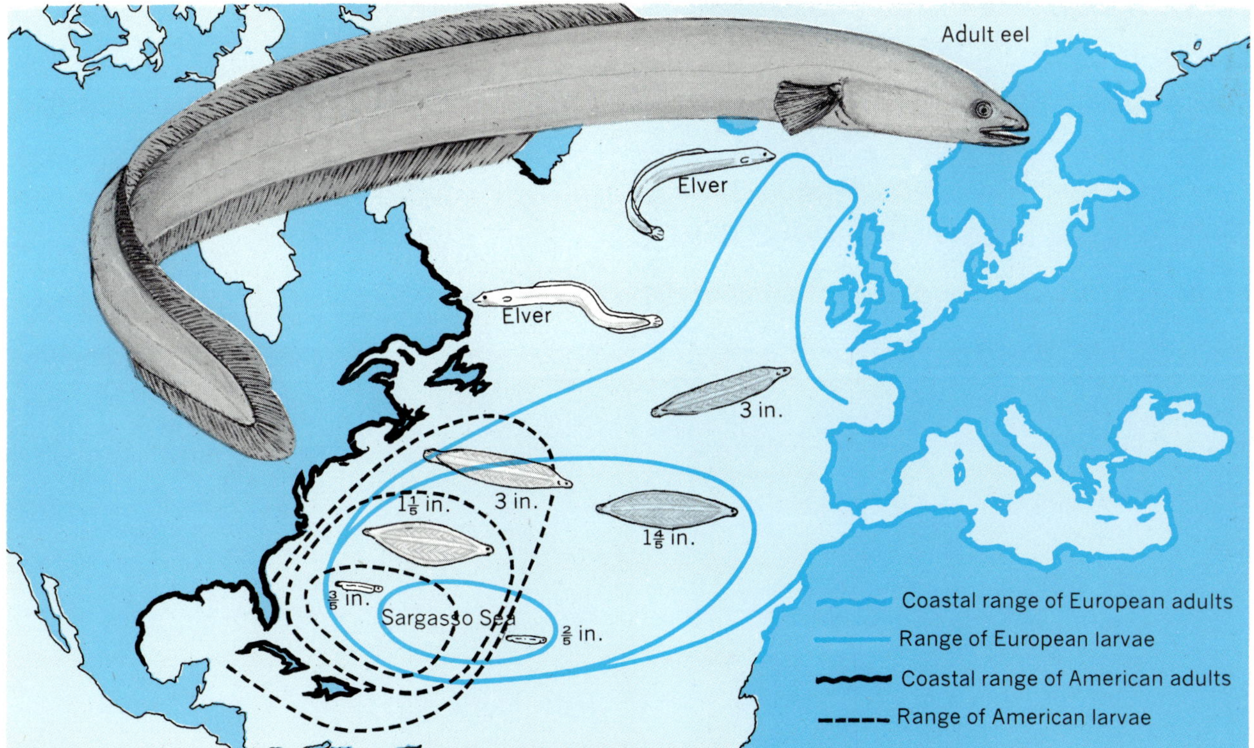

ASCENSION ISLAND VOYAGERS

Marine turtles are among the great ocean travellers. Like tunas, these animals are superbly adapted to life in the ocean. With front limbs modified into oar-like flippers, they scull across hundreds of miles of open water. Awkward on land, they are as graceful in water as birds are in air.

The smallest sea turtle, the Atlantic ridley, reaches a shell length of 2 ft; the biggest, the leatherback, reaches a shell length of $7\frac{1}{2}$ ft, a flipper span of 12 ft, and a weight of 1500 lb. Between these two are the green turtle, the loggerhead and the hawksbill.

The green turtle is the best known, largely due to the work of Dr Archie Carr, a zoologist at the University of Florida. In 1955, Dr Carr and his associates began an intensive study of green turtles in the Atlantic Ocean, with special emphasis on their breeding and migrating behaviour.

Green turtles feed in warm, shallow, continental waters where there are extensive beds of turtle and manatee grass. They breed, however, on remote islands and shorelines that are often separated from their feeding grounds by vast stretches of open ocean. One green-turtle group grazes on manatee grass along the Brazilian coast and migrates all the way to Ascension Island, an island only 7 miles long and 6 miles wide, midway between Africa and South America.

Around December, a number of the turtles leave Brazil and head for Ascension Island, 1400 miles away. All the turtles do not migrate every year. Dr Carr's tagging studies show that some breed at two-year intervals; others wait three years.

Upon reaching Ascension, the females come ashore four or more times, at 12-day intervals. Each time they dig nests in the sand and deposit about a hundred eggs, each approximately the size of a golf ball. Between trips to the beach, the females mate with the males who remain just beyond the surf. This mating fertilises eggs they will lay two or three years later. No lush pastures of marine grass exist off Ascension, so when the urge to reproduce is satisfied, the turtles set out again for Brazil.

Some two months later, the eggs hatch in the sun-warmed sand, and the baby turtles dig their way to the surface. They invariably head for open water, even when the nest is out of sight of the sea. The baby turtles presumably eat plankton and drift with the current until reaching suitable coastal feeding grounds.

Adult turtles return to the same breeding sites year after year, and Dr Carr believes that it is likely that they return to the very beaches where they were hatched, just as salmon and eels return to their native streams. Proof of this is lacking however – primarily because a practical means of permanently marking or tagging the 2-in. long turtle hatchlings has not yet been devised.

HOW TURTLES MATE AND REPRODUCE

MATING *Air-breathing green turtles churn the waters violently as they struggle to stay at the surface while mating*

NESTING *The female turtle flails her front flippers as she digs a flask-shaped nesting chamber with her hind legs*

The breeding migrations of the Brazilian green turtles pose an intriguing and exceedingly difficult question. Why do they travel all the way to Ascension Island when there appear to be suitable breeding sites much closer to their feeding grounds? Dr Carr suggests that the theory of continental drift may provide an answer. Millions of years ago, when the ancestors of the Brazilian green turtles were evolving their migrating patterns, Africa and South America may have been much closer than they are now; some turtles may have fed off South American shores and bred on African beaches. As the continents drifted apart, successive generations of turtles had to swim greater distances. Eventually they used Ascension Island as a resting-place, and later for the sites where they mated and laid eggs.

A mystery still unsolved

Even if the Brazilian turtles' migration patterns can be explained by the theory of continental drift, how do these animals navigate across 1400 miles of open sea and find an island only 7 miles long and 6 miles wide? The question remains unanswered.

Theories abound regarding the homing abilities of not only the green turtles of Ascension Island but also those of other green turtles and any ocean animal that must find its way to distant destinations over long stretches of open ocean. The senses of taste and smell may play a part. Hearing, too, may contribute; turtles cannot make echo-producing sounds of their own, as can bats, but noises made by other animals – such as the loud snappings of schools of shrimps – may

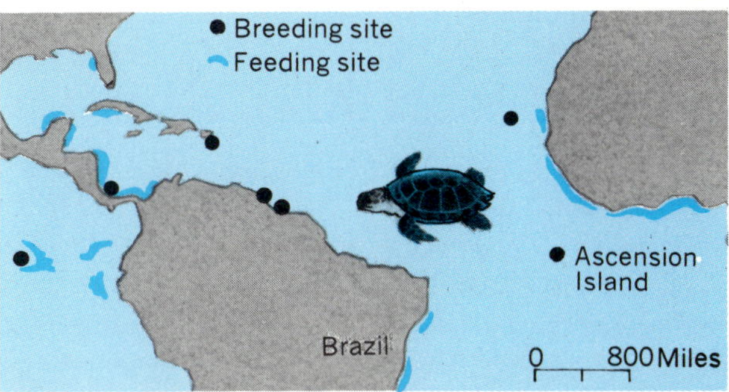

MIGRATION ROUTES *Green turtles migrate up to 1400 miles from coastal feeding grounds to the remote beaches where they were hatched. Nesting areas were once more numerous, but scientists have established that when the turtles using any breeding site die out, no new turtle takes over the hatching area*

serve as guides. Perhaps some animals are sensitive to the earth's magnetic field, or to the Coriolis force – the force created by the earth's rotation that deflects moving bodies to the right or left when they are moving north or south. Possibly some animal navigators possess an 'inertial sense' – they might detect and record all changes in their speed and direction throughout a journey. In returning to a breeding site, they would 'remember' all the twists and turns they took since leaving it, then follow them in reverse.

Dr Carr believes that the green turtles and some other marine animals navigate by the sun – that they possess biological equivalents of the sextants, compasses, clocks and maps with which human navigators chart their position and plot their course.

LAYING EGGS *About 100 eggs are laid each time the female turtle waddles ashore. She makes at least four such trips*

HATCHING *A clutch of young green turtles scurry together for the water after incubating in the nest for 60 days*

WAITING OUT THE STORM *Grey Heermann's gulls and grey-and-white Western gulls weather a Pacific storm on Monterey peninsula. Most Heermann's gulls breed on Isla Raza in the Gulf of California, then fly to British Columbia in summer*

GOLDEN PLOVER *From northern Europe, the European golden plover migrates to the Mediterranean in winter*

MIGRANTS IN THE OCEAN SKIES

Nature's long-distance wanderers

Tirelessly beating their wings and often fasting until they reach their goal, some birds fly as far as 12,000 miles over the ocean between their winter and summer homes at speeds of up to 60 mph. Probably the 'champion' of these seabird voyagers is the Arctic tern. Some of these terns enjoy two summers and almost perpetual daylight by flying between breeding grounds in the Arctic Circle and their feeding waters in the Antarctic seas. The further north they breed, the further south they migrate. Migration is triggered by changes in temperature or by the length of the days or by some kind of biological 'clock' in the birds themselves. Even flightless penguins migrate, swimming and waddling across ice-strewn Antarctic waters, to and from their rookeries.

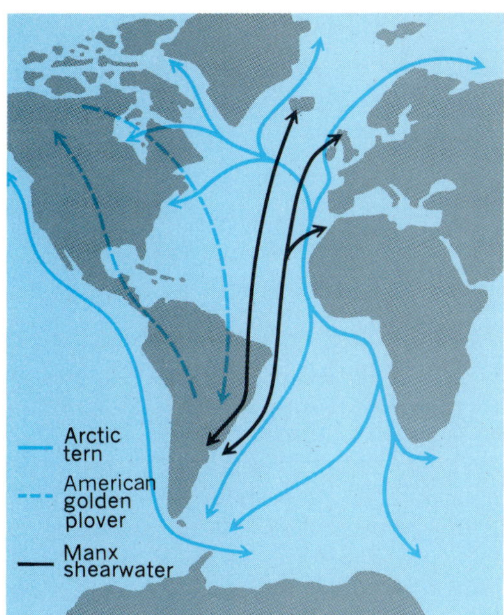

THREE WANDERERS *The Arctic tern, American golden plover and the Manx shearwater all migrate thousands of miles each year*

LONGEST JOURNEY *The further north an Arctic tern (right) nests, the further south it migrates. Terns that nest at the Arctic Circle, for example, fly to the opposite pole for the summer; terns that nest in the US fly only to northern South America*

PREDATORY SKUAS *A pair of skuas eat an unguarded egg. Skuas nest in the Antarctic regions near Adélie penguins, stealing both their eggs and their chicks. Some migrate to the north Pacific where they live by stealing fish from other birds*

INVESTIGATING THE PATH OF THE TUNA

The great journeys of swift-swimming tunas and their relatives, marlin and sailfish, emphasise our ignorance about how marine animals navigate across the trackless stretches of the sea. Tunas regularly cross both the Atlantic and Pacific oceans. Biologists do not know for certain what natural signposts the tunas follow, but it is suspected that they navigate by the sun. These strong, streamlined fish are truly oceanic travellers, for they never enter fresh water.

In May and June, large tuna schools move north between Florida and the Bahamas. In the autumn, tuna schools are seen off New York, New England and Newfoundland. These schools were once believed to be a single stock of fish that spawned in southern waters in early spring, then migrated north in summer and autumn. But it now appears that two separate stocks of these fish roam off the east coast of North America.

In 1954, Frank Mather of the Woods Hole Oceanographic Institution on Cape Cod, Massachusetts, started a programme in which sport and commercial fishermen attached metal identification tags to bluefins and other tunas. Of the first 7000 tagged, more than 1700 were recaught. Plotting the dates and locations of the recaptures, Mather made the first maps of Atlantic tuna migrations. Surprisingly, none of the 670 bluefins tagged off the Bahamas was recaptured off the north-eastern United States. It was therefore concluded that another spawning ground exists in the northern Gulf Stream, supplying the stock for the schools found along the New York, New England and Newfoundland coasts.

Mather's project also uncovered several remarkable long-distance journeys. Six bluefins tagged near Florida were later recaptured near Norway. One giant tuna made the voyage in the incredible time of 50 days. More than 30 smaller tunas travelled from the American coast between Maryland and Cape Cod to the Bay of Biscay. But these are probably not regular migration patterns followed by large numbers of fish each year.

The number of tunas crossing the Atlantic may increase with the strength of westerly winds blowing from North America to Europe. Strong westerlies mean a strong North Atlantic Current (the eastern part of the Gulf Stream that reaches European shores). At such times, a large number of fish may swim to Europe, or be carried there inadvertently. No tunas tagged off Norway or the Bay of Biscay have been recaptured off the United States. Apparently they do not return, but instead join European schools regularly migrating between Norway and Spain.

Bluefins cross the Pacific in both directions. It is believed that young fish are born in equatorial waters in the central Pacific, then join adult schools moving north in the warm currents off Japan. Later they come south to California coastal waters, where they feed for two or three years. As they grow older, they head back to the mid-ocean feeding grounds and to the coast of Japan.

Tagging has also revealed regular migration patterns among marlin. White marlin spend the summer off the coast of New England, then head out to sea in late summer and early autumn. They swing clockwise down towards the West Indies and South America. Some skirt the coast of Venezuela and follow currents around the western tip of Cuba, through the Florida Straits, and up the east coast again. Others turn north after reaching Haiti and Puerto Rico and move up past the Bahamas.

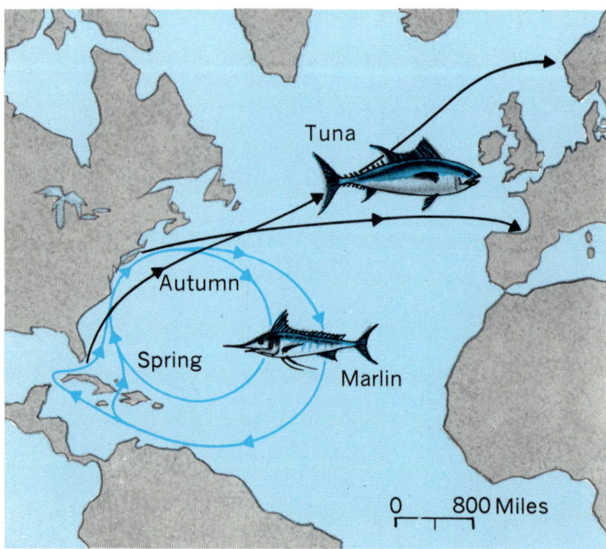

ATLANTIC WANDERERS *The bluefin tuna and white marlin both make impressive journeys. Marlin migrate clockwise around the western Atlantic and Caribbean; tunas from North America travel as far as Norway and France, but do not return to their home waters after the breeding season*

HOW SCIENTISTS STUDY ANIMAL MIGRATION

To study the fish, reptiles, birds and mammals that migrate every year, scientists have developed numerous devices which record how far – and sometimes how fast – the chosen migrants travel. The most efficient equipment – a radio transmitter – enables scientists to track a migrant's entire route

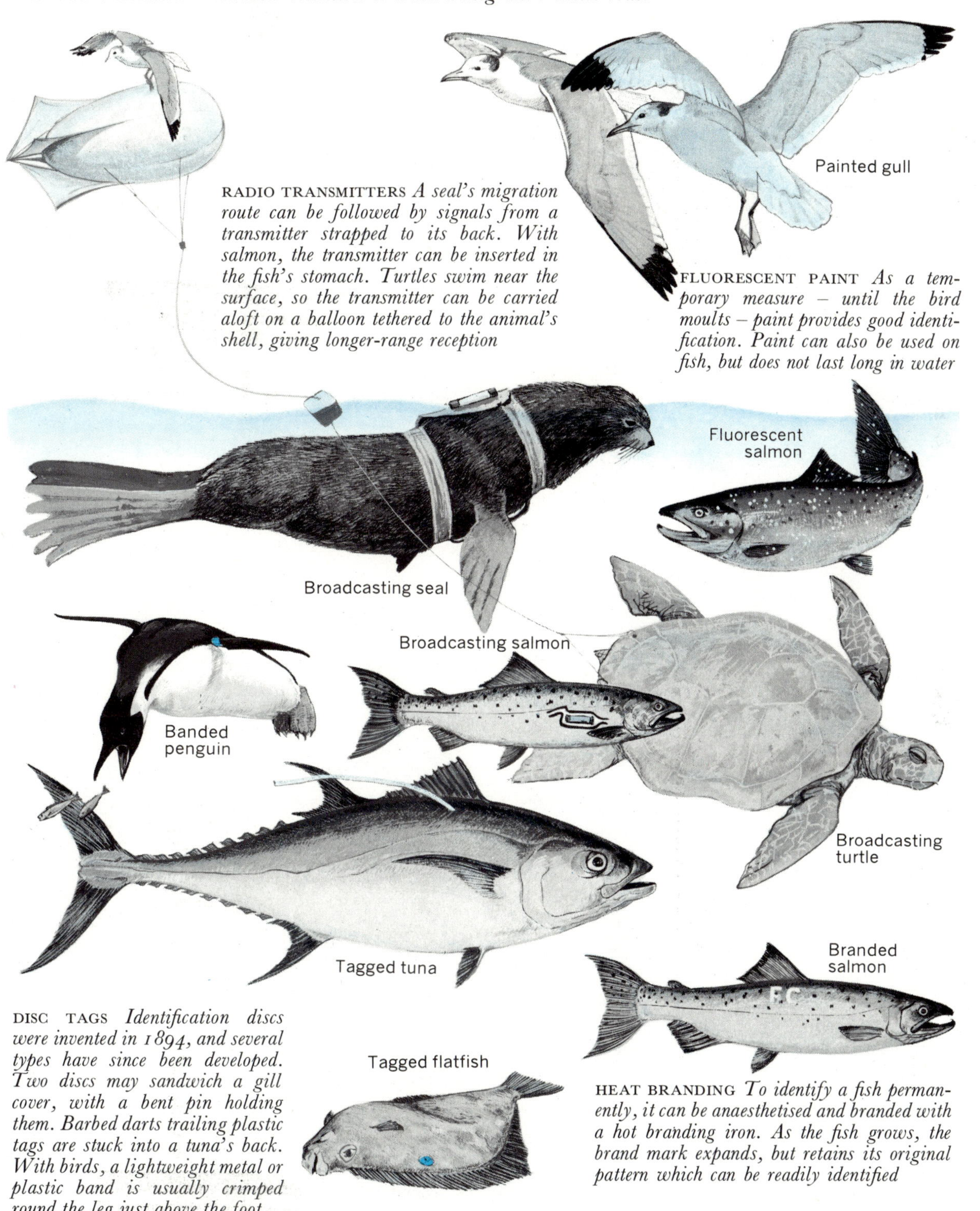

RADIO TRANSMITTERS *A seal's migration route can be followed by signals from a transmitter strapped to its back. With salmon, the transmitter can be inserted in the fish's stomach. Turtles swim near the surface, so the transmitter can be carried aloft on a balloon tethered to the animal's shell, giving longer-range reception*

FLUORESCENT PAINT *As a temporary measure – until the bird moults – paint provides good identification. Paint can also be used on fish, but does not last long in water*

DISC TAGS *Identification discs were invented in 1894, and several types have since been developed. Two discs may sandwich a gill cover, with a bent pin holding them. Barbed darts trailing plastic tags are stuck into a tuna's back. With birds, a lightweight metal or plastic band is usually crimped round the leg just above the foot*

HEAT BRANDING *To identify a fish permanently, it can be anaesthetised and branded with a hot branding iron. As the fish grows, the brand mark expands, but retains its original pattern which can be readily identified*

The backboned conquerors

Written in the rocks of the earth is one of the greatest adventure stories of all time: the rise to supremacy of the animals with backbones. Their astonishing story begins in the sea some 2000 million years ago

Life on earth began in shallow coastal seas more than 2000 million years ago. The earliest living things, probably one-celled organisms of some sort, grew by absorbing chemicals from the sea and reproduced by dividing in half. With the passing of millennia, the tiny floating cells became differentiated into plants and animals. All organisms evolved from these simple predecessors.

By 600 million years ago, the seas were populated with representatives of nearly all the major groups of organisms that dwell there today. Only vertebrates – among them fish, turtles, birds, seals and whales – had not yet appeared. Although vertebrates are latecomers to the marine community – and, indeed, to the earth itself – they have attained a mastery of their environment unmatched by animals without backbones.

Vertebrates tend to be large animals. Their size both increases the variety of organisms that they can feed on and reduces the number that can prey on them. They are strong enough to move about freely and to travel widely in search of food. Their highly developed senses provide a steady stream of information about the surrounding world; their complex nervous systems utilise this information to help them survive. Their brains, usually larger than those of invertebrates, have a greater capacity for learning.

By studying fossils, scientists can trace the rise and spread of the backboned latecomers through most of their history. Where gaps do exist in the fossil record, many essential details can be filled in by examining living animals, especially in their embryonic stages. But the origin of vertebrates will perhaps always remain something of a mystery, for their soft-bodied ancestors left few fossils in the rocks.

Most biologists believe that one present-day animal serves as a reasonably accurate 'model' of the ancient, long-extinct invertebrate from which all backboned creatures arose. This animal is the lancelet, also known as amphioxus, found on the floors of shallow, temperate seas.

An adult lancelet looks like a translucent 2-in. minnow. It spends most of its time buried in the sand with only its head protruding. When disturbed, it wriggles vigorously through the water for a short distance, then dives into the sand.

The lancelet feeds by pumping water into its mouth and out through its gill slits. Bands of sticky mucus in its throat catch food particles, and cilia push them back into the intestine. The little creature has no bones or cartilage, no fins or limbs, no jaws, no brain and, except for some light-sensitive spots, no sense organs.

RETURN TO THE SEA *The Californian sea lion spends much of its life in water and comes ashore only to bask and breed. Like all marine mammals, the sea lion, with its streamlined body and powerful swimming flippers, is descended from ancient ancestors that left their original ocean home and lived on dry land*

THE EVOLUTION OF FISH

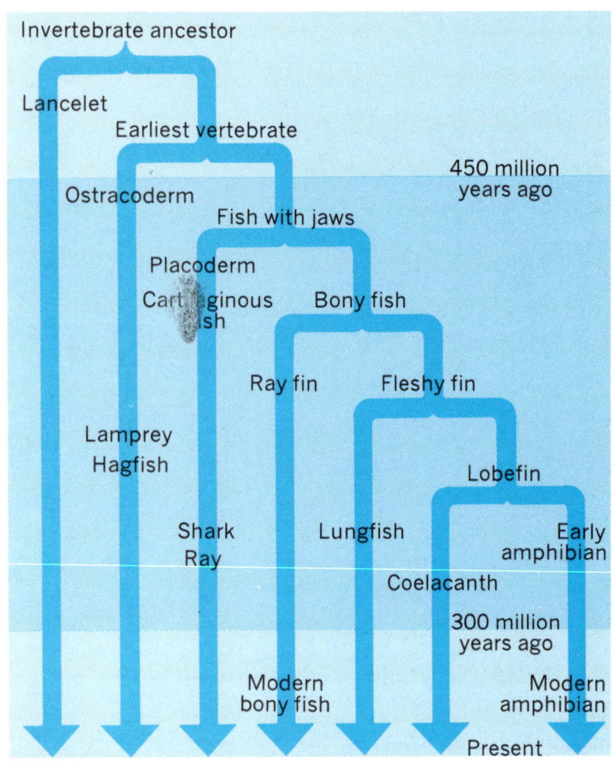

All fish are thought to have evolved from some long-extinct primitive lancelet-like creature. Most of their development took place between 300 and 450 million years ago, by which time recognisable modern species were well established

EVIDENCE OF EVOLUTION *The modern lancelet is an invertebrate and not a fish, in spite of its minnow-like appearance. But it could be a close relation of the unknown, long-extinct common ancestor of all backboned animals on earth*

The lowly lancelet seems an unlikely candidate to be closely related to the progenitor of dinosaurs, birds, whales and man himself. However, it combines characteristics found in most vertebrates. This combination makes it very probable that some ancient lancelet-like animal was the ancestor of the backboned animals.

Running the length of a lancelet's body is a stiff yet flexible rod called a notochord. A notochord is undoubtedly the evolutionary starting point for the sturdier, jointed backbone of all vertebrates. Also, a hollow nerve cord extends along a lancelet's back above the notochord. Just such a nerve bundle, its front end elaborated into a brain, is present in every vertebrate. Finally, there are the gill slits. In the lancelet itself these are no more than feeding mechanisms, but with further evolution they could develop into internal respiratory organs similar to the gills of the modern fish.

Ancestors of modern fish

Fossils of the oldest known vertebrates have been found in rocks laid down as sediments some 450 million years ago. These fish-like animals, called ostracoderms, were in several ways more advanced than the lancelets of today. Their name, which means 'shell-skinned', refers to their external armour of bony plates, which may have protected them against water scorpions or other predators. Openings for eyes and a single nostril pierced the armour. Scientists believe that ostracoderms also possessed an internal skeleton of either bone or cartilage.

Less than a foot long, ostracoderms were far less agile than modern fish. Like their living relatives, lampreys and hagfish, ostracoderms lacked paired fins and jaws. They apparently swam along the bottom by wriggling their tails tadpole-fashion and got nourishment by filtering edible material from the mud.

Another group of armoured fish, the placoderms, appear slightly later than ostracoderms in the fossil record. Placoderms did possess jaws and paired fins, adaptations that enabled them to pursue and catch prey. They usually grew longer than ostracoderms – up to 30 ft. Placoderms are known to have flourished for some 50 million years; they then died out approximately 350 million years ago.

At some time during their existence, placoderms probably gave rise to a line of primitive fish that eventually evolved into modern sharks and rays – fish with skeletons of cartilage, or hard

CAUGHT IN THE ACT *A ray-finned fish and the herring on which it probably choked some 50 million years ago. These fossilised remains, found in rocks in Wyoming, prove that parts of America's mid-west were formerly covered by oceans*

gristle. Because cartilage does not fossilise, the pedigree of sharks and their relatives is difficult to trace. The meagre fossil evidence – mostly teeth – indicates that sharks became numerous about 350 million years ago; since then they have probably changed little in appearance.

The bony fish

From the same ancestral line that produced placoderms, another group of jawed, paired-fin fish developed: the bony fish. Scientists are not certain whether bony fish arose in fresh water or salt water. The fossils that might provide details of their origin have either been destroyed by earth-forming processes or remain hidden in the ground.

Conditions on earth 400 million years ago were different from those today. The climate was far more tropical. Periods of torrential rain were often followed by sustained droughts, when rivers and streams dwindled to chains of stagnant pools devoid of oxygen. Fish that could obtain oxygen only by extracting it from water flowing over their gills died, but those that could use oxygen from the

air survived. Gradually, these ancient fish developed a throat pouch, richly supplied with blood vessels, that evolved into lungs. In all but a few kinds of modern bony fish, these lungs are no longer used in respiration; they have become the swim bladder, a gas-filled organ that helps to regulate their buoyancy in the water.

Early in their history, bony fish divided into two groups: the ray-finned fish, with paired fins of thin skin stretched between rigid spines; and the fleshy-finned fish, with stumpy, 'meatier' paired fins. The ray fins became outstandingly successful. By about 230 million years ago, they had spread from the rivers into all oceans. Of the 25,000 species of fish now living in salt and fresh water, all but about 300 are ray fins.

The fleshy-finned fish were clearly the losers among fish in the conquest of water. They now include only six species of freshwater lungfish and a single marine species, the 'living-fossil' coelacanth. But it is these same fleshy-finned fish – or, rather, their descendants – that have achieved lasting evolutionary success by becoming the vertebrate invaders of the land.

INVADERS OF THE LAND

Some fleshy-finned fish known as lobefins coped with life in stagnant swamp waters with the aid of lungs, breathing atmospheric oxygen when oxygen in the water was in short supply. Never accomplished swimmers, they moved over the floors of shallow pools on muscular fins strengthened by bones.

The refinement of these characteristics made it possible for the lobefins to invade the land, becoming the progenitors of the first backboned creatures of the land, the amphibians. Their lungs became primary breathing organs; their fins became legs.

Life on land presented problems. Prolonged exposure to air could result in fatal loss of moisture. For the first time, too, they were subject to the full effects of gravity; movement by pushing with body and tail was far less efficient on land than in water. With these difficulties, the transition took many millions of years, and the amphibians themselves never completely adapted to life on land. Today their living descendants – frogs, toads and salamanders – seldom venture far from ponds, streams or moist places. All but a specialised few return to the water to lay and fertilise their eggs.

But, limited in their own achievement, the amphibians triumphed in their successors. They opened up an evolutionary avenue for the reptiles and, through them, for the birds and the mammals that were eventually to become supreme.

Rise of the reptiles

About 300 million years ago an advanced type of amphibian began to develop the 'secret weapon' which was to make possible the Age of Reptiles – a tough-shelled egg which could hatch in a dry place. This egg, known as the amniote egg, provided the developing reptile with a miniature fortress.

Because their eggs contained their own water supply in which the embryo could develop, the reptiles were able to stay on land near their food supply instead of having to return to the sea to spawn. They developed, moreover, a tough horny skin that preserved the moisture of their own body.

With these advantages, reptiles underwent an explosion of varieties. For 170 million years, they spread across the land, where amphibians offered only feeble competition; they also conquered the air, becoming the ancestors of birds; and some of them returned to the deep, where they challenged the supremacy of fish.

The age of the great marine reptiles was an adventurous re-invasion of a once-familar element by animals which now had the advantage of 60 million years of evolution on land. They developed into an awe-inspiring range of sea monsters, thriving in all coastal seas. All remained air breathers, and most came ashore to lay their eggs. But a few became nearly as independent of land as fish.

The earliest of these aquatic reptiles probably lived in fresh water, but some of them, such as the crocodiles, later developed marine forms. Some of the early crocodile species which thrived in the seas in the Age of Reptiles were only about 12 in. long. But at least one species of giant crocodile evolved in river estuaries during this age of extraordinary giantism. A cast of its skull, kept at the Natural History Museum in London, is 6 ft long and 4 ft across. This terrifying carnivorous creature must have been at least 50 ft from head to tail.

The most prominent ancient ocean reptiles were ichthyosaurs and plesiosaurs. Both apparently took to the sea soon after early reptiles developed from amphibians. Fast and agile swimmers, the streamlined ichthyosaurs bore a remarkable resemblance both to sharks, which had developed earlier, and to porpoises, which were to come later. Well-preserved fossils, some containing embryos, reveal that ichthyosaurs did not have to come ashore to reproduce. The female retained eggs within her body until they hatched, and the young emerged live and swimming.

The plesiosaurs – less streamlined than ichthyosaurs, with turtle-like bodies and paddle-shaped limbs – were probably slow swimmers. Their long necks may have compensated for their lack of speed by allowing the head to dart and snap up prey.

About 65 million years ago, the Age of Reptiles ended in a great wave of mass extinction. The nearly simultaneous disappearance of so many creatures has never been adequately explained. The great marine reptiles faded away with their distant cousins who had ruled the land. Now only remnants of the scaly clan remain – among them crocodiles, turtles and sea snakes.

KILLER ON THE BEACH *A dangerous man-eater that can grow up to 25 ft long, the saltwater crocodile is highly aggressive and attacks human beings without provocation; but in its turn it is hunted by them for its valuable hide. Apart from their diet of fish, these crocodiles seize birds and mammals on the shore and drag them under water*

Modern survivors

Of the reptiles that have survived to modern times, crocodilians face the most doubtful future. Their two dozen species are near extinction and are being hastened in that direction by the activities of man, who kills them and destroys their habitat.

Only one species of crocodilian is fully adapted to life in the sea: the saltwater crocodile, still fairly common in coastal waters from India to Australia. It favours estuaries, but sometimes undertakes ocean journeys of several hundred miles or more.

Marine turtles have achieved much more success as invaders of the ocean than crocodilians. They remain air breathers, but their front legs have been elongated and flattened into powerful flippers for swimming. Their paddle-shaped hind feet are used for steering and for digging nests in the beach. Like open-sea fish, some marine turtles bear the camouflaging coloration of dark above, light below. The bony shell of the 'true' sea turtles – the green, hawksbill, ridley and loggerhead – is reduced in size and weight as an adaptation for floating and swimming. The leatherback, whose 1200 lb. weight makes it the world's largest turtle, retains only small, bony plates embedded in a tough, thick hide; it has no shell.

In one important respect, sea turtles are still tied to their terrestrial heritage. The females must come ashore to lay their eggs. This is the time when these massive animals are vulnerable to the predations of man; the young, too, when they hatch have to run the gauntlet with their natural enemies, seabirds, to reach the safety of the water.

Venomous latecomers

Snakes evolved relatively recently from the lizard line of reptilian development; the first snakes occur in the fossil record long after the heyday of the reptiles. In effect, snakes are lizards that have lost their limbs, perhaps in adapting to a now-abandoned burrowing way of life. Venomous fangs developed even more recently, appearing only within the last 50 million years.

Although most modern snakes dwell on land, some 50 kinds – each bearing venomous fangs – live in the sea. Less tied to land than marine turtles, only about half of these species come ashore to lay their eggs. Like ancient ichthyosaurs, the other kinds produce eggs that hatch within the female body. The young emerge able to swim and feed. All sea snakes breathe oxygen from the air, but they can remain under water several hours at a time. Valves prevent water from flooding the sea-snake's nostrils.

Most sea snakes stay close to shore, but the yellow-bellied sea snake has been sighted hundreds of miles from land. This snake has migrated east and west from the coast of Asia and is found off Africa and the Americas. Most other species confine themselves to warm coastal waters from the Persian Gulf to Japan and east to Samoa. Sea snakes sometimes congregate in enormous numbers. In 1932, a huge mass of writhing snakes '10 ft wide and fully 60 miles long' was reported near Indonesia.

Sea snakes lie quietly at the surface or on the sea floor waiting for prey – usually fish – to come within striking distance. Then their movements are sudden and rapid. Like their relatives on dry land, they can swallow meals more than twice their own diameter. When two sea snakes begin consuming the same prey at opposite ends, the smaller snake may disappear into the larger.

BLACK-BANDED SEA SNAKE *Some types of sea snake – for example, the black-banded species – slow down their metabolism so that they can remain under water for hours; it is thought by scientists that some types can even absorb oxygen from the water when they are submerged*

LOGGERHEAD TURTLE *Huge, aggressive – but graceful – the loggerhead has suffered least of all the turtle species from man's depredations. However, building developments on the North American coast and Caribbean poachers now threaten its nesting sites*

REPTILES TAKE TO THE SKIES

About 175 million years ago, reptiles took to the air. Among the first backboned flyers were the pterosaurs – clawed and scaly creatures with wings made of flaps of skin, somewhat similar to the wings of bats. They ranged from sparrow-size to giants with wing spans of 25 ft.

Pterosaurs never fully mastered the air. Their weakly supported wings could not lift them directly off the ground, so they must have launched themselves from trees and cliffs. Once aloft, these flying reptiles glided on winds and air currents, occasionally flapping their wings to gain altitude.

About the time the first pterosaurs were struggling into the air, another group of reptiles also began to fly. Unlike pterosaurs, which vanished with the dinosaurs 60 million years ago, these other flyers did not die out. They became the ancestors of birds. Birds are the most poorly represented vertebrates in the fossil record, but four specimens of a very early type, *Archaeopteryx*, have been discovered in a Bavarian limestone formation about 140 million years old. These crow-sized creatures retain many reptilian characteristics, including teeth and a slender, bony tail. They are indisputably birds, however, for the limestone shows clear imprints of feathers.

Two other primitive birds have been discovered in west-central North American rocks about 100 million years old. Both are ocean species that exhibit evidence of much evolutionary progress. One resembles today's terns. The other, about 5–6 ft long, appears to have abandoned flight in favour of swimming and diving; it has small wings and powerful webbed feet.

Beyond these clues, little is known about the early history of birds. But it is possible to speculate about how birds adapted to the demands of flight. Sustained flight requires quick responses and well co-ordinated muscles. Such activity is beyond the small brain and sluggish metabolism of reptiles. In mastering them, the birds evolved an array of adaptations that speeded up the entire tempo of their lives.

EVIDENCE OF THE EARLIEST BIRD

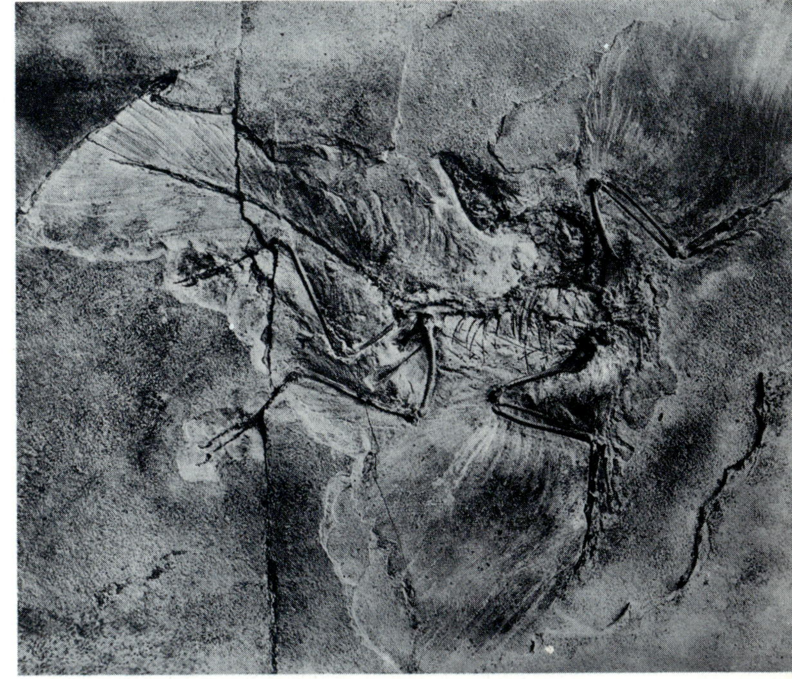

The earliest bird, Archaeopteryx (ancient wing), flew over what is now northern Europe some 140 million years ago. The reconstruction (left) shows the position of the bird in the fossil record (right) now in a Berlin museum

WANDERERS OF THE OCEAN SKIES

Only about 250 of the 8500 varieties of modern birds are adapted for life in and around the ocean. But if seabirds are less varied than land birds, they are nevertheless extremely numerous. Indeed, the most abundant bird in the world is probably a small marine species, Wilson's petrel.

Seabirds are similar to marine reptiles in their dependence on both sea and land. They usually feed in salt water, spending months out of sight of land. But they must return to shore to breed.

The solitary albatross

The wandering albatross, largest of all flying seabirds, travels tens of thousands of miles without touching down on land. It is found only flying over the wind-whipped oceans of the far southern latitudes, where it may circle Antarctica several times a year. The albatross flies in gales that would drive most other creatures to shelter. It banks and wheels just above the crests of the waves, seldom rising more than 50 ft into the air. Now and then it settles on the surface.

Aloft, the wandering albatross holds its long, narrow wings – almost 12 ft in span but only about 10 in. wide – nearly motionless. It may soar for hours, making only slight adjustments to take advantage of the wind's shifting gusts and eddies. In calm weather, the bird waits on the surface for the return of the winds.

The wandering albatrosses are solitary travellers during their long journeys. Every year or two, however, they gather in breeding colonies on remote islands scattered between Antarctica and the southern coasts of Africa, Australia and South America. After an elaborate courtship ceremony and mating, female albatrosses lay a single egg in a large grass nest. The incubation period is unusually long – about 11 weeks. After the young bird hatches, it spends nine months in or near the nest, then takes up the solitary life of an adult. Before returning to the colony, a young wandering albatross may spend eight or nine years at sea; when it is an adult, it returns to the colony to breed every other year.

Far-roaming petrels

Petrels are small relatives of albatrosses and, like them, come ashore only for reproduction. These agile flyers skim between the waves, snatching up shrimps and fish with their hooked beaks. They range in size from the giant petrel, which is as large as a small albatross, to the storm petrel, which is as small as a starling.

Petrels may travel enormous distances to reach breeding grounds which are sometimes shared with albatrosses. Wilson's petrels breed on Antarctica and nearby islands in the southern summer, then fly as far north as California and Labrador during warm months in the Northern Hemisphere. These birds thus enjoy the most congenial seasons in both hemispheres, but commute as much as 20,000 miles a year to do so. Wilson's petrels are frequently seen on moonlit nights moving in large flocks close to the waves and uttering melancholy squeals. When feeding in calm weather, they stretch out their wings and hop or run across the water on their webbed feet.

ALBATROSS BECALMED *A wandering albatross, becalmed on the ocean surface, awaits the wind that will lift it again into the air. The birds breed on remote southern islands, each pair producing a single chick (right) every second year. It is believed that the adults leave the young bird to spend its first winter alone in the nest, sustained only by the fat stored in its body*

WORLD'S SMALLEST SEABIRD *A Wilson's petrel, only $6\frac{1}{2}$ in. long, skips and flutters across the waves, snapping up fish and other food. Its annual migration takes it from the Antarctic Circle to the North Atlantic*

TOBOGGANING PENGUINS *Using their short wings partly for balance and partly for support from time to time, penguins can toboggan along flat ground at a surprising speed. They thrust themselves forward by pushing out with their strong legs*

BIRDS OF THE SOUTHERN SEAS

Curiously, the far-flying albatrosses, petrels and their relatives may share a common ancestor with the totally flightless penguins. Some evidence, at least, suggests that these two different groups are linked. One of the oldest fossil penguins has a skull and bill similar to that of the wandering albatross, and some young penguins develop tubular openings in their nostrils, a characteristic of petrels, albatrosses and related birds. The two lines must have diverged long ago – fossils found in New Zealand indicate that penguins closely resembling modern species were alive 70 million years ago.

Although penguins are traditionally associated with Antarctica, only two modern species, the emperor and the Adélie, live there. Fourteen other species dwell on barren, rocky islands in the Antarctic Ocean and along the southern shores of South Africa, South America and Australia; one species, the Galapagos penguin, is found as far north as the Equator, where it can survive because of the cold, food-rich Humboldt Current flowing up along the western coast of South America.

On land, penguins hop or waddle awkwardly on their stubby legs, or toboggan along on their bellies by pushing against the ice with their legs or wings. In the sea, they display a degree of speed and agility surpassing that of many fish. Penguins appear to 'fly' in the water with powerful strokes of their wings. They can swim at 15 mph over long distances and can go twice as fast for short periods when pursuing prey or escaping from enemies. Their flat, webbed feet stretch out behind them

COMING UP FOR AIR *Adélie penguins breathe as they skim above the surface, then duck back into the sea a moment later with flippers whirring. Thus the need to breathe does not impede their progress, which is about 15 mph*

and help to steer their torpedo-shaped bodies. To breathe, penguins break the surface periodically, arch through the air like porpoises, and disappear again with scarcely a ripple. To come ashore, most species shoot vertically out of the water and land standing on their feet.

Penguins feed on the fish, squids and krill which abound in the cold southern seas. In turn, they are the prey of leopard seals and occasionally killer whales. On land, adult penguins are virtually without enemies, but skuas, gull-like hunters that breed on the outskirts of penguin colonies, take a heavy toll of the eggs and young chicks.

Antarctic breeding

Adélie and emperor penguins breed on the Antarctic continent. The Adélies lay their eggs in summer when temperatures rise above freezing. But the 3-ft emperors lay their eggs in the middle of the frigid Antarctic winter, when the temperature drops to −62°C (−80°F) and blizzard winds blow across the waste land at 80–100 mph.

Soon after a female emperor penguin has produced her single egg, she departs for feeding grounds in open water – a journey that may take her across vast stretches of ice. For two months, the male incubates the egg on his feet. A fold of skin on his lower abdomen covers the egg and warms the embryo inside. If the egg hatches before the female returns, the chick's first food is a secretion from the crop of the father, who has fasted throughout the incubation.

When the mother returns, her crop filled with several pounds of food, the parents reverse roles. The female takes charge of the chick, and the emaciated male sets forth on the long trek to open water.

After the growing chick is about six weeks old, it is fed by both parents. With the arrival of summer, young emperors are mature enough to reach the water, which is many miles closer because of the break-up of the sea ice.

LEAPING FROM THE SEA *Adélie penguins can catapult them-selves several times their own height on to a rock or iceberg. They climb awkwardly, using their flippers for balance*

DEEP-SEA DIVERS *Unlike flying birds, Adélie penguins have solid bones. As a result their buoyancy is reduced and they have been known to plunge from an iceberg to a depth of 60 ft*

ADÉLIES, PENGUINS OF THE PACK ICE
Flightless and ungainly on land, they flourish in the Antarctic seas

The 30 in. tall Adélies are the best-known members of the penguin family, probably the most primitive of all living birds. Incapable of flight, ungainly on land, they are superbly adapted for life in the water – and flourish in the icy, food-rich seas of the Antarctic. Adélies spend the winter at sea, feeding on krill and small fish. In October (the Antarctic spring), they waddle back to their breeding rookeries across as much as 60 miles of pack ice. Males arrive first, establish territories, then attract mates with ritual displays. After pairs form, both sexes gather stones for the nest, often stealing them from neighbouring birds, and take turns incubating the eggs. At two-week intervals, one parent returns from feeding at sea and relieves the other, which then sets off to feed. The chicks hatch in about five weeks, then spend about a month in the nest and another month in a large group of youngsters, although parents continue to feed them. At the age of eight or nine weeks, the chicks moult and go off on their own, ready to face life in the sea.

PENGUIN PARADE *Huge flocks of Adélie penguins gather near the edge of the Antarctic pack ice. As many as 10,000 of the birds have been found together in a rookery – perhaps finding safety in numbers from skuas and other predators*

PENGUIN VICTIM *One of the few natural enemies of the penguin is the leopard seal, which gulps a victim down head first*

STONE NESTS *Male and female penguins take turns at incubating their eggs in nests of stone – the only material available*

MULTI-COLOURED BILLS *Puffins use their multi-coloured bills as a weapon in their squabbles, as a tool for digging their breeding burrows and as a form of display in courtship. The bills are brightest in summer when the birds go ashore to breed*

BIRDS OF THE NORTHERN SEAS

Penguins live only in the Southern Hemisphere. In the north they have distant but strikingly similar relatives – the alcids, a family that includes auks, puffins and guillemots. As awkward as penguins on land, the black-and-white alcids are remarkably agile in water. They, too, feed on fish, squids and shrimps, and 'fly' under water by means of stout swimming wings and webbed feet. These similarities are a striking example of parallel evolution – of two different groups independently evolving similar adaptations.

In one respect, however, all living alcids differ markedly from penguins – they can fly. The largest alcid, the 30 in. tall great auk, was the only flightless bird in the Northern Hemisphere during recent times. The great auk's flightlessness was its undoing; it was easy prey for hunters and became extinct in the mid-19th century.

In spring, the islands and coastal cliffs of northern oceans bustle with crowded colonies of alcids. Auks and guillemots nest on ledges and among jumbled boulders, puffins crouch in crevices and burrows on the bluffs above. These precarious nesting sites have resulted in a curious adaptation. The eggs of the ledge dwellers are markedly pear-shaped, making them less likely to roll over the edge than round or oval eggs.

Common guillemots typify the ledge-nesting alcids. They spend the winter in offshore waters, then gather in dense, noisy colonies on cliffs along the northern Atlantic and Pacific coasts. By May, the guillemots have paired off and staked claim to a small bit of ledge space. On the bare rock, the female lays one egg, which is cared for by both parents. Egg and hatchling mortality is high. Jostling adults knock many off the ledges; other eggs and hatchlings fall prey to voracious gulls.

In July, a young guillemot leaves its ledge and enters the ocean, although it has not yet developed its flight feathers. The bird literally 'takes the plunge'. Encouraged by its excited parents it walks to the brink and, after much hesitation, leaps into the sea. One or both parents join the newly launched young bird, and together they

160

THE AUK FAMILY

Members of the auk family stand erect on short legs set far back on their bodies. With webbed feet and small wings they are clumsy on land, but they are efficient swimmers and divers. Unlike their distant relatives, the penguins of the Antarctic, these modern Arctic birds can fly. Only the now-extinct great auk was unable to take to the air

swim out to sea, their home for the winter months.

The common puffins, easily recognised during the breeding season by the male's prominent red, yellow and blue striped bill, nest in about the same localities as the Atlantic populations of guillemots. Puffins arrive at the cliffs and rocky islands in late March or April. After courting and mating, each pair finds a crevice or excavates a burrow several feet long. The female lays one egg, and both birds take turns keeping it warm. The parent on duty leans awkwardly against the egg, tucking it under one wing. In about 40 days, the hatchling emerges – a black-and-white ball of fluffy down with an insatiable appetite. Both parents keep busy supplying the hungry offspring, sometimes carrying as many as 18 fish in one load. After six weeks, this parental care ends abruptly, and adult puffins depart for the open sea. The young puffin spends about another week alone in the burrow, then makes its way to the water. Here it learns to swim, dive and fly without the guidance or protection of older birds.

GUILLEMOT COLONY *No space is wasted in a crowded breeding colony of guillemots, with adults sitting nearly shoulder to shoulder on a narrow cliff edge. Each pair of birds produces one egg, which is pear-shaped and so is less likely to roll off the cliff if unattended. Even so, many eggs and young chicks are lost when they are jostled off the rock by adults returning from or embarking on feeding trips. In the absence of the adults, many fall victim to gulls and other hungry predators*

THE FEATHERED FISHERMEN

South of the alcids and north of the penguins, temperate and tropical seas are populated by immense numbers of seabirds. Boobies, cormorants, pelicans, frigate birds, tropic birds and skimmers live here, nesting on shore and feeding in coastal waters.

These middle-latitude birds use a wide variety of fishing techniques. Boobies plummet into the sea from heights of 60–100 ft and catch flying fish just as they re-enter the water. Cormorants swim along at the surface, periodically diving to chase fish; the birds have been found tangled in fishermen's nets from depths of 70–100 ft. Tropic birds roam across enormous distances, flying alone or in pairs, searching for fish.

Pelicans feed just under the surface, using their capacious throat pouches as effective seining nets. The brown pelican, the most fully marine of the pelicans, patrols about 30 ft above the water. When it sights a school of small fish, the bird folds back its wings and crashes into the water like a falling rock. The impact stuns nearby fish long enough for the pelican to gather a sizable catch, along with 2 or 3 gallons of water, in its pouch. The bird cannot get into the air again until all water drains out of its pouch. Even then it flaps and flails a good deal before taking off.

Agile frigate birds

In contrast to the diving birds, the frigate birds feed without wetting more than their bills. They skim along just above the waves, snapping up jellyfish, squids, fish, young sea turtles and bits of carrion. Particularly fond of flying fish, the frigate birds sometimes catch them in mid-flight.

Frigate birds are nearly as capable in flight as

WAITING FOR FOOD *With a wing span of up to 5½ ft, and more than 2 ft long, the masked boobies are the largest of the booby family. They feed mainly on flying fish, which they catch either by snapping in mid-flight or by diving under water in pursuit*

albatrosses. With exceptionally light bodies and 7-ft wing span, the birds can swoop and soar on air currents for hours before returning to land. But if a frigate bird is forced to settle on the ocean's surface, it is unlikely to rise again; it has no oil-producing glands to waterproof its plumage. Unless a strong gust of wind quickly lifts it aloft, the bird will probably drown.

Frigate birds are accomplished pirates. A flying booby that has just swallowed a fish is no match for these swift, noisy marauders. The frigate birds pummel and jostle their unfortunate victim until it regurgitates its catch, then they grab the morsel before it disappears into the ocean.

During the breeding season, frigate birds build loose platform nests in trees and shrubs on isolated islands or mangrove tangles. The birds gather feathers, bones and other nesting materials from the ground, steal from other nesters, and break branches off trees and bushes. One parent always remains at the nest, for frigate birds regard the eggs and chicks of their neighbours as fair prey.

RED BADGE OF COURTSHIP *A scarlet throat sac distinguishes the male frigate bird. During courtship, the sac is puffed up to attract females. The bird's wing span is up to 7 ft*

BIRD THAT CANNOT STAND *The red-billed tropic bird, found in warm coastal seas around the world, can fly, swim and dive – but it cannot stand or walk. To move about on land the bird rests on its belly and pushes itself along with its feet*

MARINE MAMMALS

The early history of mammals generally parallels that of birds. Like birds, mammals evolved from reptiles, became warm-blooded, and developed an insulating body cover (hair instead of feathers), keen senses and a larger brain.

Unlike birds, however, nearly all mammals abandoned the shelled-egg method of reproduction. The earliest mammals, which may have resembled the modern platypus and spiny anteater, probably laid eggs. But most modern mammalian young develop within the mother's body and are linked by a placenta to her circulatory system. After birth, the young are nourished by the mother's milk and receive parental care for an extended period.

When many of the mammals' reptilian competitors became extinct between 60 and 65 million years ago, the pace of mammalian evolution accelerated tremendously. Their numbers, variety and size rapidly increased, and they spread from dry land into rivers and seas.

The first mammals to enter the sea were probably the ancestors of the whales. Next came the ancestors of dugongs and manatees. Millions of years later came the ancestors of seals, sea lions, sea otters, walruses and polar bears.

The gradual adaptation of mammals to life in the ocean is illustrated by the marine mammals alive today. From polar bears to whales, there are mammals in various stages of transition from a terrestrial to a marine way of life.

The sea-going bear

On the evolutionary time scale, the polar bear has only very recently become a part-time marine mammal; only a scant million years ago, its ancestors were land dwellers. The polar bear and the land-dwelling brown bear are so closely related that in captivity they can interbreed. But thousands of years of hunting on the pack ice and in the sea have changed the polar bear. Adapted for a life in which swimming plays an important role, it has a smaller head, a longer neck and a more slender and streamlined torso than the brown bear. The polar bear also displays a 'Roman nose' profile that cuts through the water with a minimum of resistance.

Polar bears venture hundreds of miles out to sea, but most commonly they keep to regions that offer a mixture of land, ice and open water. They

POLAR BEAR *The great white aquatic bears that are found in the polar regions are protected from cold by a 3-in. layer of fat*

THE RIGHT SHAPE *More streamlined than the brown bears, slim, long-necked polar bears can hunt on ice and in water. Their claws are short and straight and their feet are insulated against the cold sea and ice by a covering of hair*

LEARNING TO HUNT *Young polar bears (extreme left and right) stay with their mother for almost two years, learning to hunt. When their preferred food – seals – is not available the bears live on birds, fish, lichens and moss*

give birth to their young on land; they hunt for seals, their staple diet, out on the drifting floes of ice.

To catch a wary, fast-swimming seal, a polar bear relies upon stealth rather than speed. When it spots a seal dozing in the sun on an ice floe, the bear moves downwind and begins a slow, cautious approach. The seal may periodically wake up to make a quick check of its surroundings; the bear freezes motionless until the prey closes its eyes again. If the final approach must be made through water, the bear quietly eases itself in, hind feet first, and swims just beneath the surface. Occasionally it raises its head just above the water to breathe and to measure the remaining distance. Upon reaching the seal's patch of ice, the bear clambers stealthily out of the water and crushes the skull of its unsuspecting victim with a single blow of its massive forepaw.

When no seals are available, polar bears consume anything edible: fish, seabirds and their eggs, lemmings, plants and carrion. A dead whale washed ashore attracts polar bears from many miles around.

During the brief midsummer breeding period, pairs may travel together; at other times, polar bears are solitary roamers. In autumn, a pregnant female excavates a roomy den in a snow-bank and slumbers until the birth of her twins in early winter. The den shelters the mother and cubs well; temperatures inside may be 40 degrees higher than outside. The family leaves the den in April, and the cubs often remain with their mother until they are two years old. Under rather severe maternal supervision, they learn how to survive in a bleak world where starvation will be a life-long threat.

Man is the polar bear's only enemy, but a formidable one. The bears have been badly over-hunted, presently numbering less than 15,000. However, now that hunting is legally limited, it is to be hoped that the polar bear will not join the lengthening list of animals that have been wiped out by human greed or indifference.

EATING AT SEA *A sea otter devours a starfish with its powerful incisors and molars. Sea otters also eat shellfish*

THE OTTERS WHICH SLEEP AT SEA

Superb fur – and a taste for shellfish

The smallest marine mammals are the web-footed sea otters, members of the weasel family. They are well adapted to ocean life and seldom come ashore except to give birth. They live in dense North Pacific kelp beds, where they swim or sleep among the algae's fronds. Unlike most marine mammals, sea otters are not insulated by a layer of blubber beneath the skin. Instead, they are kept warm by air trapped in their thick, heavy fur. These luxurious coats nearly doomed them; by the end of the 19th century, fur trappers had almost exterminated the species. Through the efforts of conservationists, sea otters are now making a strong comeback.

ANIMALS THAT USE TOOLS *Sea otters sometimes smash open shellfish by hammering them against flat stones balanced on their chest. Because they compete with man for the meat of the abalone shellfish, they are often killed by fishermen*

PRODIGIOUS APPETITE *A sea otter, weighing about 80 lb., lies on its back to nibble a mussel. To supply the energy needed to stay alive, a sea otter must eat about 20 lb. of shellfish meat – a quarter of its own weight – every day*

BUOYANCY AND WARMTH *Sea otters keep themselves afloat and warm by pockets of air inside their dense fur. Unlike other marine mammals they do not have insulating layers of fat. They often congregate in herds – called pods – of several dozen animals*

MANATEE AND CALF *A manatee calf (top) nuzzles its mother. The manatees, related to land elephants, lost their hind legs in adapting to a life spent entirely in the water. They feed by sweeping up ocean vegetation with the bristles attached to their lip pads. Because they often nurse their young in an upright position, manatees have been mistaken by sailors for mermaids*

HERDS OF GIANTS *Walruses frequently congregate along the northern ocean shores and ice floes in herds of 100 or more animals. Bulls often reach a weight of 3700 lb. — which includes 900 lb. of blubber*

AT HOME IN THE ICE *Emerging from under the ice, a Weddell seal surfaces near Scott Base, Antarctica. These seals winter beneath the ice, after cutting breathing holes with their strong teeth. Once under water, they can find the holes they have prepared*

Fin-footed mammals at home in the water

Like sea otters, seals and their relatives still come ashore to bear their young. But in every other respect, these pinnipeds – 'fin-footed ones' – are a more fully marine group of mammals. They travel far, swim fast and dive deep in nearly all the oceans of the world. They are big animals: the smallest, the ringed seal, is $4\frac{1}{2}$ ft long and weighs 190 lb.; the largest, the male southern elephant seal, reaches 20 ft and $3\frac{1}{2}$ tons.

The pinnipeds evolved at least 50 million years ago from four-legged land dwellers. The earless or 'true' seals have progressed furthest towards a full-time oceanic life; their hind limbs have become almost tail-like and can no longer turn forwards. As a result, an earless seal on land 'humps' along somewhat in the manner of an outsized caterpillar. Eared seals – fur seals and sea lions – and walruses hobble slowly and awkwardly on all four legs.

Each year northern fur seals form the world's largest herd of mammals when about $1\frac{1}{2}$ million of them go to their breeding grounds on the Pribilof Islands in the North Pacific.

The earless seals make astonishingly long and deep dives. The Weddell seal of the Antarctic Ocean can dive to nearly 2000 ft and remain submerged for more than an hour.

When it dives, a seal conserves oxygen with a number of extraordinary mechanisms. The animal's metabolic rate drops, so that its body consumes oxygen at a much slower rate. The heartbeat slows to one-tenth its normal pace. The pattern of blood circulation changes, and the heart and brain, the organs most quickly damaged by an oxygen shortage, receive a greater share of the blood flow. Also, because their blood carries more oxygen than that of most mammals, seals can take a larger supply of the life-sustaining gas down with them.

The ability of the seals to dive far beneath the surface of the sea demonstrates the versatility and adaptability of the basic mammalian design. But even the remarkable seals have not fully exploited the possibilities for ocean life. Among the mammals, the true conquerors of the marine environment are the whales.

Of whales and men

The whale family ranges from 100-lb. porpoises to 100-ton blue whales. Their ancestors once walked on land but whales are as much at home in the water as fish. Only one enemy – man – threatens their continued survival

Whales are the mightiest creatures the world has ever known. A fully grown blue whale weighs as much as 30 elephants and could easily carry the largest dinosaur on its back.

Because whales are mammals, they must have evolved from land animals. But few fossils have been found to document their transition from a terrestrial to a marine way of life. Attempts at reconstructing their early history begin with the oldest-known whale fossils, those of the extinct archaeocetes – 'old whales'. The archaeocetes had peg-shaped front teeth similar to those of some modern whales, and triangular molars similar to those of certain primitive land-dwelling predators. The presence of molars, absent in modern whales, suggests that the archaeocetes might have been descendants of small terrestrial meat-eaters.

The first step in that long, mysterious evolution may have come as these land mammals began wading into the water along shores of lakes and shallow seas some 70 million years ago. During the next 15–20 million years, their descendants ventured deeper and deeper and stayed longer and longer. Perhaps food was more plentiful, or enemies scarcer in the water. Whatever the reasons for leaving land, the ancestors of whales adapted to ocean life mainly in this period.

The fossil trail begins early in the great age of mammals, about 50 million years ago, with the appearance of the archaeocetes, which closely resembled modern whales. The skulls of these ancient whales were elongated; the nostrils had moved from the snout towards the top of the head – the ultimate location of the blowhole. During the 20 million years of whale evolution unrecorded in fossil history, the mammalian hind legs disappeared, except for internal remnants. Forelegs became flat flippers, used for steering and maintaining balance, while the tails assumed a dominant role in locomotion. The tail vertebrae increased in size to become the largest in the spinal column, permitting the attachment of powerful swimming muscles.

The archaeocetes reached their peak of development with a group of giants. The largest had slender 70-ft bodies and were proportioned more like the sea serpents described in works of fiction and seamen's legends than like the whale species that still exist today. About 25 million years ago, the last of the archaeocetes vanished, but not before the appearance of ancestors of the two groups of living whales – toothed whales and baleen whales. Toothed whales first appear in the fossil record about 50 million years ago, baleen types about 30 million years ago.

SEA-DWELLING MAMMAL *All modern whales, including the white 800-lb. beluga (left), are the descendants of land mammals that took to the sea some 70 million years ago. Beluga whales gather in small herds in coastal waters within the Arctic Circle and swim south when the pack ice becomes too thick*

A MARVEL OF EVOLUTION

Seventy million years of evolution have gone into the refinement of one of nature's most awe-inspiring marvels – the giant mammal that has triumphantly conquered the seas.

As whales evolved, they developed shapes strikingly similar to those of fish. Indeed, fish and whale shapes differ in only one major respect. Fish tails are vertical, whale tails horizontal. Fish swim by undulating their bodies and beating their tails from side to side; whales swim by fanning their broad tails up and down.

'In the tail,' wrote Herman Melville in *Moby Dick*, 'the confluent measureless force of the whole whale seems concentrated.' With vertical sweeps of its great tail, a blue whale thrusts its ponderous bulk through the water at 22 mph for short sprints. The sperm whale can match this swift pace when alarmed, or swim on long journeys at 6 to 7 mph. Even porpoises and dolphins, which are small whales, can outrun a ship steaming at 16 knots (18 mph) by flexing their tails three or four times each second. Sometimes these whales 'hitch' rides in the bow waves of sizable ships. With a burst of speed, they swim into the waves and glide along effortlessly just beneath the surface.

Streamlining contributes to the whales' swimming ability. Shorn of the coat of hair that covered their land-dwelling ancestors, smooth-skinned whales slip through the water with little friction. The tender outer layer of skin covers a somewhat firmer but pliant inner layer. The layers wrinkle and flex together to keep water flowing smoothly past the body.

Just before diving, sperm, right, grey and humpback whales often toss their enormous flukes – the fleshy tail lobes – into the air in what appears to be a magnificent salute. Since whales are buoyed up by their thick blubber, perhaps this gesture helps them to dive. Sometimes whales flip their tails out of the water and bring the heavy flukes down on the surface in a tremendous slap, creating a thunderous boom audible for miles. Whalers have dubbed this behaviour 'lobtailing', and it may be an act of sheer playfulness, rather than a necessary contribution to movement.

Like seals and walruses, whales are accomplished divers. Dolphins and porpoises dive to

SPOUTING STALE AIR *When whales surface from the ocean depths they exhale the stale air in their lungs in mighty geysers, up to 11 ft high. These two spouting whales are Californian greys – probably a mother (on the left) with her calf*

650 ft. Baleen whales, such as fin whales, dive to at least 1500 ft when frightened. One harpooned fin whale was still sounding – diving straight down – when it broke its neck by striking bottom at 1600 ft. The deepest divers, sperm whales, are known to plunge to 3600 ft and stay under water for well over an hour.

When a whale submerges, its pulse rate drops. A dolphin's heart beats 90 times a minute at the surface, but only 21 times under water. Blood circulation to skin and blubber diminishes as the whale dives, conserving both body warmth and oxygen. As the animal plunges, blood retreats from limbs and muscle layers, and the life-sustaining oxygen supply is saved for the brain and heart. Additional oxygen in the blubber and muscles passes to the bloodstream.

When the whale finally surfaces, it exhales stale air from its lungs in great spouts. Geysers spurt up from the puffing animal. In *Paradise Lost*, Milton compared the whale to 'a moving land that draws in, and at his breath pours out a sea'. Poets and whalers alike perpetuated the misconception that whales spout water. But the warm, fishy-smelling spout is actually exhaled air mixed with mucus and oils from the respiratory system and sea-water from outside the blowhole, condensed into a visible cloud of water droplets.

HITCHING A RIDE *Dolphins, and other members of the whale family, frequently ride along like surfers in the bow waves of passing ships. Their low-friction, flexible skins and stream-lined bodies help them to reach high speeds over short distances*

LARGEST CREATURES THAT EVER LIVED

In spite of their size, baleen whales are among the world's least aggressive creatures. These sea-dwelling mammals, which include the 100-ton blue whale, the most massive monster the earth has ever seen, feed only on minute plankton which they strain from the sea through rows of horny plates – called whalebone or baleen. The baleen whales differ greatly in size: the largest, the blue, may be 100 ft long; the fin usually no more than 80 ft; the black right whale, 60 ft; the humpback, 50 ft; and the minke whale, 30 ft. Several species migrate from polar feeding grounds, where they have been hunted almost to extinction, to breed in more hospitable, temperate waters

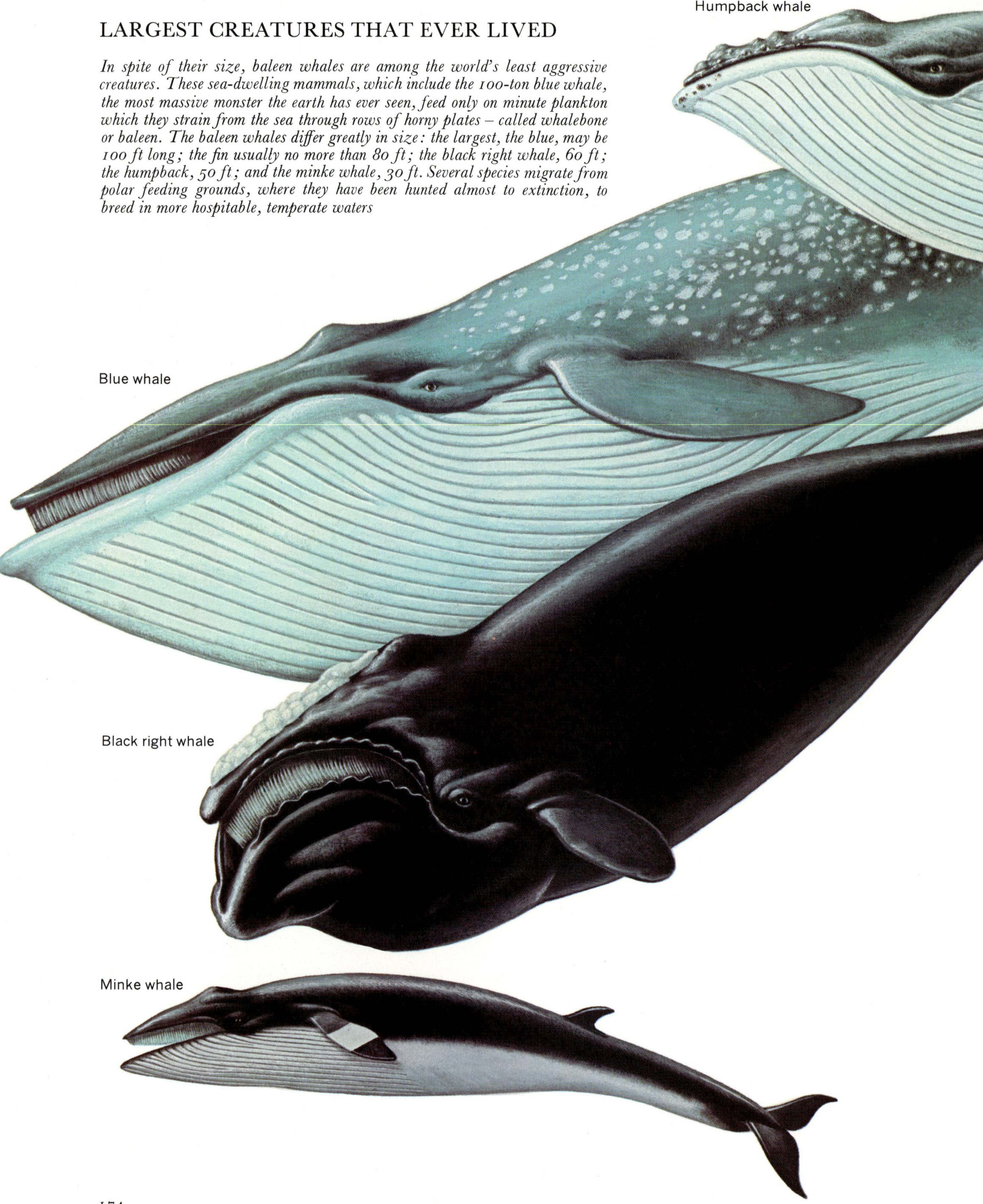

BALEEN WHALES, THE GENTLE MONSTERS

Baleen whales are named after the thin, triangular plates that hang from the roofs of their mouths. These tough, resilient baleen (or whalebone) plates are made of the same horny substance as fingernails and horses' hoofs. From the middle of the 17th century, baleen whales were hunted, both for the oil from their blubber, and for these plates, which were used as corset stays, umbrella ribs, carriage whips, sledge runners, and other items requiring both stiffness and flexibility.

A baleen whale's mouth contains 250–400 baleen plates hanging one behind another, less than $\frac{1}{2}$ in. apart. Long, hairy fringes on their inner edges strain food from sea-water. To eat, a baleen whale lets sea-water pour into its mouth. Then it closes its mouth and squeezes the water through the baleen and out of the corners of its mouth by moving its tongue or contracting its throat muscles. Krill and other small organisms suspended in the sea-water tangle in the baleen threads. The massive tongue, which may weigh as much as a fully grown elephant, licks off the food and pushes it down the throat. Thus the largest creatures in the ocean feed on some of the smallest, but in prodigious quantities. A fully grown blue whale consumes as much as 4 tons of krill a day.

The largest baleen whale – and the largest creature that ever lived – is the blue whale. This species reaches a length of nearly 100 ft and weighs up to 135 tons. Named after its bluish back, it is also known as the 'sulphur bottom' because a coating of yellowish diatoms, the minute single-celled plants, lives on the blue whale's flanks and belly when it feeds in polar waters.

For all its size, the blue whale is remarkably streamlined. Its body is relatively long and slender, its head is fish-like and flat on top. The jutting lower jaw, together with the throat and chest, creates a bowl-shaped reservoir that holds a small lake of salt water and krill. The blue whale, like the fin, sei and humpback, has grooves or pleats running along its undersurface. The function of these pleats is not known. They may expand and allow the tongue to move back, thereby increasing the mouth's capacity. They may function as stabilisers for swimming, or as some kind of braking device.

Right whales, which grow as long as 60 ft, are less streamlined. Their rather thick, blunt shape makes them slow swimmers, and they rarely move

WHALES IN COURTSHIP *Two grey whales frolic off the coast of California. Before mating, the whales swim side by side and nuzzle each other for several hours. Mating, when the whales face each other, may take only a few seconds*

faster than 5 mph. Their backs are finless, their bellies pleatless. Their enormous arched heads contain up to a ton of baleen plates.

Before the days of fast boats and high-powered harpoon guns, these animals were the 'right' whales to hunt. Their blubber contains so much oil that they float after being harpooned. Other large baleen whales sink unless they are quickly inflated with compressed air.

12,000-mile migrants

Baleen whales populate temperate and polar seas in both the Northern and Southern hemispheres. Blue whales and several other species are believed to feed only in the warm months, fasting and breeding in the winter. This means that northern herds are breeding when southern herds are feeding, and vice versa, so that whales from the Northern and Southern hemispheres do not mix.

Whales nearly always travel in groups, called herds or schools. They are sociable and apparently loyal; if one whale is in danger, others will gather round to help it.

Spring blooms of diatoms and krill attract blue, fin, sei and humpback whales to the Arctic and Antarctic, and grey whales and other northern species to the Arctic. As sunlight wanes with the passing of summer, plant production dwindles and autumn ice begins to cover the feeding grounds. With the exception of Greenland right whales, baleen whale herds move out of polar seas and begin the annual migration to subtropical breeding areas.

In late autumn, grey whales set off on one of the longest migrations undertaken by any mammal – a 6000-mile journey from feeding grounds in the Bering Sea and the Arctic Ocean. By December or January, they reach the coast of southern California. The pregnant cows arrive first, followed by other females and the males.

The long voyage ends in shallow breeding lagoons on the coast of Baja California, Mexico. There, 1500-lb. babies up to 14 ft long are born. In March, April and May, adults and young begin moving at a steady pace up the California coast on the return leg of their 12,000-mile round trip. Thousands of these barnacle-blotched creatures, some almost 45 ft long, swim northwards to their Arctic summer home.

Courtship—with slaps

Scientists believe grey whales and other baleen species pair for life. Courtship play, which sometimes lasts for hours before mating, takes a variety of forms. A pair of whales may rub one another with their flippers, bite, nuzzle, roll over and over in an embrace, or leap together from the water. Humpback whales are most demonstrative. Ugly in appearance, with unsightly lumps on their heads, chins and jaws, they act like tender lovers. E. J. Slijper, a Dutch zoologist, reported that these 45-ft giants gently stroke their partners with their exceptionally long flippers before mating. They also use the flippers to give their partners slaps that can be heard for miles.

Female baleen whales usually produce a single calf every other year. In a lifetime of 15–30 years, a female bears from six to 15 young. The newborn baleen calf enters the world tail first, eyes wide open, and usually with enough stamina to swim past its mother to the surface for its first breath. If the calf cannot do so, the mother and other adults rush to push the infant upwards.

The calf is about one-third adult size at birth. A newborn blue whale may measure 25 ft and weigh $2\frac{1}{2}$ tons. The calf wastes no time looking for food; it begins to suck milk immediately after finding its mother's teats, concealed in two slits near the tail. Suckling is speeded up by the mother pumping milk into the calf's mouth. On the fat-rich milk, a young blue whale gains up to 200 lb. a day. By seven months, it is 50 ft long and can become independent of its mother's milk.

STANDING ON ITS TAIL *A grey whale, straining plankton from its baleen plates into the first of four chambers in its stomach, 'stands' in the water. The animal, which has surfaced to breathe, can maintain the standing position for up to a minute*

MIGRATING IN HERDS *Grey whales usually migrate to their breeding grounds in herds or pods – often as many as 40 whales swimming together. On the return journey northwards, when the whales are not searching for a single destination, they scatter more widely*

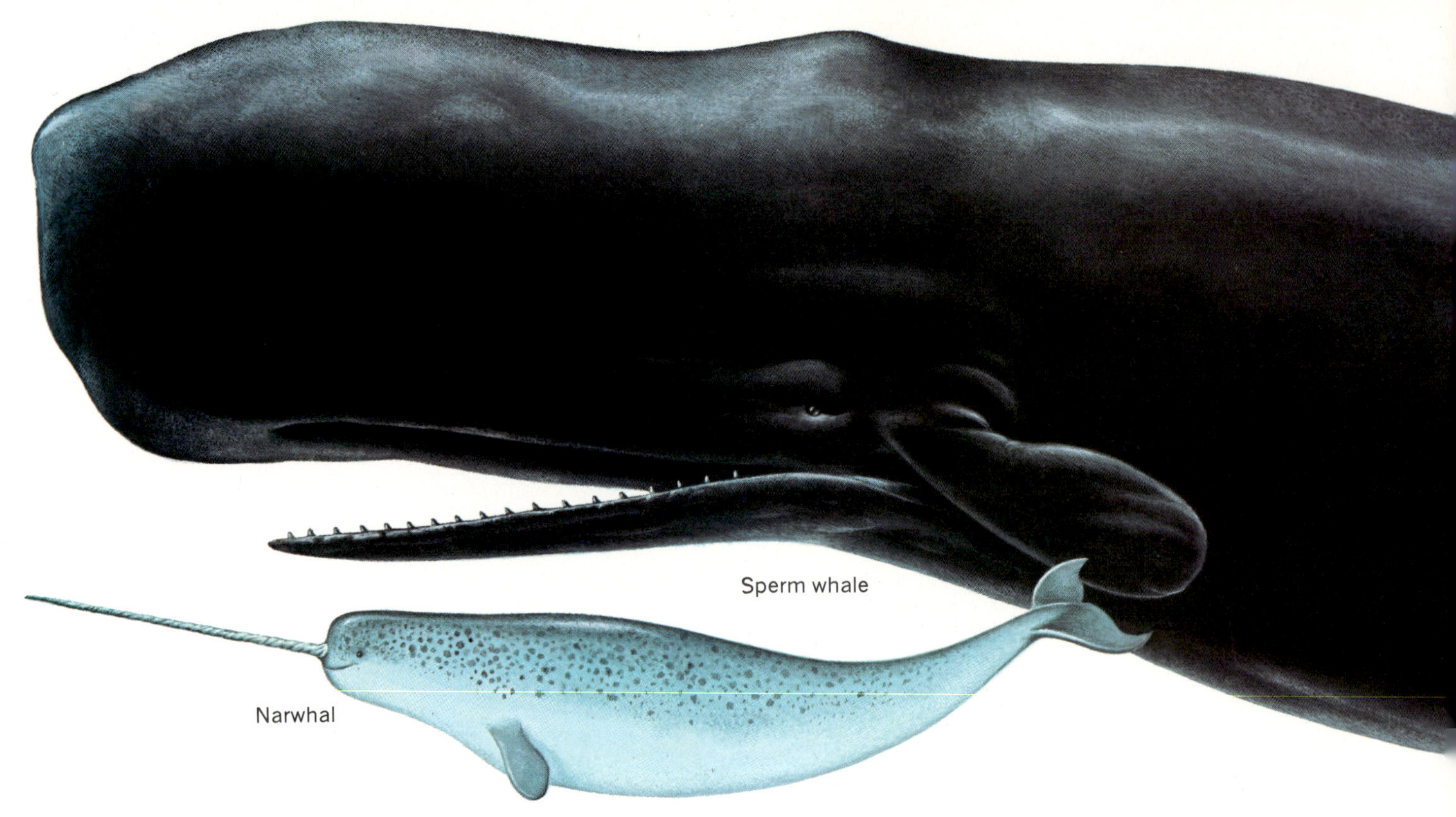

THE WHALES WITH TEETH

None of the toothed whales is as big as the blue whale, but the biggest toothed animal is nevertheless a whale. Toothed whales use their teeth not for chewing but for grasping prey, which they swallow whole. Fish use their teeth in the same way, but land animals do not.

Seventy-five species of toothed whales survive, compared with ten baleen species. Among the toothed kinds are sperm whales, porpoises, dolphins, narwhals, belugas, killer whales, false killer whales and pilot whales.

The giant of the toothed whales is the sperm whale. A 60-ft bull sperm whale can weigh more than 50 tons. One-third of an adult's entire length is its formidable head, which gives the animal a majestic aspect. Herman Melville likened the flat front of a sperm whale's head to a battering ram. The eyes are low on the sides, just above and behind the lower jaw.

Compared with the rest of the head, the lower jaw seems absurdly small. It is a long, narrow boom, studded with 36–60 conical teeth, each weighing up to a pound. These the sperm whale uses to grab and hold giant squids and other large prey. The remains of a squid 34 ft long were found on one occasion in a sperm whale's stomach.

Behind the huge forehead is a large oil-filled reservoir. Whalers mistook the oil for the animal's semen – hence the name 'sperm whale'. Chilled sperm oil yields spermaceti, a waxy substance once in great demand for candles and now used chiefly in the manufacture of ointments. Refined sperm oil makes an excellent lubricant for watches and precision machinery.

The sperm whale also produces ambergris, a grey substance found in the intestine. Purified ambergris is an ingredient of expensive perfumes, although synthetics now frequently replace it.

Most toothed whales do not migrate. Sperm whales, however, do make seasonal journeys between breeding and feeding grounds. During the mating season, a sperm bull lives in a harem, a group of about 30 cows and calves. In summer, some bulls leave their family groups in near-equatorial waters and swim as far as both polar seas. The migration may be connected with the movements of cuttlefish and other prey. Remaining in warmer waters, cows carry their young for about 16 months and produce a calf about once every four years.

MIGHTY HUNTERS
OF THE SEAS

Toothed whales – from the fierce 50-ton sperm whales to the 8-ft dolphin – are carnivorous animals that prey on fish and molluscs in the oceans. Although they have terrified human beings from biblical times, even the voracious 30-ft killer whale seldom attacks men; it feeds mainly on fish, seals and other whales. With its distinctive shark-like dorsal fin, it roams most of the world's oceans. The 35-ft bottle-nosed whale, and the 20-ft pilot, are found in northern waters, as is the 17-ft narwhal, easily recognisable by its 8-ft tusk

PLAYFUL BUT TREACHEROUS *Although killer whales can be trained to entertain, they are dangerous predators with up to 50 large teeth. They feed on a variety of warm-blooded animals and have been known to topple ice floes to catch penguins*

The terrifying killers that hunt in packs

Baleen and sperm whales have been ruthlessly slaughtered by centuries of whaling, but smaller toothed whales have generally escaped this fate. They make difficult targets and yield much less blubber than larger whales.

Killer whales are highly social creatures of considerable speed, power and intelligence, with no natural enemies. They prey primarily on fish and squids, but also devour seals, sea lions, young walruses, porpoises and dolphins. They also attack large baleen whales. Reliable reports tell of grey whales so paralysed with fear in the presence of killer whales that they roll over on their backs, as if helplessly awaiting the end. Not even the enormous blue whales are safe. When killers attack blues, rights and other baleen whales, they tear away only the tongues, leaving the rest of the victim's body for other predators.

Killers in packs of two to 40 roam all oceans. Easy to recognise, they have jet-black backs, white bellies and throats, and triangular dorsal fins as high as 6 ft on males. Females seldom grow longer than 15 ft, which is about half the length of the biggest bulls. Both sexes have about 50 large, strong teeth.

For many years, killer whales were considered the most vicious predators in the sea. This impression was reinforced when a 21-ft individual was found with the remains of 13 porpoises and 14 seals in its stomach. But in 1964, a killer was captured alive and studied closely for the first time. Named Moby Doll, it proved to be an intelligent, friendly, playful animal that allowed men to swim in its pen, liked to have its belly scratched, and ate fish out of a person's hand. Since then, several captive killers have been trained to perform in aquariums. Swimmers put their arms – and even their heads – into the whales' tooth-filled mouths, pull the animals around by the blowholes, and ride on their backs. The whales seem to enjoy leaping out of the water, and can be quickly trained to do so on command. Trainers and scientists agree that these once-maligned animals exhibit a gentleness and high intelligence matched only by the more familiar entertainers – dolphins and porpoises.

HUNTING IN PACKS *Male and female killer whales hunt together. Unlike the baleen whales, where the female is always larger than the male, the female killer whale is never more than about half the size of her partner*

STREAMLINED FOR SPEED *With powerful tail flukes propelling it through the water, the streamlined dolphin can reach speeds of 30 knots. Just before diving the dolphin inhales a 15-minute supply of air through its single blowhole, seen open above*

BOTTLE-NOSED DOLPHIN AND CALF *Dolphins are born under water, but they surface immediately to take their first breath of air. If a calf does not do so, it is usually nudged upwards by its mother or other members of the herd*

DOLPHINS AT PLAY *Dolphins frequently gather in great herds, often many hundreds strong, to cavort off the coastline in warm seas. Leaping and nudging one another near the bows of ships, they are also a common sight for ocean travellers*

The playful and mischievous bottle-nosed dolphins

There are 50 species of dolphins and porpoises, and much confusion about which is which. Strictly speaking, porpoises are smaller than dolphins and have rounded snouts instead of beaks. But for most Americans and Europeans, 'dolphin' or 'porpoise' means the bottle-nosed dolphin found in temperate waters of the Atlantic and Pacific. Named after its prominent beak, the bottle-nosed dolphin can grow to 12 ft and weigh 500–600 lb.

In the wild, bottle-nosed dolphins leap out of the water, frolic around ships, and push floating objects, sometimes tossing them into the air and catching them. Tales of dolphins and porpoises coming to the aid of drowning people date back to ancient times. These whales have undoubtedly pushed floundering swimmers ashore more than once, but there is no reason to believe that they deliberately try to save human lives. Rather, they seem to have an instinctive urge to push objects they find floating in the sea.

Extremely gregarious, dolphins live in groups of from six to several hundred individuals. Large groups usually consist of a number of smaller family units. They take excellent care of their young and are quick to help one another. A blast of dynamite once stunned a member of a dolphin school. Instead of immediately fleeing, the others took turns holding their injured companion at the surface until it revived. Only when it could escape did the rest of the pack speed to safety.

In captivity, bottle-nosed dolphins are playful and mischievous. They tease fish by pulling their tails, take unwilling turtles for rides, throw objects out of their tanks, and squirt water at spectators. When rewarded with fish, dolphins can be trained to perform a variety of tricks. They jump through hoops, leap as high as 30 ft, toss rubber balls with their beaks, and dance on their tails. But captive dolphins are not all fun. They slap and bite one another in fights over females and food, and they attack human beings when antagonised or handled roughly. Several scientists and trainers have been rammed or bitten during experiments and training sessions.

Some of the most fascinating stories about wild dolphins involve instances where human beings have been given rides. The best known modern case refers to an 8-ft female bottle-nosed dolphin named Opo. She regularly visited a beach near Opononi, New Zealand, and allowed children to ride on her back. Such behaviour is by no means common among wild or even captive animals. Dolphin rides are evidently rare, special events involving exceptional animals. But rare or not, they show that dolphins can develop personal relationships with man – at least when no other dolphins are near by.

NAVIGATION BY SOUND WAVES

FINDING THEIR WAY *Relying primarily on echo-location and on a sense of hearing that is second only to that of bats, dolphins and other toothed whales are superb underwater navigators. In most cases, their vision is not keen*

For centuries, marine creatures were thought to be mute and the oceans silent. During the Second World War, naval technicians lowered microphones into the water to listen for submarines. To their amazement, they heard a bedlam of sounds produced by invertebrates, fish and whales.

Toothed whales most commonly produce staccato bursts of ultrasonic clicks. One of the first to discover the significance of these sounds was Arthur F. McBride, the first curator of Marineland in Florida. In 1947, he was trying to capture some dolphins, but every time he drove them toward nets, they would stop a good distance short. It made no difference whether he pursued the dolphins during the day or at night; they never even ventured close enough to glimpse the nets in the murky water. McBride rightly assumed that somehow the dolphins used sound to sense the obstruction in their path.

HOW WHALES NAVIGATE BY EAR

When a toothed whale – for example a dolphin – swims under water, air circulates under pressure through its nasal passages to produce clicks or similar sounds. The whale's melon – a round mass of blubber between the blowhole and the front of the head – directs the sounds forwards to hit fish and other objects in its way. It has been found, however, that the sound cannot be directed to objects below the line of the jaw, perhaps explaining why otherwise skilful navigators sometimes become stranded on gently sloping beaches. When the sound signal hits a fish it is reflected back to the whale and is picked up by sensitive areas in the lower jaw which transmit it to the animal's inner ear. Since lower-pitched sounds have longer wavelengths, the whales probably use them for long-distance location; higher-pitched clicks are thought to give them information about their immediate surroundings

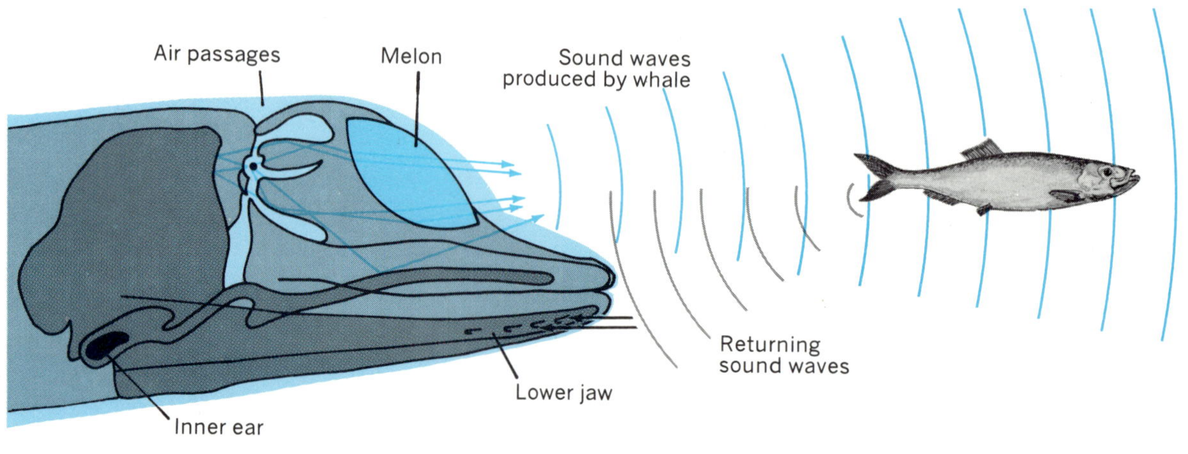

In the early 1950's, Winthrop N. Kellogg of Florida State University showed that dolphins use their clicking sounds just as men use underwater echo-sounding devices. The clicking sounds travel through the water and bounce off anything solid. Dolphins determine the exact distance to an object by measuring the interval between an outgoing click and its echo. They can also tell a great deal about the size of an object. In tests, blindfolded dolphins have distinguished between $2\frac{1}{2}$-in. and $2\frac{1}{4}$-in. ball bearings from as far away as 10 ft. Therefore, avoiding rocks on a dark night or catching something as small as a 4-in. herring in turbid water should not be the least bit difficult for any healthy dolphin.

The ability to navigate by sound is not confined to dolphins. All toothed whales produce high-pitched clicks. Baleen whales, on the other hand, apparently produce relatively low-pitched sounds, which bend around obstructions instead of bouncing back to the source. Grey whales moan and make bubbling noises, but emit no clicks. One grey whale blundered into a barrier of sound-reflecting aluminium tubes stretched across a Mexican lagoon, became entangled, freed itself, and again swam into the barrier. During the same experiment, a group of bottle-nosed dolphins detected the barrier from a considerable distance, inspected it, and retreated – all the while emitting their echo-sounding clicks.

Despite their ability to use sound signals, toothed whales periodically run aground and perish in shallow water. About 200 false killer whales – smaller than true killer whales and all black – beached themselves at Fort Pierce, Florida, in January 1970. Similar disasters are known to have struck pilot whales, killer whales, sperm whales, dolphins and porpoises.

The composition of the sea floor may be to blame. Some mud floors absorb sound instead of bouncing it back, and some gently inclined sea floors may reflect sound waves past the whales rather than back to them. Certain hard floors produce a confusion of echoes. Any of these situations can produce disorientation and panic, and whale strandings may be much like cattle stampedes. Whales have a strong instinct to stay with the herd, and if several run aground, the rest follow. Even whales pulled into deep water swim back to their stranded companions.

DEATH-TRAP BEACH *With a navigation system that apparently cannot detect gentle slopes, these false killer whales are stranded on a Florida beach. They cannot move because of their weight and they die from dehydration and sunburn*

DOLPHINS HELP SCIENCE *Experiments with bottle-nosed dolphins have helped scientists to understand how whales navigate and communicate with one another. Sophisticated equipment is used to study the functions of their bodies*

THE LANGUAGE OF WHALES

Both baleen whales and toothed whales use low-pitched sounds for communication. Bottle-nosed dolphins have 32 distinct whistling sounds. If each represents a complete expression, these whales are limited to 32 signals. Even this number is quite an accomplishment, but if each whistle pattern represents a symbol or 'word' the sounds could be combined into what amounts to an animal language.

A bottle-nosed dolphin in distress utters a short, sharp whistle over and over again. A captive female bottle-nosed dolphin fell ill in her tank and could not rise to the surface to breathe. With the distress call, she summoned two companions; they held her up for four days. The treatment then continued intermittently for two weeks until the sick dolphin recovered. Such rescues are always accompanied by frequent whistling.

When two male bottle-nosed dolphins are reunited after a separation, they carry on dialogues that sound like old friends meeting. But vocalisations may be more than just calls of greeting or distress. Scientists studying sea noises in a Gulf of California lagoon kept watch on a group of bottle-nosed dolphins. When the dolphins found their way partially blocked by a line of buoys, they immediately held a whistle conference. First one scout and then another left the group to examine the barrier. Each time a scout returned, another conference took place. Finally, the group moved off together and cautiously passed under the buoys.

Gregarious dolphins are among the most loquacious whales, but they are by no means the only ones known to 'talk' frequently. Killer whales in pens adjoining open water carry on whistle exchanges with free killer whales outside the enclosures. Eskimos report that killer whales whistle to each other before launching group attacks. The false killer whales rescued from the 1970 stranding on the Florida coast communicated with those trapped on the beach.

Humpback whales sing during their northward spring migration, possibly to help keep the herd

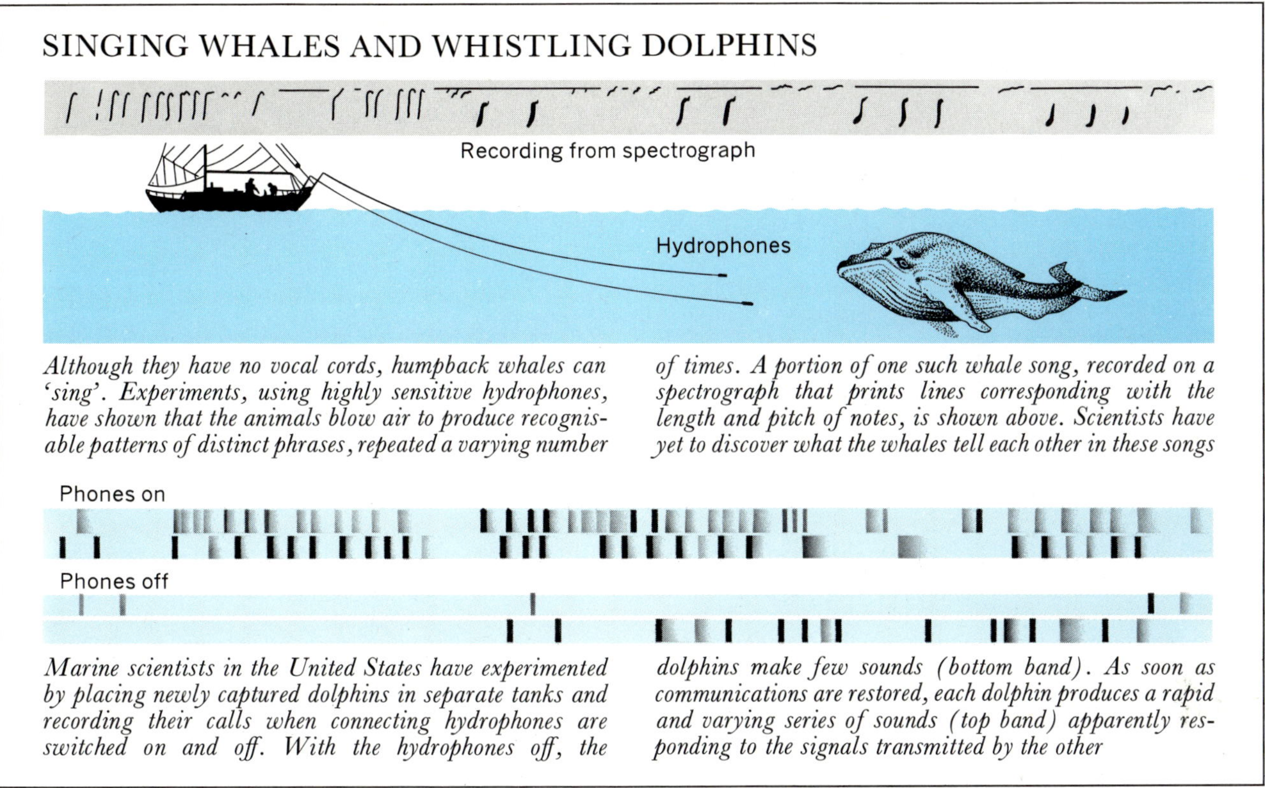

Although they have no vocal cords, humpback whales can 'sing'. Experiments, using highly sensitive hydrophones, have shown that the animals blow air to produce recognisable patterns of distinct phrases, repeated a varying number of times. A portion of one such whale song, recorded on a spectrograph that prints lines corresponding with the length and pitch of notes, is shown above. Scientists have yet to discover what the whales tell each other in these songs

Marine scientists in the United States have experimented by placing newly captured dolphins in separate tanks and recording their calls when connecting hydrophones are switched on and off. With the hydrophones off, the dolphins make few sounds (bottom band). As soon as communications are restored, each dolphin produces a rapid and varying series of sounds (top band) apparently responding to the signals transmitted by the other

together. Roger Payne, of Rockefeller University in New York City, has recorded these calls of humpback whales, which he describes as haunting, warbling notes. Lower in pitch than those of birds, the calls are repeated from seven to as long as 30 minutes at a time.

No one understands exactly how whales make sounds, for they have no vocal cords. Evidently the sounds result from air forced past valves and tongue-like flaps in sacs below the blowholes. Differences in the way these structures vibrate probably account for the very noticeable variations in the 'voices' of individual whales.

Even less is known about how whales 'hear'. Certainly their ear holes, seldom more than $\frac{1}{2}$ in. in diameter, are important. They channel sounds into the inner ears, linked by nerves to the brain. Bones are excellent conductors of sound, and some whales may use their skulls as sounding boards. Two narrow channels in the jaws of bottle-nosed dolphins connect to the inner ear and may pick up echoes of the high-frequency clicks.

Scientists still question how much meaning whales convey by their vocalisations. Are their calls like the chirping and singing of birds and the barks and growls of dogs, or are they more like the human whistle languages of Mexico, Turkey and the Canary Islands? Arguing for the latter idea, some scientists point out that a dolphin brain compares in size and complexity with that of a human adult.

Several attempts have been made to communicate with dolphins. One group of scientists tried to match recordings of dolphin sounds with films of their actions. A computer was programmed to classify the sounds so that people could learn their whistle 'language', but the idea did not work out as planned.

John C. Lilly, who founded the Communications Research Institute in the Virgin Islands, tried to teach whales to talk English. He first succeeded in getting bottle-nosed dolphins to make sounds in air instead of under water, then to repeat a series of as many as 12 nonsense syllables. The dolphins accurately mimicked the number and duration of the sounds. Lilly taught a dolphin named Elvar to repeat such phrases as 'More, Elvar', 'Stop it', and 'All right, let's go'. But other scientists who listened to recordings of Elvar did not agree that the utterances sounded like human speech. Few scientists believe that men will ever 'converse' with dolphins or with other whales.

AN ANCIENT AND BRUTAL BUSINESS

Whales and whalers face an uncertain future

Stone Age Scandinavians pursued whales; so did Eskimos 3500 years ago. For centuries men put out to sea in flimsy boats to challenge these giants. The opponents were not well matched: both history and fiction tell of smashed boats and drowned sailors. As man's technology improved, however, the balance shifted; and for almost a century now the whales have been losing the battle. They are no match for grenade-tipped harpoons and diesel-powered whaleboats. But international agreements are at last restricting the catches of some whale species to levels which present stocks can support. Continued control may allow their numbers to increase again and provide a brighter future for both the whales and the whalers.

EXPLOSIVE HARPOONS *Men once stood in the bows of small fragile boats and threw their harpoons at whales; now they use lethal cannons to fire harpoons fitted with grenades which explode and kill the whale within seconds of contact*

INFLATED TO PREVENT SINKING *When a whale is harpooned from a catcher ship it is usually inflated by the whalers to ensure that its massive body cannot sink during the journey to the fleet's factory ship, which may be up to 150 miles away*

TRADITIONAL WHALING *Fishermen from the Faeroe Islands still brave the North Atlantic to hunt whales from small wooden boats. Here two threshing whales are killed in shallow water by the fishermen's hand-flung lances*

WORK ON THE FACTORY SHIP *When the whale's body is delivered to the factory ship, strips of blubber are removed and boiled to give whale oil*

WEAPONS OF THE WHALERS

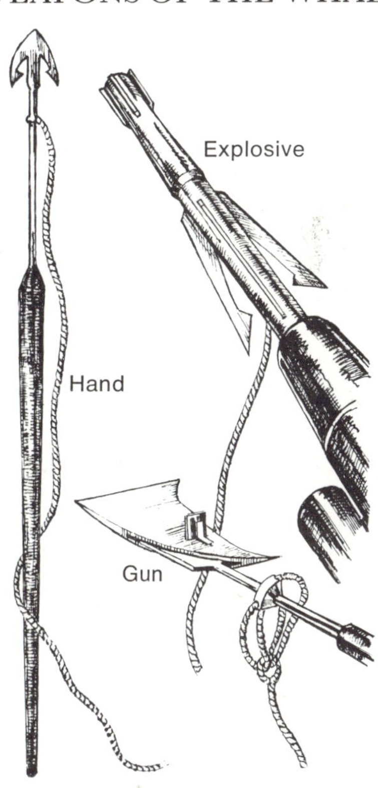

The earliest harpoons had to be flung by hand, relying entirely on the whaler's skill. Gun harpoons gradually widened the fishermen's range until today when they use cannon harpoons with explosive heads

MAN, THE WHALES' ONLY ENEMY

The common enemy of all whales – baleen and toothed, large and small – is man. Records of whale hunts go back 4000 years. Drawings scratched on rocks in Norway by Stone Age men depict two porpoises, with a hunter in a boat close behind. Whale bones unearthed among the remains of Alaskan Eskimo settlements establish beyond doubt that Eskimos were killing whales at least as early as 1500 BC.

As early as the 11th century, Basques and Spaniards sailed out to hunt black right whales in the Bay of Biscay, where these slow-moving creatures roamed in herds more than 100 strong. In the 17th and 18th centuries, large Dutch and English whaling fleets pursued right whales on a grand scale in Arctic waters; in the 19th century, Americans extended the pursuit into the South Atlantic, Pacific and Indian oceans. Right whales steadily disappeared from one area after another until they were given protection from commercial whalers by an international agreement in 1935. Some scientists think the ban may have come too late to save several species from extinction.

For hundreds of years, the sprinting speeds of large baleen whales protected them from capture. But fast steam-powered boats and the harpoon gun developed in the last part of the 19th century drastically changed the situation. In the peak year of Antarctic whaling, 1930–1, about 30,000 blue whales were killed. From then on, the catch dropped steadily: less than 7000 in 1948, and only one in 1965. With just a few thousand of these great animals left, most whaling countries placed them under protection – again, not until the species was close to extinction.

As the blues disappeared, whalers first turned to fin whales – the second largest species – and then to the smaller sei whales. Blues could no longer be hunted, but 'blue whale units' could. Under the system imposed by the International Whaling Commission, two fins or six seis equal a blue whale unit, since these whales yield an average of a half and one-sixth as much oil, respectively, as a blue whale. In 1970, the commission set a limit of 2700 units in the Antarctic whaling grounds, where most large baleen whales are hunted. This meant that 5400 fin whales, 16,200 seis, or various combinations of the two could be killed that year. This was thought to be the maximum number of whales that could be taken without destroying the future productivity of Antarctic whaling. But many conservationists believe quotas like these put too much pressure on individual species; they want limits defined in terms of num-

A MOTHER'S LOVE *A grey whale tries to rescue her calf which has been harpooned despite conservation regulations. Male greys help females under attack. Females always stay beside their calves, but they swim away from males in distress*

MAN-MADE BREATHING HOLE *A white beluga whale surfaces for air at a breathing hole made by Canadian conservationists on the Eskimo lakes. The herd of beluga had been trapped 50 miles from the sea by a sudden river freeze*

bers of each baleen species that can be caught, not in terms of the total oil yield.

The noble sperm whales, which escaped extinction in the heyday of American whaling, were threatened as stocks of baleen whales diminished. In 1970, biologists estimated that 10,000 sperms could be harvested from the North Pacific without seriously depleting the stock. The whaling nations in this area agreed on an upper limit of 13,551.

A diminishing demand

Attempts at international regulation alone probably would not save the great whales, but economic factors may. Large factory ships, with their retinues of smaller catcher boats and spotter planes, are expensive to maintain. At the same time, demand for whale products – used in paints, soaps, cosmetics, pet food and other items – declines as substitutes become readily available. The United States has banned the importation of whale products, thus eliminating about 20 per cent of the world market.

As long as any great whales remain, there is hope for their kind. The story of the grey whale offers strong encouragement. In 1846, whalers began intensive killing of grey whales. They tracked the animals to their shallow breed-ing lagoons, where hunting was easiest, and slaughtered them there. In the 1850's, an estimated 30,000 grey whales ranged the Pacific coast of North America; 40 years later, the species was nearly extinct. Whalers turned their attentions elsewhere. By the 1920's, the grey whale population had begun to build up again, and the herd was saved by an international treaty in 1937. By 1943, the population had increased to nearly 3000; by 1970, to nearly 11,000.

It is to be hoped that the other endangered species can make equally spectacular recoveries so that future generations may yet see right whales lobtailing and blue whales spouting. Yet the future of the greatest animal ever to live on earth remains in doubt. Some scientists fear that the point of no return may already have passed, and the blue whales may be unable to find mates. A factor which may have saved the grey whales from extinction is the migration instinct which brings the whole herd together in a single breeding ground each year; but the blue whales range for breeding and feeding across the open ocean from the Equator to the Antarctic, and it is possible that the few survivors may now find it too difficult to rendezvous with others of their kind in these vast ocean reaches.

PART TWO
Wonders of island life

Worlds apart

Islands are worlds in miniature. They have their own unique plants and animals which have evolved in isolation, and these have provided man with clues to the mysterious origins of all species, including his own

Islands are meeting places between sea and land. The land, rising above the waters, affords a home for plants and animals; the surrounding sea provides an equally important element: isolation. Each of the world's 500,000 islands is a world apart, a world in which isolation over thousands or millions of years has allowed the evolution of creatures who may now be altogether different from their cousins on the distant continents.

The five weeks which the British naturalist Charles Darwin spent on the Galapagos Islands in 1835 gave rise to ideas which he said haunted his life until he expounded them in his book *On the Origin of Species* almost 25 years later. These ideas, born out of the study of island life, have changed man's view of the nature of all life on earth.

But if islands have played their part in teaching us about the past, they are now, still more urgently, the subject of study for lessons they can teach for the future – lessons on which the survival of mankind may well depend. For every island has its individual system of ecology – the balance of animals and plants which have evolved in relationship to one another; and because islands often contain many fewer species than exist together on the mainland, scientists are able to work out what each contributes to the whole.

This understanding of the workings of ecologies may be vital if we are to conserve the rich resources of the world. For man now has the power to disrupt the balance of all living things in the world and even to exterminate life on a massive scale. If we are to manage our impact so as not to impair its capacity to support life we must understand how ecological systems work.

Islands are natural laboratories for this research. But some of the most beautiful and interesting of them are being disrupted and disfigured before naturalists have had time to probe their secrets. Remote islands are vulnerable. Their plants and animals are threatened with catastrophe when man brings in his domestic livestock or the pests – such as rats and mice, weeds and weevils – that travel with him.

Many of the fabled islands of the past have been ravaged beyond repair; some of the most fascinating of island creatures are lost beyond recall. But it is not too late to save something. There are signs that the peoples of the world are awakening to their danger in the nick of time. In marvelling at the wonders of island life, we are reminded that the earth is itself an island – well stocked with every conceivable variety of plants and animals, but pathetically vulnerable and utterly isolated in the surrounding oceans of Space.

ISLAND FROM THE SEA *The British colony of St Lucia in the West Indies, like many islands scattered across the world's oceans, was once nothing more than a flow of gleaming molten lava; over thousands of years it has become a lush tropical resort. The 2461-ft peak Petit Piton is one of two dramatic upthrusts on the island*

The rift that runs the length of the Mid-Atlantic Ridge is the most active volcanic segment in the worldwide rift system

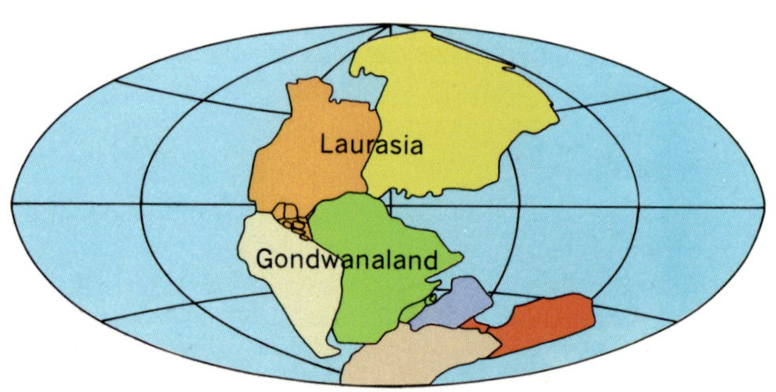

HOW THE EARTH BEGAN *Geologists believe that, more than 200 million years ago, all the land on earth was joined in one great mass, which they have called Pangaea. About 180 million years ago the northern section, Laurasia, split from the southern mass, Gondwanaland. Over millions of years these two super-continents broke up, and today the fragments — our present continents — are still moving apart*

HOW ISLANDS ARE BORN

Islands originate in one of two basic ways. *Continental* islands are fragments of land once part of a continental mainland. *Oceanic* islands emerged from the ocean floor and were never connected to other land. Some continental islands have been cut from the mainland through the erosion of a former land connection. Others owe their origins to geological forces that long ago led to major changes in the surface features of the earth. The

The Pacific and Indian oceans, unlike the younger Atlantic, are shrinking, as the Americas move westwards and Africa drifts eastwards

world has not always been as it now is; the continents have shifted about in a slow, ponderous movement that still continues.

Some 180 million years ago, scientists now believe, dry land was gathered together in two major super-continents: Gondwanaland, which gave rise to today's South America, Africa, India, Australia and Antarctica; and Laurasia, consisting of North America, Greenland, Europe and Asia all fused together. From the earth's mantle – the layer of dense, plastic rock that extends 1800 miles down from the surface – molten rock began to well up into the surface layers of the crust along enormous cracks. Giant convection currents within the mantle are believed to have produced this slow upwelling, carrying sections, or plates, of the crust away from one another as if on an enormous conveyor belt. The ocean channels widened, perhaps at a rate of a few inches in 100 years.

Riding on the plates of the crust, the land masses were eventually carried far from their original points of connection. Some of these points are evident where the contours and geographical composition of continental shorelines match those of shores far across the sea. At the same time as the breaking up of the super-continents came a separation of land into lesser fragments, which became islands at an early date. The drift of the great plates continues today.

Oceanic islands commonly form along the lines

where molten rock wells up into the ocean floors. The slow piling up of lava flow upon lava flow over thousands of years builds a volcanic mountain. Eventually the mountain may grow to such a height that its summit breaks the ocean surface.

From the moment of birth, volcanic islands are exposed to the relentless erosive action of wind, rain and waves. In warm seas, however, the life-span of volcanic islands may be considerably extended by the growth of offshore coral reefs, which form a barrier against the crashing ocean waves. But even coral reefs carry no immunity against destruction. A change in environmental conditions may pose a threat to an island that depends on living reef builders to maintain its structure against the forces of erosion.

Throughout the history of the world, the state of the polar ice caps has played a significant role in the appearance and disappearance of both continental and oceanic islands. During the Ice Ages, when more of the water of the earth was locked into massive glaciers, the seas were several hundred feet lower than they are now. Under such conditions, the number of islands must have been far greater, since many of today's underwater mountains and offshore slopes were exposed. Conversely, when the ice caps melted and the seas were higher than they are at present, many of today's islands were submerged.

It is impossible to determine how many islands have come into being on the face of the earth, only to erode away or to be covered by rising seas. But during its transitory existence, almost every island is destined to be settled by organisms arriving from continents and from other islands. And as evolutionary mechanisms come into play, each island develops its own unique community of living things, some of which are similar to life elsewhere but can never be exactly the same.

Life gains a foothold

How do plant and animal inhabitants of an island come to be there? To almost every island that man has settled or visited, he has brought – sometimes intentionally, sometimes accidentally – ornamental and food plants, domestic pets, livestock, and even such pests as mice and rats. In the study of island life, however, these introduced forms are of secondary interest. The primary concern is rather with native forms, the species whose ancestors were the original colonisers. Species that evolved in their island environments and that live nowhere else are called endemic.

A continental island, of course, already supports a community of living things at the time it becomes separated from the mainland. At first, its populations may be similar to those of the mainland. But as colonising species arrive from elsewhere, and as evolutionary processes give rise to endemic species, the composition of the island's life undergoes significant change. After long isolation, if the island is far removed from the life of its parent continent, its community of living things may come to be distinctly different from that of the mainland.

In contrast to continental islands, an oceanic island begins its existence devoid of surface life. A volcanic island is completely sterile when it emerges from the sea, and a coral island at first supports only the builders and marine inhabitants of its reef structure. The development of surface communities of plants and animals on oceanic islands thus depends entirely on the arrival of life from elsewhere.

The sea is an impassable barrier to many forms

INTRODUCED TO ISLANDS *The Indian grey mongoose has become a serious pest on several of the West Indies and Hawaiian Islands. It was originally imported by sugar-cane planters to fight rats, but as the rodent population was brought under control the agile, aggressive mongoose turned its attention instead to young pigs, goats, sheep, poultry, native birds, lizards, crabs and even fruit*

of life. Prolonged contact with salty ocean water is fatal to most species of land plants and animals, and even to frogs, salamanders and river fish, which thrive in fresh water. The plants and animals that reach an island in a condition to survive are those with special qualifications. Some are sea-feeders – seabirds, turtles and seals – for which a new island provides new breeding sites amid abundant food and remote from the predatory mammals and land birds of the mainland. But the true colonisers of an island are the organisms which manage to cross oceans to become permanent settlers on its dry land.

Such colonisation needs unusual powers of dispersal – and luck. Since islands are generally small and scattered, the process of settlement is long-drawn-out and risky. Only a few species even arrive; of these, fewer still manage also to survive, to propagate, to become established. Those species that do gain a foothold may have a whole miniature world almost to themselves.

NATIVE OF THE ISLANDS *The Japanese serow, which has evolved in isolation since Japan split from the Asian mainland, is smaller and has a woollier coat than its continental relative. Only about 3000 of these sure-footed goats survive in the wild, on forested hillsides and mountain ridges*

VISITORS TO THE ISLANDS *Southern elephant seals come ashore on remote islands in the Antarctic only to breed and moult. Here a moulting male throws sand over himself, probably in an attempt to soothe the irritation caused by his old skin and fur*

THE PIONEERS ARRIVE

The land plant species that have the best chance of arriving are the simple forms that produce microscopic airborne spores: algae, lichens and fungi. Mosses, ferns and their relatives – with somewhat bigger spores – are also good travellers.

Although seed plants are somewhat less promising candidates for long-distance dispersal, some island plants produce seeds with devices that catch the wind. Other plants grow small seeds that can be transported unharmed within the digestive tracts of birds. Some unusually successful plants produce seeds that adhere to the feet and feathers of birds. Birds that fly far out to sea may nest on heavily vegetated cliffs and in forests; others nest among ferns growing on the limbs of trees, or in dense plant growth on the ground. Such close associations almost inevitably result in the transportation of seeds and spores to distant islands.

Some plant species produce seeds that can float long distances without damage from sea-water, and some of these take root when they reach a tropical island beach. Among the best known of these species are the mangroves, the screw 'pine' (or pandanus), several kinds of morning glory, and *Scaevola*, a low shrub that forms shoreline hedges on many Pacific islands. Conifers, on the other hand, have not done well in crossing wide stretches of ocean. Their seeds are neither well protected against salt water nor likely to be borne long distances by wind or by birds.

By sea and air

The animals which have proved most successful in these long ocean voyages include not only birds, but also bats, lizards, land snails, spiders and insects. Young snails and insect eggs have been found encased in mud on the legs and plumage of wide-ranging birds. Larger animals travel on the foliage or among the roots of trees dislodged from shorelines by floods or storms and sent floating hundreds of miles out to sea. 'Rafts' of vegetation often consist of single trees or parts of trees, but sometimes they reach the proportions of small islets. Dense with grasses, and with trees rising 20–30 ft in the air, some of these rafts were even charted by early mariners, whose successors were puzzled when the 'islands' could not be found later. When such a huge raft is beached on some more permanent piece of land, it deposits a sizeable cargo of living plants and animals.

Some spider and insect species on islands are derived from passengers on rafts of vegetation, but most have descended from strays caught in strong winds and carried into the sky and across wide expanses of ocean. Although spiders do not fly, it is common for their young to spin short strands of fine webs that catch the breeze and carry the tiny spiderlings far aloft. Sometimes the sky glistens with multitudes of these little parachutists transported in the air stream.

Throughout the world, at altitudes of 10,000 ft or more, there is a sparse but discernible aerial 'plankton' composed of a variety of spiders and insects. Spitsbergen, in the Arctic Ocean, has been colonised repeatedly by insects carried by both birds and wind. Spiders have been found spinning webs in crevices of the barren St Paul's Rocks, in the equatorial Atlantic. Brazilian moths have been caught over 2000 miles away on Tristan da Cunha, most of whose insects seem to have blown in from South America.

The normal trade winds, although dependable and often quite brisk, account for only a few of the insect immigrants to tropical islands. Cyclonic storms have undoubtedly been of at least equal importance. When a tropical hurricane hits an island it may uproot palm trees, and an insect or other small animal, whipped up with forest litter, may be transported a great distance. Newly arrived butterflies, locusts, flies, beetles and other insects have been found on Pacific islands hundreds of miles from the nearest land, but almost always shortly after a storm. Birds and amphibians – even small fish – have crossed salt water on storm winds.

In a world of their own

Despite the hazards and improbabilities of transport, the single greatest challenge in colonising an island is not getting there but becoming established. An animal or plant suddenly arriving – often exhausted and half-starved – may have to make a rapid and difficult adjustment. The new environment may be too dry or wet, too cold or warm; it may lack the kind of soil, vegetation or – most important – food to which an organism is accustomed. It is no wonder that many invaders fail to survive. Success in becoming established is reserved for those colonising organisms able to become part of the island's living conditions.

Isolated though it may be, an island is related to the world about it through tides and currents and through the winds that blow above. Yet each island is a separate entity, an ecological system unduplicated anywhere else and embodying the relationships of all its plant and animal inhabitants to one another and to their physical environment. Time, geological change, biological opportunity and the variations of living things mould this ecological system into a distinctive entity.

Every plant and animal plays its own role in the functioning of its ecological system and fills a specific place in the interdependent lives of other plants and animals. The ecological niche occupied by any organism is determined partly by its physical make-up and partly by its specialised way of living: where it lives, how it secures nourishment, how it copes with competition from other organisms for food and living space, how it may supply other organisms with food or shelter, and how in life and death it affects the physical conditions of the environment.

Similar, but not the same

The distinctions between the niches filled by different species in an island's ecology may be extremely subtle. Two or more species living in the same area and looking much alike may actually live quite differently from one another. For a long time, it was thought that all the grazing animals of the African plains ate the same food. But biologists have recently learnt that each species has its plant preferences, and that when two species favour the same plant, one may eat only the top leaves and the other only the lower portions.

An ecological role which exists in many parts of the world need not be performed everywhere by precisely similar plants or animals. In Australia, which has no antelopes, the role of grazer has been filled by the large kangaroos. Similarly, in the absence of the leafy trees common on continents, forests on some of the Galapagos Islands are composed of tree-like sunflowers and cactuses.

As new species arrive or evolve, competition in an island environment increases. In the relatively limited space of an island, any competition may be short-lived, for the weaker species has nowhere to which it can retreat. The less successful organism either takes on a new lifestyle or is eliminated. If large numbers of antelopes were introduced to Australia and competed with kangaroos for a specific kind of grass, and if neither species could change its food preferences, only the species which proved more successful in foraging would survive.

THE TOOL-USER

A Galapagos woodpecker finch digs a hole in a tree trunk, probes with a cactus spine and extracts a grub for a meal. Members of the true woodpecker family have evolved a long tongue to reach insects, but the Galapagos woodpecker finch is not so well equipped; it has evolved instead into a tool-user

A USEFUL NOSE *A male proboscis monkey uses its bulbous nose to honk its distinctive warning to its troop in the Borneo jungle. Females and young males have normal noses: only in the fully grown male is the nose specially developed*

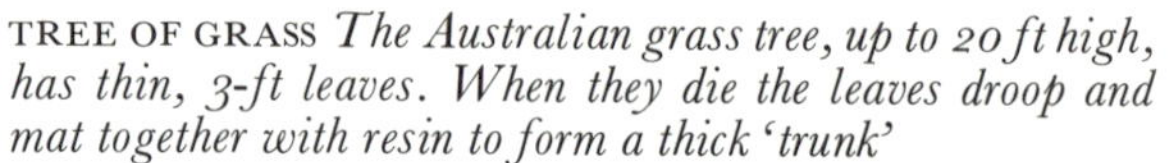

TREE OF GRASS *The Australian grass tree, up to 20 ft high, has thin, 3-ft leaves. When they die the leaves droop and mat together with resin to form a thick 'trunk'*

GIANT CACTUS *The prickly pear tree, found on the Galapagos Islands, is a giant cactus. Pads at the top of the trees collect water during the rainy season. When they fall they provide food and moisture for the islands' giant tortoises*

ODDITIES OF THE ISLANDS
A strange world of dwarfs and giants

Isolated island dwellers have evolved some unusual adaptations to their physical environment and to one another. Shapes or habits may appear which are startlingly different from those of the ancestral stock – or a plant or animal may survive unchanged for millions of years. The largest lizard on earth (the Komodo dragon), the longest-lived animal (the giant tortoise) and also the smallest primate (the mouse lemur) all live on islands. Such island oddities gave Darwin the idea which transformed man's view of himself and his world. For Darwin's discovery that island forms of life are merely variations on a mainland theme compelled him and later researchers to the conclusion that all animals belonging to the same group must have evolved from a single ancestor.

THE LARGEST LIZARD *Many island reptiles have become unusually large – perhaps because of the absence of competing mammals. The 9 ft long Komodo dragon, a monitor lizard found only on the Indonesian island of Komodo, preys on monkeys and deer*

UNIQUE TUSKS *The upper tusks of Indonesia's babirusa boar protrude not between the lips as in other pigs, but through the skin between the eyes. Curved backwards, they are sexual adornments, not weapons*

THE SMALLEST PRIMATE *The 4 in. long mouse lemurs of Madagascar are the smallest of all primates. They are also among the most primitive. Mouse lemurs spend the day hiding in foliage and come out at night to feed on insects*

FORCING-GROUNDS OF EVOLUTION

A visitor to a tropical Pacific island relatively unchanged by man is often overwhelmed by the abundance of living things, such as the scuttling droves of land crabs, and by the variety of shapes, sizes and colours.

With time, all living things undergo change, most of them evolving into forms markedly different from their remote ancestors. Every individual organism carries a unique combination of traits inherited from its ancestors, which make it different from all other individuals. The parts of living cells carrying these traits are called genes; all of an organism's genes together form the genetic make-up. Every organism with two parents has received half of its genes from each one, and its genetic make-up is different from that of either parent. Genes are reshuffled again and again in successive generations, with each individual offspring inheriting genes in unique combination from its parents and exhibiting traits that make it more or less adapted to survive.

The pool of genes available to an island species is likely to be severely restricted, for even a large island population has usually developed from a very few individuals that made a fortunate landfall in days long past. The first immigrants did not carry in their genetic make-up all the characteristics of their entire race. They were a chance sample, no more typical of the homeland population than any other group of a similar size. The newly established island population is therefore immediately different from the parent stock.

The kind of change called evolution begins when a random accident occurs to the molecules

SNAILS THAT WENT THEIR OWN SEPARATE WAYS

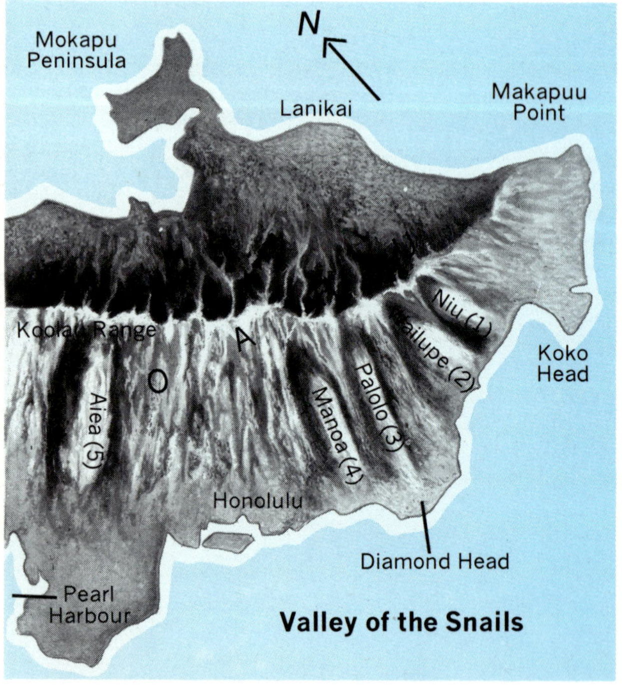

Five neighbouring valleys on Oahu, one of the Hawaiian Islands, have different species of snails. Scientists have found that all the snails, although now easily distinguished, are descended from a common ancestor that was taken to the *island long ago either by birds or on a raft of vegetation. As time passed, rivers and gorges separated the early populations of snails and each group began to evolve in its own way to suit its isolated environment*

AN ISLAND 'FAMILY' EVOLVES

These five species of vanga shrikes, descendants of one common ancestor, show how birds diversify to meet the needs of different environments. The 12 vanga shrike species found on Madagascar differ widely in diet, shape, colour and size of beak

carrying the genetic make-up. Such accidental alterations, or mutations, of the genes are caused by a variety of environmental influences, including ultra-violet radiation from the sun and cosmic rays from far out in Space. Heat, chemicals and other physical factors may also alter a single molecule within a sperm or egg cell.

The vast majority of mutations are harmful and often fatal to the organisms in which they occur. The infinitesimally small percentage of mutations that are beneficial, however, improves the ability of the new organisms to adapt to environmental conditions. With enhanced powers of survival, such organisms are likely to pass on the mutant genes to their offspring.

Only a few members of a species are affected when a favourable mutation occurs and is inherited by the next generation. But if the change persists through successive generations, the species gradually comes to consist of two distinct and diverging groups of organisms: those with the mutant genes and those without.

As additional mutations reinforce earlier ones, the process of diversification is fostered within the mutant group. Natural selection becomes increasingly influential. The individuals that are most successful in a given set of conditions – plants with roots long enough to reach scarce water, birds with beaks strong enough to crush seeds, predators or prey with a good camouflage coloration –

multiply and flourish. But organisms that are not so fit in one respect may, instead of dying off, become specialists in another direction and survive as another branch of the family tree. In time, both groups may take forms strikingly different from their pioneering ancestors, as well as from their neighbouring cousins, and relationships may be nearly or quite impossible to trace.

A multitude of such specialised forms may come from just a few hardy colonising species. The 12 kinds of vanga shrikes on Madagascar apparently evolved from a single species; the more than 3700 insect specimens on Hawaii radiated from only 250. Isolated from the competition of mainland life, islands have produced or preserved oddities such as the flightless kiwi of New Zealand and the egg-laying mammals of Australia. Sometimes giants have developed: the Galapagos Islands harbour centipedes up to 12 in. long, huge tortoises weighing as much as 600 lb. and sunflower trees 50 ft tall.

The rate of mutations is not less on continents than it is on islands, but mutant individuals on a continent must face intense competition from an abundance of other well-entrenched individuals, and the new characteristic is more likely to be lost in the larger gene pool. In an island environment, a mutant organism that could well have met quick destruction on the mainland may become a successful experiment.

COLOUR BY THE SHORE *The long-legged scarlet ibises are the most colourful birds of Trinidad. Until they are mature the birds have brownish feathers. Anhingas and a black-crowned heron can be seen in the small tree on the right*

THE COLOURFUL LIFE OF CARIBBEAN ISLANDS

These remnants of extinct volcanoes have become a paradise of exotic plants and animals

THE CARIBBEAN *With a total land area of 91,000 square miles, the many islands of the West Indies are scattered over 971,400 square miles of open ocean*

There are about 7000 lush islands and coral reefs in the Caribbean, stretching from Florida to South America. A very few are continental in origin, but the great majority are the result of volcanic action long ago. Most of the volcanoes that formed them are extinct, but in 1902 the eruption of Mount Pelée on Martinique wiped out 30,000 inhabitants. Probably all the land animals arrived on floating debris from the Americas, and most of the exotic birds of the islands are also found in South America. But the West Indies have one unique bird family, the tiny, brilliantly coloured todies. They are often seen sitting on low branches with their oddly flattened bills pointing upwards at a 45-degree angle, watching for insects.

A UNIQUE PARROT *Many of the larger islands in the West Indies have species of parrots which are found nowhere else. This red-necked parrot lives only in the mountain forests of Dominica*

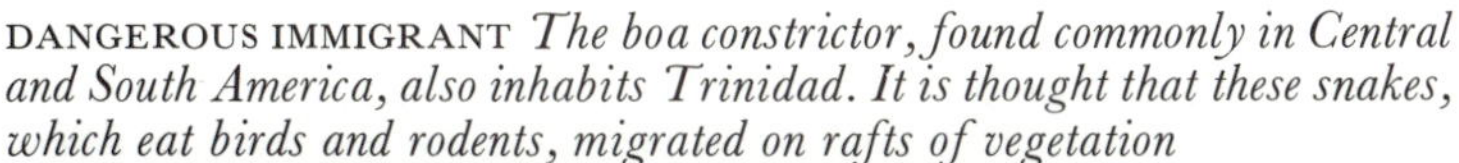

DANGEROUS IMMIGRANT *The boa constrictor, found commonly in Central and South America, also inhabits Trinidad. It is thought that these snakes, which eat birds and rodents, migrated on rafts of vegetation*

TOWERS OF COLOUR *Tall yellow agaves bring colour to Puerto Rico's semi-arid hills. Most species grow in North and Central America*

TINY INSECT-EATER *The 4-in. Jamaican tody catches its insect prey by snapping its long bill like a kingfisher. These small, colourful birds are found only in the islands of the West Indies*

IN DANGER OF EXTINCTION *The shrew-like Solenodon of Cuba and Haiti is an insect-eater with poisonous saliva to kill its prey. This 12-in. mammal, closely resembling animals alive in the age of reptiles, is now in danger of extinction*

HOW TIME HAS STOOD STILL

As immigrants arrived, both they and the original inhabitants were usually forced to adapt or perish. Most of the island species of today are highly modified; some have changed so radically that it is difficult to trace their origins. But occasionally a plant or animal was able to survive in its isolated environment without much change. Such a plant or animal is called a relict.

Among the more interesting animal relicts are those found on the ancient continental islands of New Zealand, Australia, the Philippine Islands, Madagascar, Cuba and Hispaniola.

Possibly the most archaic of the larger land animals still living is the tuatara, a lizard-like reptile of New Zealand. Externally the tuatara superficially resembles some modern lizards, but its primitive skeleton identifies it as belonging to a family of reptiles widespread 200 million years ago. As dinosaurs came to rule the earth, this

group of small reptiles died out everywhere but in New Zealand; there, isolated from enemies, the tuatara managed to survive.

Three closely related kinds of New Zealand frogs are also very primitive, with muscular characteristics like those of their long-tailed salamander ancestors. In Australia, three freshwater fish – a lungfish, barramundi and blackfish – have persisted in isolation for millions of years, the only representatives of their kinds still in existence.

Small insect-eating mammals first appeared in large numbers over 100 million years ago, during the reign of the dinosaurs. They remained largely unspecialised and insignificant until most of the giant reptiles had vanished. Many of these furry little creatures probably lived inconspicuously in the thick forest undergrowth, where their insect prey was plentiful, although some, as indicated by the grasping structure of their feet, evidently climbed trees. The only surviving descendants that strongly resemble them are the several species of insect-eating tenrecs found exclusively in Madagascar, and two species of another insectivore found only in Cuba and Hispaniola.

Cuba was once heavily forested and populated by a wide array of animals. Its coastline, over 2000 miles long, surrounds an essentially low country with few mountains. Its equable temperature and year-round, plentiful rainfall make it a semi-tropical paradise except when hurricanes roar over it. Today, most of the forests are gone, replaced by farms. A few native hardwoods, some pines and many introduced trees still grow on the hills, and the shores remain heavily fringed by several types of mangroves. In the shallow coastal sea, life remains almost unchanged.

Cuba has approximately 30 mammals but more than two-thirds of these are bats which, of course, are able to cross the inter-island straits without difficulty. Most of the others are several kinds of rodents. But it is the remaining two mammals that excite the greatest interest. They are insect-eaters extinct elsewhere but related to the tenrecs of Madagascar. They belong to the genus *Solenodon*, and they resemble large, 12 in. long shrews. *Solenodon* is rare in both Cuba and Hispaniola (where it lives primarily in Haiti). It has scent glands, but the odour, though unpleasant, is not strong. Its many, rather unspecialised teeth are an archaic feature, although they are efficient in shredding insect prey. Its eyes are small, and its nostrils are directed to the sides, possibly an adaptation to its snuffling, searching way of life.

The mysterious tarsier

Another interesting island relict is the tarsier, a curious little primate with a strange mixture of primitive and advanced traits. It survives today only in the Philippines and on other islands of the region such as Sumatra, Borneo and Celebes. The tarsier is not easily observed: small and nocturnal, it dwells high in trees.

The most startling feature of a tarsier's appearance is its pair of large, staring eyes, so huge that the rest of its face is pinched and reduced. Such eyes indicate that it feeds by night. Like most primates, a tarsier has a poor sense of smell, but its hearing is unusually acute. Sound is channelled into its ear canals by big, almost bat-like ears that move sensitively back and forth to identify the source of the slightest noise. The tarsier probably has the most mobile neck among present-day mammals: without twisting or moving its body,

ANCESTOR OF MAN? *The Philippine tarsier, only 4–6 in. long, can make a 6-ft jump from branch to branch, holding tight with pads on its fingers. The first primate – ancestor of lemurs, monkeys, apes and men – may have resembled a tarsier*

the gnome-like little creature can turn its head nearly 180 degrees in either direction to inspect an object directly behind it.

With its long, slender hind legs, the tarsier can make prodigious leaps from branch to branch. The tip of each finger and toe is a bulbous pad of flesh, with ridges and grooves that almost completely eliminate the chance of slipping. When a tarsier lands on a branch, it hangs on tightly, hunched up, ready for another springing leap.

It is difficult to place the tarsier in the long sequence of primate evolution. Its brain, teeth, digestive tract and fingers are essentially rather primitive, yet the tarsier displays a number of advanced, monkey-like characteristics: the structure of its retina; the rounded head, reduced face and mobile lips; and certain features of its reproductive system.

All primates living today – the lemurs of Madagascar, tarsiers, monkeys, apes and man – are of course modern creatures, and none is an ancestor of any other. Possibly some tarsier-like animal of the past was the ancestor of all the primates.

Most of the islands of the Philippine archipelago are small and without human settlement. Some of the larger islands are mountainous, and a few volcanoes are still active. With a 20,000-mile shoreline, and a climate extending from humid warmth at sea level to the chill of peaks over 9000 ft high, the Philippines have presented a wide variety of homes to plants and animals.

In the lowlands, fruit trees, broad-leaved timber trees and flowering plants – especially orchids – grow in abundance. Monkeys, large fruit bats and most of the carnivorous mammals live in the lowlands. Some monkeys are also found on the lower slopes, and birds, small mammals, lizards, snakes and even freshwater crabs have invaded the higher elevations as well. The beautifully marked leopard cat – only 2 ft long – is a fierce predator in mountain forests.

Among the great variety of spiders inhabiting the islands are large, tarantula-like ground spiders that live in burrows, smaller tree-dwelling species, and mimics that live in populous ant colonies and are almost indistinguishable from ordinary worker ants. Some of the large tree-dwelling (and house-dwelling) huntress spiders spend the day hanging, completely relaxed, from the four legs on one side; three of the legs on the other side dangle, while the last leg forms a triangular brace to support the heavy abdomen.

'Ceremony' of the lizards

Folklore everywhere in the world is full of myths that are really embellishments of natural phenomena. One tradition of Philippine folklore is the belief that small climbing lizards descend to the ground late every afternoon or early evening to kiss the earth, the implication being whatever human imagination wishes to make of it. Actually, the lizards do just that, but the reason is clear only to a careful observer. In the tropics, as the sun goes down after a hot and rather dry day, the air cools suddenly, and the moisture it held in the heat of the day quickly condenses on stones, soil and leaf litter. The lizards descend to lap the surface moisture while there is still enough light for them to see.

One mountain on Luzon Island supports over a dozen species of rats found nowhere else. This forested elevation with thick undergrowth is an example of the environmental isolation that can occur within an island that has itself been separated from the mainland for millennia.

THE HARSH LIFE OF A POLAR ISLAND

By comparison with islands in tropical and temperate regions, life is sparse on polar islands. Occasionally such islands support large populations, but they lack a wide variety of types. Cold in itself limits the number of species, and an ice cover is an even greater handicap.

WAITING FOR ITS PARENTS *A wandering albatross chick awaits its parents on snowy South Georgia Island in the South Atlantic.*

No matter how well adapted to the cold an animal may be, an ice cover almost assures starvation. Grazers cannot reach the few green things that are able to grow, and carnivores therefore have no animals on which they can prey.

The great ice sheet covering Greenland, one of the world's largest islands, is almost devoid of life. The ice-free fringes support only stunted vegetation. Insects are the most common animals. There are no cold-blooded land vertebrates – amphibians or reptiles. Only four land birds – the raven, snowy owl, redpoll and ptarmigan – winter commonly in Greenland, although seabirds and waterfowl visit the coasts and breed abundantly in summer. Greenland has only eight land mammals, ranging in size from the large caribou, musk ox and polar bear to the tiny lemming. Most of them came originally from North America.

Antarctic islands, including Antarctica itself, support even sparser populations on their moss and lichen vegetation: the largest true land animal is a tiny midge. Seals, and about 15 species of seabird, including five kinds of penguin, are the best-known animals.

Because the adult birds spend 11 months hatching their eggs and caring for their young they breed only every second year. As the chick grows, it is fed less and less frequently; one adult, known to be still feeding its young, was found 2640 miles away

DOUBLE COAT FOR WINTER *Arctic musk-oxen huddle together in battle formation to protect their young. Their double coat of soft wool overlaid by tough hair reaching to the ground is impervious to cold and wet. They shed the wool in spring*

CHANGE OF DIET *The Alaskan brown bear, which normally gorges itself with grass, lichens and bilberries, takes advantage of the season and scoops a salmon trying to get upstream to breed. The five-month-old cubs are too young to catch fish*

WEIGHTY ANIMALS KEEP WARM

Animals with large bodies have a smaller surface area in proportion to their total bulk, and so dissipate less heat. For this reason the larger animals of any species are usually found in cold conditions – for example the largest bear, the Alaskan brown, and the largest penguins, the emperor and the king. The smallest bear, the sun bear and the smaller penguins, the blue and the fairy, live in warmer climates

GUARDING THEIR EGG *A pair of 9-lb. macaroni penguins guard their single egg in a sub-Antarctic nest of mud and grass. The female usually lays two eggs, but hatches only one. If the second, smaller egg is hatched, the penguin chick will be a dwarf. Macaroni penguins hatch their eggs in December, and look after the young until March, when they are able to take to the water and fend for themselves*

How animals survive in the polar cold

There is evidence that Greenland and Antarctica have not always been as frigid as they are now: they experienced completely different conditions in the remote past. Coal beds on Greenland indicate that it had a temperate or subtropical climate some 125 million years ago. Antarctica has fossil skeletons of amphibians and reptiles which could not have lived in a polar climate.

Warm-blooded creatures of the far north and south tend to be large, a characteristic that allows them to conserve heat because their bodies possess a small surface in proportion to their bulk. The largest land carnivore in the world, the 9 ft long Alaskan brown bear, lives on the mainland and on Aleutian Islands, while one of the smallest bears, the Malayan sun bear, is tropical. Puffins – fish-eating birds found as far south as the Balearic Islands in the Mediterranean Sea – increase in size steadily northwards through the British Isles and Iceland to the Arctic islands of Spitsbergen.

A related phenomenon is the reduced size of ears, legs and tails of animals living in the polar latitudes. Small ears and short legs and tails lessen the amount of heat loss and the danger of freezing. The stocky musk-ox and the short-eared Arctic hare are quite different from their relatives far to the south: and the lanky Arizona jackrabbit offers an extreme contrast to the Arctic hare. Many Arctic and Antarctic species have developed a dual heat-balancing circulation system, unknown in similar animals in temperate countries. Warm blood from the heart heats cold blood returning from the animals' extremities and the returning cold blood cools the warm blood flowing to the extremities.

Island life around the world is characterised by its diversity. Each island, after long isolation, shows its own unique set of plant and animal adaptations. It is this uniqueness – born of chance arrivals, nourished by isolation, and expressed through a multitude of bizarre and beautiful forms – that makes islands endlessly fascinating.

How island
life evolves

All islands – whether they have separated from a continent or sprung up from the bed of the ocean, remote from any other land – depend on the continents for their plants and animals

Ancient and remote islands are more likely to have unique plants and animals than newer islands which are close to continents. The passing of millions of years allows the few species that succeed in establishing themselves on a remote island to evolve new shapes, colours and characteristics to suit their new environment. Remoteness ensures the other main condition for the evolution of species – isolation.

Continental islands, such as Japan and the British Isles, have few unique species – because they are not sufficiently isolated from the mainland and also because their existence as separate islands is not particularly ancient. Oceanic islands, such as the Galapagos Islands and the Hawaiian archipelago, have great numbers of unique creatures because they are both ancient and isolated.

The British Isles are separated from the mainland of Europe by only 20 miles of shallow water. As sea levels fluctuated during the Ice Ages the islands and the mainland were alternately linked and isolated again, and the most recent separation took place only 7000 years ago. Some British plants and animals may have survived the last period of ice and tundra along the Atlantic seaboard, but most are later immigrants from Europe and are indistinguishable from the mainland species.

The area stayed cool for several thousand years after the ice began to recede. Although there was a land bridge to the Continent, the sub-polar conditions did not suit many species, such as cold-blooded amphibians and reptiles which are unable to regulate their body temperatures. They moved northwards slowly as the climate warmed, but water from the melting ice caps flooded the Channel before many reached Britain.

Because the Irish Sea was flooded before the English Channel, Ireland has no snakes and only one type of lizard, one toad and one newt. Britain has three types of snakes (one venomous), three lizards, and six amphibians including newts, toads and one native frog. A similar pattern applies to freshwater fish. Ireland has half as many native species as Britain, which in turn has only half as many as Belgium.

Warm-blooded mammals, although they colonised polar regions successfully, also have only half as many species in Britain as in mainland Europe, and Ireland again has only half the British number. But poverty in species does not mean that the total number of animals is less in Britain: only that there are fewer kinds. Of 28 British mammals, 11 show some endemic features.

UNIQUE IN ISOLATION *A grey heron, standing 4–5 ft high, surveys a colony of the Galapagos Islands' unique marine iguanas. Descended probably from North American land iguanas, the Galapagos species provide living evidence of how even remote islands are colonised by continental animals which then adapt in isolation*

Birds can easily cross the sea, and the British species closely resemble those of Europe. But there are some native races, for example the red grouse. The jays, dippers and coal tits of Ireland and the hedge sparrows and thrushes of the Outer Hebrides have distinctive colours, and there are different forms of wren on the remoter islands, such as St Kilda, 40 miles out in the Atlantic.

These differences may have come about partly because the birds evolved along different lines in isolation, like the spectacular species of the Galapagos and Hawaii. But the story may be even more complicated. Washed by the warm waters of the Gulf Stream, the British Isles have an equable, oceanic climate – in the same way that much of Japan benefits from the warmth of the Kuroshio Current. As in western and south-eastern Japan, the warmest and coldest months in western Britain vary little in temperature. Frost is rare at sea level and even in the driest areas of southern England, the winters are warmer and the summers less hot and dry than on the European mainland. The result is that such islands offer far less varied habitats than the widely different expanses of a great continental mainland.

It is possible, therefore, that some of the many species that flourish on the nearby continental mainlands may be absent from Britain and Japan not because they could not get there but because the islands do not offer a suitable habitat.

Mountainous lands

Both Japan and Britain are rugged lands, with evidence of considerable early volcanic activity. Britain's volcanoes are long since extinct, but most of the 40 Japanese craters are only dormant. The major contribution of mountain ranges in both countries is to divide the islands into distinct geographical regions.

Occasionally, even continental islands may have species that have died out on their nearby parent mainlands. Northern Japan, for example, harbours a few unusual plants not found elsewhere on the Asian mainland – just as Kent has some 54 plants that do not grow around Calais in France, only 20 miles away. The katsura tree, which for long ages grew across all the continents of the northern hemisphere, survives today only in Japan. The magnificent cryptomeria, which creates the mystical blue-shadowed atmosphere of deep Japanese forests, is almost gone from the rest of Asia.

THE INFLUENCE OF WARM OCEAN CURRENTS

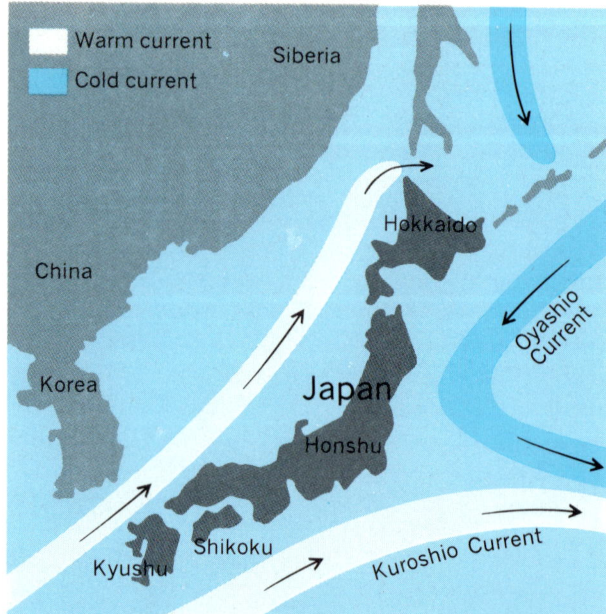

In the seas around Japan, cold waters flowing from the Arctic in the Oyashio Current are offset by the warming influence of the northbound Kuroshio Current. North-east Japan has snowy highlands; in the south and west, the climate is mild

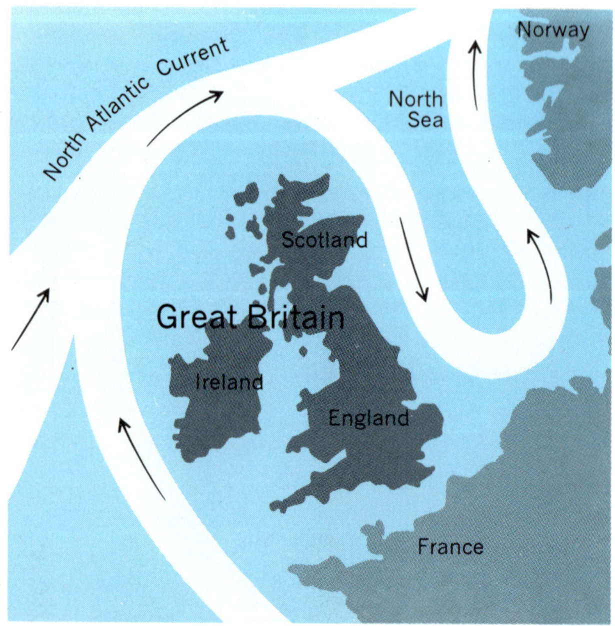

The British Isles have a temperate climate, although they lie in cold northern latitudes, because they benefit from the warm waters of the Gulf Stream, or North Atlantic Current. They have milder winters, but cooler summers, than northern Europe

SACRED MOUNTAIN *The 12,388-ft peak of Japan's sacred mountain, Fujiyama (Mount Fuji), is always snow-capped; but the foothills of the volcano provide many different environments for wildlife, ranging from deep forests to lakeside grasslands*

LINKS WITH THE ASIAN MAINLAND

The animals of Japan are a little more specialised than those of Britain because the archipelago was not heavily glaciated and was separated from its mainland millions, rather than thousands, of years ago. The islands have received colonisers from all their neighbours: Siberia in the north, China in the south and Korea in the west.

Japan's mammals, once abundant, today offer few surprises. The bears hunted and kept in captivity by the Ainus (the original inhabitants of the archipelago) are much the same as those of Siberia. Wolves, foxes, badgers, otters, mink and other carnivores scarcely differ from their Asian and American relatives. Large mammals are no longer common, but small species still live on forested slopes that have few human inhabitants.

Perhaps the most unexpected mammal in Japan is a light-coloured monkey – a macaque – red-faced and cloaked in long fur that insulates it against the bitter cold of winter. Except for man, it is the only primate to inhabit such cold latitudes; and, like man, it takes advantage of unusual local sources of warmth. In winter, the monkeys immerse themselves in hot springs. A few other colonies of the same species of monkey inhabit much warmer areas of Japan.

The birds of Japan, more than 400 species in all, are like those of the Asian mainland. Many of them – storks, herons, pheasants, ducks and small songbirds – are frequently depicted in Japanese prints, paintings and sculpture. Since no point on the islands is far from water, aquatic birds are especially numerous.

The archipelago possesses many species of freshwater fish – sure proof of previous continental connections. Japan also has freshwater crayfish and freshwater crabs similar to common Chinese species, found on the mainland.

Amphibians, absent from distant oceanic islands, appear in variety on all the Japanese islands. Some are large, but one minute green tree frog can easily sit on a child's fingertip. The giant among Japan's amphibians is the world's largest salamander, *Megalobatrachus*, a 5 ft long creature that lives in the cool mountain streams of southern Honshu and Kyushu. Like the American axolotl, it matures without ever losing its external gills and so is still able to breathe under water. Because it has been one of man's favourite delicacies for centuries, it is now rare.

Japan has comparatively few venomous snakes, but much feared is the mamushi, with heat-sensitive pits beneath its eyes. Pit vipers are not common. Most poisonous snakes, such as the cobra, have venoms that attack the nervous system; pit vipers' venom destroys blood cells.

More common are Japan's non-venomous snakes, which include rat snakes up to 5 ft long. Among its other reptiles are lizards and two kinds of turtle, one of which is endemic to the islands. A turtle living in rivers in central and southern Japan is often represented in the country's art; a long train of algae cascading down its shell lends grace and colour to what would otherwise be a rather solid and angular animal.

A significant snail

The Inland Sea, or Seto Nai Kai, is famous for its clear beauty, its multitude of picturesque islands, and the plentiful life of its waters. Among the marine animals found here, one green mollusc stands out. It is a snail, but the adult is difficult to recognise as such; encased in a pair of delicate shells hinged at the top, it is an almost perfect replica of a small bivalve (two-shelled) clam. The snail grazes on green algae in shallow water. If it is disturbed, it immediately withdraws, clapping its shells together for safety.

When the larva hatches, it looks like any other snail larva about to grow a typical spiral shell. But the juvenile shell soon divides into two parts, which gradually become more symmetrical, and the spiral is all but lost. A hinged ligament holds the two valves together, and a single muscle closes them. Two muscles close a clam's shell.

The little green snail is scientifically important because it provides an opportunity to study how such highly specialised bivalve forms as the clam may have evolved in the distant past.

Even a casual survey of the animals and plants of the British Isles and Japan demonstrates that isolation on islands brings about changes in form and behaviour as the few survivors settle into their new, more restricted environments. The animals of Japan have become more diverse and specialised than those of Britain, because they have had more time to do so. Far more complicated examples of the specialising effects of isolation appear on other remoter islands, much further from the continental land masses of the world.

MONKEYS THAT CAN STAND THE COLD

FINDING FOOD IN WINTER *When Japan's climate changed from tropical to temperate some 35 million years ago, its monkey population managed to adapt to long winters. Their present-day descendants, the macaques, live on bark in winter*

THE FAMILY UNIT *Macaque mothers care for their infants for two years at least, but the young females stay near their parents for longer. The babies are born six months after the winter breeding season and are suckled for a year*

TAKING A HOT BATH *Beating the winter cold, a troop of macaques – snow monkeys – huddle in a hot-springs pool. Younger monkeys are more adventurous than their elders. At the coast they frolic in the ocean while the adults stay ashore*

Islands that inspired Darwin's theory

Around the world, many long archipelagos, consisting of hundreds of islands, string out from their parent continents, or arise quite independently in the ocean; or several islands may cluster far from any other land. The plants and animals living in such clusters or chains of islands reveal much about the slow sequence of colonisation.

The earliest study of island specialities, and still one of the best, was made by Charles Darwin in 1835, during his trip around the world in the survey vessel, HMS *Beagle*. Darwin had already spent nearly four years studying South America when the *Beagle* arrived at a group of rocky islands 500 miles off Ecuador – the Galapagos Islands.

For the first time, Darwin had a chance to see the effects of isolation on a number of familiar animals: birds, lizards and tortoises. It is clear that at the time of his visit, he did not realise the importance of what he saw, but he made careful collections. Years later, as the germ of his evolutionary theory grew in his mind, Galapagos life provided his most important evidence.

THE GALAPAGOS ISLANDS

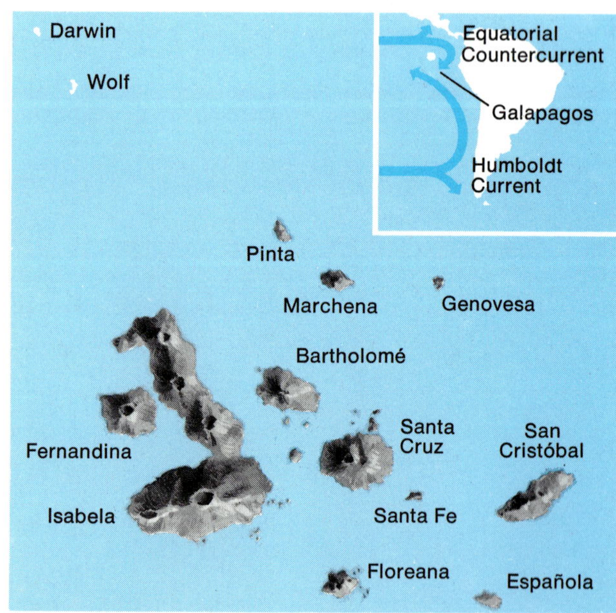

The Galapagos group comprises 16 major islands and numerous tiny ones, some of them still active volcanoes. The islands' first living things may have been carried by the waters of the Humboldt Current from the shores of South America

MOST NORTHERLY PENGUIN *A Galapagos penguin chick waddles across a lava shore. The species, which nests on several of the islands, is the only one found in the tropics. Females lay eggs in holes in the rocks a few feet above high water*

SHARED ACCOMMODATION *Galapagos sea lions, which are related to species living on islands off California and Japan, bask on the rocks used for nesting by swallow-tailed gulls from Peru. Like the gulls, the sea lions live mainly on fish and squid from the surface waters of the Humboldt Current around the islands. Sometimes, however, they may also attack the small native penguins. Each bull sea lion usually has as many as four or five cows in his harem*

In the geological time scale, the islands are of recent origin; the oldest fossils found there are perhaps no more than 2–3 million years old. Both the Hawaiian Islands and the Galapagos are the exposed tops of great volcanoes with huge bases, 10–20 miles across, resting on the sea floor far beneath. The islands are strewn with jumbled masses of lava blocks.

The communities of shallow-water marine creatures along the shores resemble those along the tropical and sub-tropical American mainland. Because the nutrient-laden waters from the deep sea are forced upwards against the shores of the Galapagos, the shallow coastal seas are exceedingly rich in living things.

Other volcanic islands in the equatorial Pacific are densely covered with vegetation, but not the Galapagos. They lie in the path of the great Humboldt Current, which brings sub-Antarctic waters up the west coast of South America: in consequence they are one of the coolest equatorial regions in the world. The south-east trade wind blows for three-quarters of the year and brings little rain, except to the highest peaks. The dry volcanic terrain does not easily weather to make soil in which plants can grow. On the hot plains only vegetation suited to an arid climate thrives.

Because the Galapagos rose from the sea with no previous connection to the mainland, their life consists only of what was able to arrive by sea and air. There are now five large and 11 small islands, with a total area of slightly less than 3000 square miles. Geologists have worked out only part of their history, but much information comes from biology. The most dramatic evidence of change is the small lizards and snakes that cannot swim or fly across the channels but whose distribution, resemblances and differences provide clues to the separation of the islands.

How did terrestrial life arrive? Off Ecuador, the cold Humboldt Current is deflected towards the Galapagos Islands and merges with the Equatorial Current to become a reliable conveyor belt from South and Central America.

Penguins and fur seals from the south and sea lions from the north originally swam in on the currents. Other creatures, seeds and living plants were carried from America aboard natural rafts, on winds, or on the feet and feathers of seabirds. Of all the animals now inhabiting the Galapagos, only one – a land snail – seems likely to have come from the broad Pacific to the west.

HOME OF THE GREAT TORTOISES

Apart from seabirds and seals, the first successful settlers of the Galapagos must have been plants, since land animals could not have survived without some sources of plant food. The progressive enrichment of the vegetation with new colonists was essential for the development of a varied plant community.

Each island differs from the others in area, in altitude and in exposure to the rare rain-laden winds. The distinct zones of vegetation include mangrove stands along the shore; tough cactus and shrubs, leafless and flowerless except after the brief rainy season, in arid coastal lava fields; and scrub forests and lush meadows of grasses and ferns on high slopes where moisture condenses from clouds and encourages their growth. At still higher elevations (2000 ft or more), exposed to constant cool winds, the dominant plants are grasses, low shrubs, mosses and lichens.

Prickly giants

Several trees of the Galapagos are derived from small ancestors. The prickly pear cactus, common in the Northern Hemisphere, usually grows close to the ground or is only a few feet high. The flattened pads (actually modified stems) of this cactus are buoyant, and they can propagate new plants; it is likely such pads drifted to the Galapagos from Central America a long time ago. Prickly pears apparently became established in the Galapagos at least twice, for two basic kinds of specialisation are found among the several major groups of their modern descendants. Some species that closely resemble their ancestors cling to rocky cliffs. Others produce exceptionally long pads that thicken into stout trunks more than 3 ft in diameter; dense clusters of more conventional spiny pads crown the tops, perhaps 15 ft high.

Today this cactus and another tree-like species – the candelabra cactus – play major roles in the biological development of the islands. The cactuses propagate regularly and abundantly enough to sustain the islands' iguanas and tortoises, which eat some of the young plants as they appear above the ground. A significant percentage of cactuses, however, escape being eaten; once they produce massive trunks, the cactus trees are proof against attack. Indeed, they seem to grow largest on the

DISTINCT VEGETATION ZONES *On larger islands, vegetation grows in distinct zones determined by rainfall, temperature and height above sea level. The plants range from cactuses on the arid coast to grass and ferns in the wetter heights*

islands where tortoises live or have lived in the past, while much smaller prickly pear bushes grow in areas never inhabited by tortoises.

When rainfall encourages the growth of less formidable plants, tortoises prefer to eat them, but in the dry season cactus pads are almost their only source of food and moisture. The pads, which have grown in profusion at the top of the tree during the beginning of the rainy season, become swollen with water, and are often so heavy that they fall to the ground. The tortoises then forage for the fallen pads, which have relatively small, soft spines.

A tortoise is not merely a consumer, however. The tough cactus seeds it swallows with the pads and fruits pass undamaged through its digestive system and are deposited, together with fertilising waste matter, in most parts of the island, where many germinate and grow. In this way, the reptiles have widened the distribution of their essential food source.

The seeds or pollen of many Galapagos plants are scattered by other animals besides tortoises: birds, flies, carpenter bees and hawkmoths are among the most important. At the same time, these isolated islands support many self-pollinating plant species, more than are usually found in continental environments, which have numerous varieties of insects that serve as pollinating agents for plants and flowers.

Side by side with cactus trees, another tree, 30–40 ft tall, has evolved from the weed-like sunflower family. On the Galapagos Islands the sunflower family has given rise to several species. Sunflower trees make up the forests of the middle slopes, taking over where the cactus trees begin to die out. Further up, guava and other smaller trees become dominant.

The sailors' prey

It is from the giant land tortoises that the islands take their name – *galápago* is the Spanish word for 'tortoise'. Darwin had been told that each island supported a different kind of tortoise and he was surprised to find it true. The ten or more different kinds that are or have been present are descended from one species that must have arrived by raft from South America. Smaller relatives still live there and in the West Indies.

Long ago giant tortoises were abundant on at least ten of the islands. At mating time, the hoarse cries of the males resounded over hundreds of yards. But early in the history of human exploration, the islands became ports of call for sea-going

GIANT IN DANGER *A giant tortoise, now in danger of extinction, plods through a Galapagos cactus forest. The reptiles, which weigh only 3 oz. at birth, may take some 30 years to reach a weight of 600 lb. They are thought to live for 100 years*

ships. Even a medium-sized vessel could stow away 300 tortoises in the hold, where they could be kept alive, without food or water, for as long as a year to provide meat and fat for the crew. Later, during the early 19th century, whaling vessels were still removing thousands of the huge, completely fearless and defenceless reptiles.

Today there are self-sustaining colonies of tortoises on only three of the islands; the tortoises of the other islands are either gone or – because their reproductive rate does not keep pace with their declining numbers – in danger of extinction. The eggs and the young of the tortoises are often eaten by wild pigs, wild dogs and rats, whose ancestors were brought in by man. On some of the islands, even the once-extensive plant food of the tortoises has been gradually destroyed over the years by pigs and imported goats.

WORLD'S ONLY MARINE IGUANAS

The large and conspicuous marine iguanas of the Galapagos are descendants of green vegetarian iguanas similar to those that still inhabit mainland forests, where they feed on the abundant leaves. At some time in the distant past, a section of an American estuary bank may have broken away, with the plants and animals living on it, and drifted hundreds of miles to the islands. If, over many thousands of years, this happened repeatedly, it would have brought not only the iguanas but also smaller land animals.

The Galapagos are easier to reach in this way than many island groups. For most of the year winds and current are from the mainland, and with their aid the journey would take about two weeks. Any land iguanas cast ashore would then face the problem of finding suitable food.

Establishment would be easiest in the rainy season when the arid coastal plains have a brief flush of greenery. Wandering inland, iguanas, fortunate in their unspecialised feeding habits, would find a range of fruit, flowers, seeds and water-filled cactus pads. The fissured lava offers crevices giving escape from the mid-day heat.

One may guess that today's marine iguanas, with their unique specialisations, evolved from ancestors established on land in the coastal zone. The thick mat of brown seaweed (mostly a kind of Sargassum weed) exposed on the rocks at low tide

or cast up in storm debris is nutritious, and once the animals had adapted to it, could support many more individuals than the scanty land vegetation.

Today the 3 ft long marine iguana is very different from its ancestral stock on the mainland. It is dark rather than green: some races are black, others are a mottled red and black and show a trace of green only during the breeding season. The dark colour may make it possible for these cold-blooded reptiles to absorb heat. The rather long, leaf-gathering snout of the forest iguana has become a short, heavy, bulldog type, suited for browsing on the dense mats of seaweed hugging the rocks. Biting first with one side of its jaws and then the other, the big lizard tears at the tough plants, all the while clinging to the slippery rocks with powerful clawed toes.

Unlike the land iguanas, which can obtain water from plant food, a marine iguana is obliged to swallow salt water when it feeds on seaweed at low tide. When it eats under water – a feat made possible by the swallowing of small stones, which lessen its buoyancy and enable it to cling securely to coral heads – the iguana takes in even greater quantities of salt water. So much salt in its body would be fatal except for the specialised tear glands that secrete salt.

The marine iguana is a good swimmer; with legs folded to the side, it undulates its body and tail. It

224

avoids heavy surf, and does not venture out more than a few dozen yards. Sharks are numerous in Galapagos waters and are known to prey on the lizards. An iguana rides in on a swell and crawls up the rough lava. If it does not find a home immediately, it wanders about, touching the rocks with its tongue, which carries scent to a special sensory organ in the mouth. Once the scent of other iguanas is found, the iguana rests safely above high tide, almost motionless, even when bright red crabs scurry over its scaly body, stopping occasionally to pick off ticks.

Marine iguanas seldom travel more than 10 yds inland. Lying in great swarms along the dark lava shore, they soak up the sun until the tide recedes and it is again time to eat. Each male occupies a small area of territory, usually in the company of one or two females. If an intruder wanders past, it is attacked at once and bitten ferociously; it usually retreats in haste.

Most fights result from the deliberate challenge of an invader. Such contests begin with much threatening and bluffing: the males show their bright red mouths, rise high on their legs, and may squirt twin sprays of moisture from their nostrils. Each combatant, butting with its heavy, flattened head, tries to shove the other away. If the battle is decisive, the loser lies down in a deflated attitude of submission, and the victor stands high and nods vigorously. Since no serious injury results from such a fight, the colony is not depleted, and the population remains unaffected.

While females are excavating nests in the all-too-scarce volcanic sand, they too may come into conflict with one another. If the 12 in. deep burrows have not yet been covered when the fight begins, the pair of eggs exposed in each burrow may be destroyed by island mockingbirds, which dart in to feed on the nutritious yolks.

Three-eyed lizards

The Galapagos land iguanas are much the same size and have the same sluggish vegetarian habits as their marine relatives. There are also some seven species of smaller reptiles, called lava lizards, which display similar aggressive behaviour. In the centre of some lava lizards' heads is a vestigial third eye, not totally functionless: it senses the amount of solar radiation received by the reptile, which accordingly regulates its body temperature by moving in and out of sunshine, or between hot rocks and cooler crevices. When light and heat are excessive, there is a marked reduction in the lizard's activity.

TERRITORIAL COMBAT

A male land iguana points his snout upwards and jerks his head up and down to scare away an intruder on his territory

The competing male iguanas fight fiercely, trying to grab the skin on each other's flanks or snapping at the enemy's head

Wounded by his opponent, the intruder retreats. The iguanas rarely continue fighting after one has been injured

SHOWING OFF *A male frigate bird puffs up his scarlet throat sac in a courtship display, while his mate rests beside him. The throat sac is conspicuous only during the bird's breeding season*

ATTRACTING A MATE *Blue-footed boobies strut in an elaborate courtship dance, inflating their cheeks, cocking their tails and lifting their webbed feet high off the ground. They stay with their chosen mate for life*

TAKING HOME A GIFT *A flightless cormorant waddles past an iguana colony clutching seaweed for its mate. The bird's wings are useless for flying, but food is plentiful and they have few enemies. They use the wings to shade their young*

BIRD CLUE TO EVOLUTION

The Galapagos are among the few islands in the world where large bird populations share their environment with large numbers of reptiles. Over 100 bird species, many of them migratory seabirds, nest on the islands. The land birds, probably blown in by storms, have become highly specialised. A few species of birds are almost extinct: there are only 200 Galapagos flamingoes left. The flightless cormorants and the Galapagos penguins are abundant; there are hordes of boobies, and about 2000 pairs of Galapagos albatrosses.

Darwin's finches

The most famous birds of the islands are the finches that Darwin collected, described and studied for two decades. Forty years after their first mention by a Captain James Colnett, Darwin's more detailed account of them in his journal, *The Voyage of the Beagle*, gave no indication that they would become basic evidence in the development of his theory of evolution. But he noted: 'Seeing this gradation and diversity of structure in one small, intimately related group of birds, one might really fancy that from an original paucity of birds in this archipelago, one species had been taken and modified for different ends.'

The ancestor of the Galapagos finches is not known, but probably it was one of the common seed-eaters of the mainland. What Darwin finally realised is that different specialities may arise from a common ancestor. From the original finches arriving on the Galapagos thousands of years ago, there are now 14 species of birds. All are still essentially finch-like, mostly brown, short-tailed birds that construct roofed nests in which they lay pink-spotted white eggs. But in size, beak structure and habits they are widely different.

The shape of a finch's bill suggests its food preference. Some bills can tear at fruits, others are grain-crushers; some are curved and can reach deep into flowers such as those of the coral tree. There are straight bills that bore into wood, a fly-catcher type, and the usual finch shape. Once Darwin recognised that these different shapes were all modifications of a single original bill design, the conclusion was inevitable: given time and opportunity, species change. Undoubtedly the Galapagos finches are still in the process of specialising even further.

The Galapagos Islands have been studied repeatedly since Darwin's day, and now have a permanent research station. Despite the ravages of early visitors and of the once-domesticated animals they left behind, the natural life of the islands is still more nearly intact than that of many other islands. From the Galapagos scientists continue to learn the basic principles of how plants and animals have developed over millions of years.

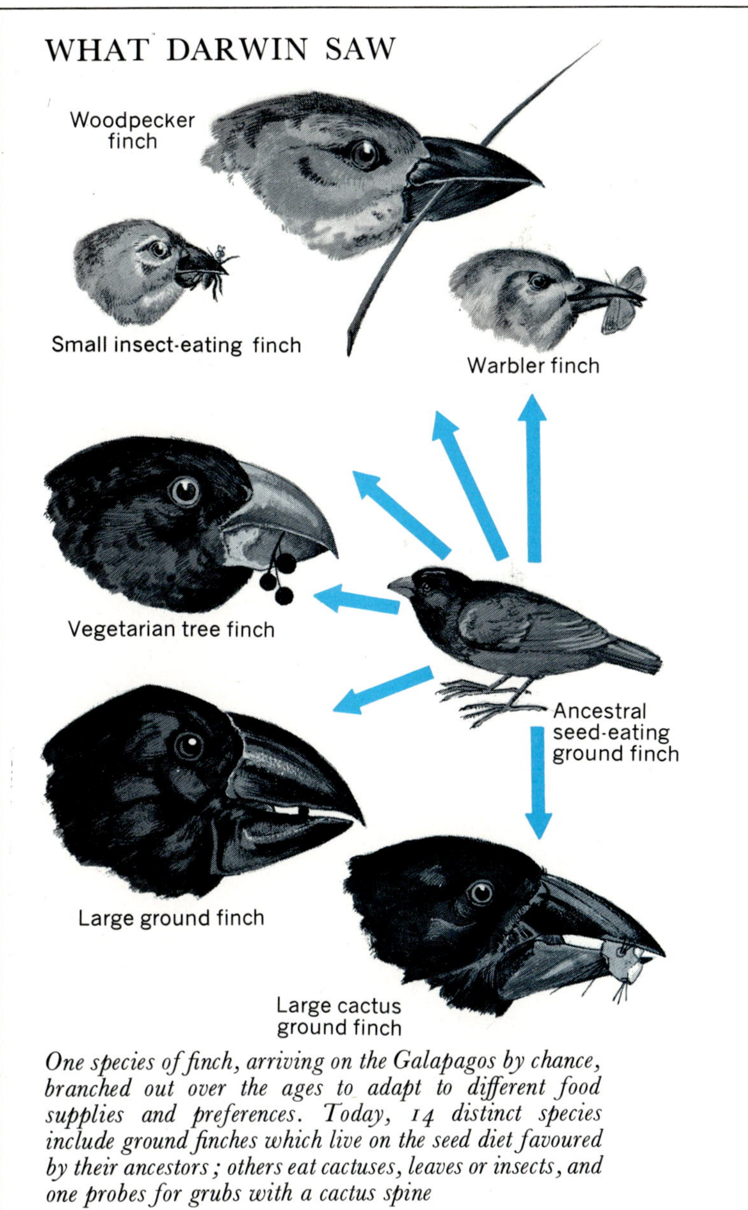

One species of finch, arriving on the Galapagos by chance, branched out over the ages to adapt to different food supplies and preferences. Today, 14 distinct species include ground finches which live on the seed diet favoured by their ancestors; others eat cactuses, leaves or insects, and one probes for grubs with a cactus spine

Isolated in the world's largest ocean

Although the Hawaiian Islands have been influenced by early Polynesian and more recent settlers, they still offer some of the most dramatic examples of island life and of changing species from one island to the next. There are even clear distinctions within a single island.

The area of the Pacific Ocean, the world's largest body of water, exceeds that of all continents and islands put together; it stretches nearly 9500 miles in one direction and 10,000 miles in the other. The Hawaiian Islands, in the middle of this vast expanse, are more isolated than most islands; the land nearest the archipelago is little Johnston Island, 600 miles to the south-west.

Yet animals and plants have succeeded in colonising the Hawaiian group as they have most Pacific islands. There are no native amphibians; the present freshwater fish evolved from marine species. The ancestors of the islands' geckos and skinks may have accompanied early Polynesian settlers rather than making the trip on their own. Except for one kind of bat, no mammals existed on the islands before the arrival of man.

Whereas Galapagos life is distinctly descended from that of South and Central America, the Hawaiian Islands are largely populated by species from the Pacific and Indonesian regions to the south-west, though the birds come mainly from Asia and America. Oceanic and atmospheric currents have changed somewhat over the ages, and many stepping-stone islands have disappeared through erosion or submergence, and so the pathways taken by the first colonisers are obscure. Only occasionally scientists are able to dredge fossil evidence from sunken islands.

The islands are volcanic and have never been connected with any continent. Four of the major islands were formed by the union of smaller ones, and deep water now separates some small islands that were once joined. The north-west islands are several million years old, today little more than slight protuberances above the sea. Sixteen hundred miles to the south-east, the great island of Hawaii, about a million years old, continues to grow through volcanic action.

The Hawaiian Islands offer many more examples of development from single ancestors than do the Galapagos; it is a pity that Darwin did not visit them or even know of their plant and animal variations, which have progressed much further along the evolutionary path than his examples on the barren islands off Ecuador.

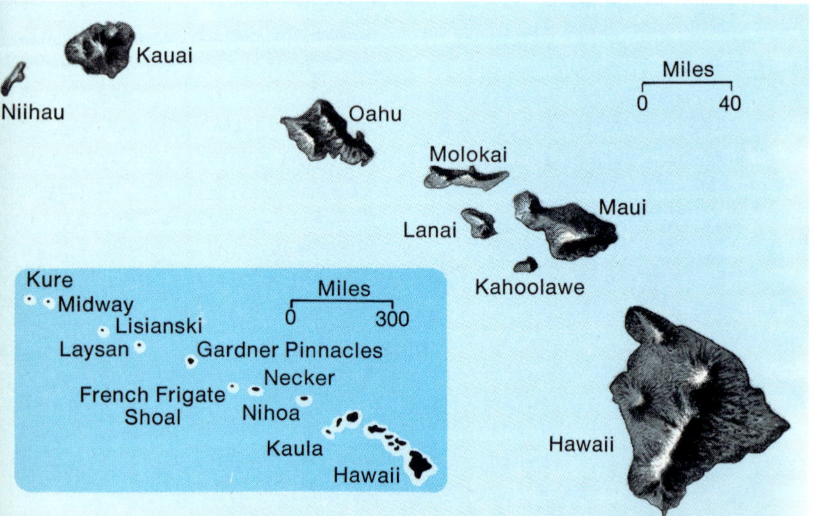

VOLCANIC CHAIN *The Hawaiian Islands chain was formed along a great crack in the ocean floor, beginning at the north-west end. The eastern islands are the youngest and largest; those at the other end, 1600 miles away, are greatly eroded*

THE PLANT PIONEERS

Colonisation of the volcanic Hawaiian soil had to begin with hardy pioneer plants. Only after plants were well established could vegetarian animals survive, and of course neither predators nor parasites could live without an earlier resident animal population.

Before man arrived, the plants of the islands consisted of many grasses, tree-size lobelias, shrub violets, 12 ft high geraniums, and a host of other seed plants, ferns and lesser forms. More than 1700 kinds of seed plants were derived from perhaps 272 original colonisers, most of them from the Indo-Pacific region. There were no mangroves or pines, and very few orchids or palms; all of these are plentiful today. There were enormous

hardwood forests forming an unbroken cloak of vegetation from shore to timberline. Man has changed all that – first the Polynesians, then settlers from modern Asia and America.

The original forests were composed of a wide variety of plants, but three trees peculiar to the islands are especially well known: the great koa, which grows on mountainsides; the widely distributed ohia lehua with its dense wood; and the fragrant Hawaiian sandalwood, a partial parasite that taps the roots of other forest trees.

Some of the rarest and strangest plants on any islands are those of the alpine zone on Maui and Hawaii, a region beginning at about 9000 ft and rising to almost 14,000 ft. Here are found 13 species that exist nowhere else in the world, while nearly two dozen rare species can be found in the islands' adjoining sub-alpine zone.

The spectacular silversword, a distant relative of the sunflower, is known as one of the oddest plants in the world. Its dense spherical cluster of pointed, silvery leaves is an adaptation to the severe dryness of volcanic cinders and the intense sunlight of high altitudes. After as much as 20 years of growth, it produces a towering complex of several hundred flowers, and then dies. The flowering stalk is thickly covered with sticky hairs that permit only flying insects to reach the pollen and nectar. The silversword has several relatives on the islands, including the greensword, but since nothing like it exists anywhere else in the world, its evolutionary history is obscure, although it is related to the tarweeds of California and Chile.

NATIVE BIRDS OF HAWAII

Colonisers from North America were the ancestors of some of the bird species endemic to the archipelago: the Hawaiian crow, two ducks, a hawk and the nene (Hawaiian goose). All have developed in isolation for so long that they now differ perceptibly from their cousins on the mainland some 2200 miles away.

The nene is obviously descended from the Canada goose, but so much changed that it has become a distinct species. No longer associated with water, it lives on the high, desolate mountainsides of Maui and Hawaii. Here it builds its nests and raises its young amid scrubby vegetation on the massive lava deposits. Although the nene can fly, in recorded history it apparently has inhabited only the high mountains of these two islands. It is rare today, but at times it can be seen walking slowly over the lava clinkers, its neck curved into a U shape as it searches for food. The nene's feet

have lost much of their webbing, its legs are extended, and its waddle is gone. On the ground the nene is rather quiet, making only short, creaking calls; its honk in flight has a higher pitch than that of mainland geese.

The Hawaiian duck, or koloa, apparently evolved from the American mallard, but the males fail to develop the resplendent and precisely delineated breeding plumage of the mallard. North America also provided the ancestors of the Laysan duck, a coot and a gallinule, none of which has changed much. A resident black-crowned night heron does not differ in the least from its North American relatives. The Hawaiian stilt belongs to the same species as the North American black-necked stilt. The honeyeaters seem to be descended from Australian forms that could have come by a series of stepping-stone islands. The Hawaiian owl has many close relatives in the

SAVED FROM EXTINCTION *Pairs of Hawaiian nene geese, once on the point of extinction, have recently been bred at Slimbridge Wildfowl Trust in England. The birds released back on Haleakala Crater still bear identification tags on their legs*

Northern Hemisphere, where such owls have been seen flying 1000 miles from land, but this owl is active by day. The ancestry of the Hawaiian and Laysan flightless rails is unclear.

The forbears of most of the birds found only on the islands were accidentally deposited long ago, probably blown off course by cyclones. Such chance arrivals still occur. A few years ago, a pair of North American kingfishers appeared on Hawaii, 2200 miles away; a Chinese cuckoo was sighted on Wake Island, 3000 miles from home.

Unlike seabirds and regularly migrating species, none of the native birds leaves the islands voluntarily despite their migratory ancestry, although a Hawaiian hawk has been recorded flying over California. The birds no longer follow the old patterns of seasonal migrations, and breeding times may be prolonged or repeated within a year. Several strong-flying birds remain in restricted territories on only one or two islands of the chain. The hawk lives and nests on Hawaii and visits neighbouring Maui, but is unknown on the other islands unless it is blown to them.

THE MOUSE-CATCHER *The Hawaiian hawk ranges in colour from tawny to nearly black. Its main food is mice, but it also eats eggs, insects and spiders. The bird builds a circular nest in high trees on Hawaii's open parklands*

STANDING ON WATER *The elongated toes of the Hawaiian gallinule or moorhen, allow it to tread across lily pads looking for tiny clams and plants. The shy bird rarely ventures far from cover. It swims well but flies poorly*

DIPPING FOR NECTAR *One of the commonest honeycreepers, the apapane, perches on a cluster of broussaisia flowers which are native to Hawaii. The bird draws nectar from blossoms with a brush-tipped, tubular tongue. It also eats caterpillars*

NECTAR EATER *The 6-in. crested honeycreeper is nearly extinct and now lives only in remote forests on the north-east slopes of Haleakala on Maui. Its chief food is the nectar of the ohia lehua tree, but insects are also part of its diet*

LARGEST OF THE HONEYCREEPERS *The 7-in. Laysan finch is larger than other honeycreepers. Nearly extinct and found now only on Laysan and Nihoa, it can crush tough seeds with its parrot-like beak*

BILLS TO SUIT FLOWERS *The long bills of the i'iwis can reach nectar in even deep-belled flowers, and are also useful for snapping up insects. These rare birds of Kauai and Hawaii nest in shallow cups of dry plant material placed in the trees*

THE ADAPTABLE HONEYCREEPERS

Among the most interesting of Hawaiian birds are the spectacular honeycreepers. They surpass the Galapagos finches in diversity; a wider variety of feeding grounds has been available to them, and they have had more time in which to evolve. Their ancestors probably came from the American tropics, but scientists cannot be certain because the birds have changed so dramatically since they settled on the Hawaiian Islands.

Some species of honeycreepers are extinct, but the several surviving species inhabit the high forests of the larger islands. Two, the apapane and the i'iwi, are brilliant red; others are yellow, green or olive. The apapane and i'iwi are often seen sipping nectar from the ohia lehua tree.

Some honeycreepers have large, powerful, parrot-like beaks well suited to crushing tough seeds, and short, thick tongues with which they efficiently manipulate hard seeds. Some have beaks specialised for tearing into fleshy fruits. Others have long, slender bills that fit the curve of native lobelia flowers, and long tongues that, when rolled into tubes, become organs for sucking up the sweet nectar. A woodpecker-like honeycreeper opens crevices in trees with its short under bill, then probes for grubs with its much longer upper bill.

Although the first honeycreepers on the islands were probably nectar-feeding birds, some of their early descendants adopted the ways of seed-eating finches. They then branched out to fill the niches usually occupied, on continents, by other species of birds. The different kinds of honeycreepers, which evidently evolved from a single species, illustrate the tendency of island species to proliferate. Once a particular plant or animal has invaded an island, isolation and new opportunities usually evoke a rapid evolutionary response.

233

Islands from the past

The creation of Madagascar and the islands of New Zealand dates back to the breaking-up of the super-continents. Isolated for tens of millions of years, they have become 'arks' of unique plants and animals

Madagascar, lying in the Indian Ocean 250 miles off the east coast of Africa, is one of the world's most fascinating and puzzling islands. It is known as the land of the extinct elephant bird, the primitive lemur and the curious 'traveller's palm'. It is a huge island – 1000 miles long and almost 250,000 square miles in area.

On a map, Madagascar looks exactly like what it is – a broken-off piece of Africa. Evidence that dinosaurs once lived there shows that Madagascar was long ago part of the great super-continent of Gondwanaland. During the drifting movements and eventual fragmentation of the super-continents, Gondwanaland broke up to form Africa, Antarctica, India, Australia and South America. Among the many smaller fragments, which became what we call 'continental islands', were New Zealand, New Caledonia, Cuba, Hispaniola and Madagascar. Thus Madagascar's birth dates back at least 70 million years.

Madagascar has had time and opportunity to develop a richly varied flora and fauna of its own. Its long history as part of a continent, its relatively early isolation, and its prolonged exposure to invasion by outside forms of life have produced some truly extraordinary endemic species. Too little is known about them, and discoveries that are still occurring raise as many questions about the island's history as they answer.

All the backboned animals found on the island, apart from species brought by man, are descended from a few creatures which flew in or were washed ashore on 'rafts' of logs or vegetation – probably between 30 million and 60 million years ago. This limited stock of comparatively primitive creatures has developed over millions of years into a marvellous array of specialised forms of life to fit the conditions offered by the island's extraordinarily wide range of climate and vegetation.

Madagascar is now the home of 70 species of land mammals and 118 species of birds which exist nowhere else in the world – in addition to a large number of unique reptiles and amphibians.

The mammals include all the world's lemurs and tenrecs; the reptiles include the grotesque and fascinating chameleon. Such creatures make Madagascar an island treasure house. No other island of its size has so many examples of backboned animals and seed plants that have adapted to suit their environment. Much remains to be learnt about Madagascar's wildlife, much that will help to explain how living things develop in isolation. Unfortunately, man's activities are destroying many of the most interesting species of Madagascar faster than they can be studied.

MONKEY SUBSTITUTE *Monkeys and apes had not yet developed when Madagascar broke away from Africa; some of the island's lemurs gradually evolved the way of life characteristic of monkeys elsewhere. Mongoose lemurs, the size of cats, are now rare and in danger of extinction. They eat fruits and seeds*

ISLAND OF CONTRASTS

Because of its size, its north-south orientation and its central ridge of mountains, Madagascar offers a variety of environments: wet rain forests, dry upland forests, savannahs, grasslands and near-deserts. These conditions have made it an island laboratory for the development of its own unique collection of plants and animals.

Some of Madagascar's odd plant adaptations seem almost to be imitations of similarly specialised but quite unrelated plants in distant countries.

An example is the family of didiereas, plants unique to Madagascar, which grow in the island's more arid regions. Tall, slender and drooping, the didiereas have spines like those of cacti, yet they evolved from small, herb-like ancestors not even distantly related to the spiny plants of American deserts. The stems of the didiereas rise in clusters 30 ft or more in the air, each stem armoured with spines and, in season, adorned with leaves and yellow, ivory, green or bright red flowers.

Other plants growing in the dry areas of Madagascar are even more cactus-like. One species of kalanchoe has broad, fleshy, water-storing leaves 2 ft long. If these plants grew on the African mainland, they would be succulent fare for herbivores, but Madagascar had no indigenous large grazing animals to devour them or to threaten their survival. In contrast, the leaves of a vine-like species of kalanchoe are so narrow and curled that they resemble rings or hooks. Serving as tendrils, they curl around neighbouring stems and leaves during the plant's upward growth.

Members of the sunflower family on Madagascar, as on many other islands in warm climates, have evolved into forest trees, far exceeding the development of their continental counterparts. The island also supports many ancient forms of flowering plants that survive elsewhere only on islands near by or – strangely – one-third of the way across the world, in New Guinea.

One of these Madagascan oddities, the 'traveller's palm', is partly misnamed, since it is not a palm at all but a cousin of South Africa's spectacular bird-of-paradise flower. But the other part of the name is perhaps justified: the tubular base of each leaf forming the plant's fan-shaped crown traps rainwater with which the parched traveller may conveniently slake his thirst.

The baobab tree is certainly one of the world's most bizarre plants. Some kinds of baobab grow in Africa and Australia; one of Madagascar's varieties achieves a grotesqueness approaching the absurd. Fully grown, the base of the tree trunk is a great bloated cylinder 30–40 ft in circumference, its naked bark stretched smooth by the spongy, water-filled tissues inside. The gross trunk rises tapering and branchless to a narrowed point, from which all the branches and foliage grow in a flattened and disorderly array. The whole effect is that of a child's drawing of a tree. Although the plant appears to be firmly based and sturdily built, its roots are shallow and the tree can be

MADAGASCAR'S VEGETATION

Madagascar's east coast receives heavy rainfall, but the high interior of the island, with one peak of 9468 ft, prevents much moisture from reaching the west coast, parts of which are semi-desert. Lush tropical rain forest originally flourished on the east coast; tropical thorn forest and tropical grasslands on the arid west coast; temperate vegetation in the cooler central highlands. Today, forests and grasslands have been so disturbed by man that only fragments of that pattern remain

easily toppled to the ground. Once down, it soon crumples and rots quickly like a ripe fruit.

Animal oddities

Many of Madagascar's animals are as unusual as its plants. Spiders in great variety populate deserts and forests alike. Termites inhabit the towering, rock-hard hills that dot the arid plains, each hill used century after century by successive generations of the light-shunning insects. Butterflies are abundant and dazzlingly conspicuous. They settle by the hundreds along shorelines, opening and closing their brilliant wings as they uncurl long tongues to probe for moisture.

Like their relatives in South America, Madagascar's moths are mostly day fliers, so big and colourful that they have no equals in Africa. One moth in particular has a remarkable distinction: its existence was predicted 40 years before scientists found it. In 1862, Charles Darwin described the Christmas-tree orchid from Madagascar, with a spur more than 10 in. long. No scientist had seen an insect that could reach the nectar in such a spur, but Darwin reasoned that the plant must be pollinated by a moth with a 10-in. tongue. When, finally, the moth was discovered, scientists naturally called it *praedicta*.

Marine fish, no matter how sluggish, theoretically can roam the wide oceans, so physical restrictions would not seem to be a problem for them. Yet even in the sea there are examples of geographical isolation. Living in pools along Madagascar's shoreline are some of the strangest fish known, the mudskippers. Related species are found only in western Africa, Australia, Borneo and a number of smaller Pacific islands.

Mudskippers spend much of their lives out of water, where they breathe partly through their skin, like frogs or salamanders, and return only occasionally to the water to moisten their gills and bodies. Their blunt, frog-like faces, complete with bulging eyes, and their habit of climbing up rocks and muddy beaches – even slanted tree trunks – make mudskippers droll and delightful little creatures. They scull up slopes by using jointed front fins, which are almost leg-like in structure. To keep from slipping backwards, they anchor themselves in position with other fins extending from their undersides and serving as suction discs. Pairs of mudskippers can sometimes be seen sunning themselves, and raising and lowering their dorsal fins in a form of courtship display.

ANCESTRAL HOME OF CHAMELEONS

Madagascar is the ancestral home of chameleons – gaudy, ornamented lizards with neatly coiled prehensile tails – and the island still harbours many more species than the few found in Africa, Ceylon and southern India.

The chameleon's colour does not always render it inconspicuous. At night the chameleon is white, but by day its colours seem to respond readily both to the brilliance of sunlight and to the animal's emotions – such as fear or anger.

A frightened or angry chameleon will puff out its throat; at the same time, its whole body blushes green or brown, perhaps with bluish stripes or bright yellow streaks, or it may even turn jet black. Colour cells in the chameleon's skin enable it to put on this outlandish display. The cells contract in darkness, leaving the skin essentially colourless, but expand when the animal is excited or exposed to light. A chameleon sitting in shade-dappled sunlight shows distinct patterns of colour on its scaly body where twigs or leaves have cast their shadows; this may help camouflage it.

Most chameleons have heads armoured with long horns, ridges, shield-like projections, flaps and various other devices. The 36 species on Madagascar differ widely in size, ranging from 2 ft in length to fully grown adults that are not more than $1\frac{1}{2}$ in. from nose to stump of tail – possibly the world's smallest reptiles.

The chameleon's behaviour is as bizarre as its

A RESERVE 'LEG' *A chameleon's tail, here neatly coiled, can also serve as a third 'hind leg' when it stands erect*

appearance. On Y-shaped toes, it climbs trees slowly and deliberately, wobbling a little from side to side; its scaly, turreted eyes swivel independently, one watching where it is going, the other perhaps following the movement of a nearby fly or moth. In slow motion, the chameleon approaches its victim, then stops perhaps 12 in. away, its mouth slightly open. A sudden contraction of muscles, and out darts a tubular tongue faster than the eye can see. The blunt, sticky tip unerringly hits its target. Just as quickly, the tongue snaps back, and the insect prey is drawn into the chameleon's mouth, to be slowly and methodically chewed before being finally swallowed.

Refuge of the boas

The island's other reptiles present a panorama of mixed ancestries. Most of them are of African origin, with a few from the Orient. But the large snakes, unexpectedly, are boas, found elsewhere only in the Americas; there are no pythons as there are in Africa or, indeed, any dangerously poisonous snakes, although one species reputedly has a mildly toxic bite. It is believed that boas were once world-wide and reached Madagascar from Africa, where they were later supplanted by the more successful pythons. Madagascar became their last refuge in the Old World.

Like other tropical islands, Madagascar has a population of skinks and geckos peculiarly its own. These creatures are often so camouflaged that they go unnoticed even by careful observers. One gecko not only mimics background colour and texture, but also casts no shadow when it clings to a tree because folds of skin hang down like curtains from its jaws and body to touch the bark. Even its eyes are camouflaged by overhanging flaps of skin that obscure their margins.

HOW A CHAMELEON CHANGES ITS COLOUR

A chameleon's colour changes are triggered by light, temperature or its emotions. This chameleon has its brown pigment cells expanded in the top picture. In the lower picture, the brown cells have contracted and the yellow cells have expanded

A BIRD FROM 'THE ARABIAN NIGHTS'

Madagascar once possessed several species of elephant bird, the like of which the world had never seen before and will not see again. One species, the largest and heaviest of all birds, walked the island's countryside within the last 1000 years, possibly as recently as three centuries ago. Ten feet tall and weighing up to 1000 lb., the giant elephant bird was exceeded in height, but not in weight, only by the New Zealand moa. Although the elephant bird never flew, and its wings were mere vestiges, it apparently inspired the story in *The Arabian Nights* of Sinbad the Sailor's encounter with the giant flying roc. Earlier, Marco Polo had written that the bird could carry away an elephant. He reported seeing a feather of the bird, but his description suggests to present-day researchers that what he had actually seen was a dried palm branch.

Bones of the bird are still found today, not yet fossilised. Broken shells of eggs with a capacity of 2 gallons are often discovered in sandy soil, the thick pieces so preserved that a whole egg can be reassembled easily from the fragments.

Probably the elephant bird's extinction was caused partly by man and partly by a slow change in climate that altered much forest to near-desert. In any event, men once saw this enormous creature, ate its eggs, used the sturdy shells as water containers, and perhaps hunted the young chicks, only a few months old but as tall as a man.

Like the extinct moa, as well as today's cassowary and emu, the elephant bird illustrates the characteristics of flightlessness and giantism that can develop astonishingly in island birds.

Some of the surviving families of birds restricted to the island show typical island adaptations. Several rails do not use flight as a means of escape, although they can fly. They rely upon rapid and skilful running – surely a step toward the flightlessness characteristic of isolated rails. Another group of birds, the mesites, have already lost the power of flight, although at times they may weakly and ineffectually flap their wings. The asitys, on the other hand, are able to fly but are relatively fearless, another result of long isolation in an environment with few predators.

Nearly 200 different kinds of birds still live on Madagascar; 118 of these species are found no-

Elephant bird

African ostrich

THE ELEPHANT BIRD *The largest bird ever known, the elephant bird of Madagascar, became extinct within the past 1000 years. It was the heaviest of the flightless ratites, a group which included the New Zealand moa, now also extinct, and a few surviving giants – Australia's emu and cassowary, and the largest living bird, the African ostrich. The biggest elephant bird weighed three times as much as an ostrich and laid eggs six times the size of ostrich eggs*

where else on earth. In addition to the endemic birds, there are many African ones, some of which fly back and forth in annual migrations. A smaller number of birds appears to be of fairly recent Asian origin. Strangely enough, despite the proximity of the African mainland, many African birds that could easily make the flight have failed to emigrate to Madagascar, or have never succeeded in becoming established on the island.

Brilliant flamingoes

The most dramatic and colourful of the African visitors are flamingoes. Huge flocks of two species feed in shallow lakes during morning hours, sifting out crustaceans and other tiny animals with their curious bent beaks held upside-down. Although the two species mingle, they apparently do not compete directly with one another for food, since the larger species reaches down further into the water and the smaller species feeds closer to the surface. After a busy morning of feeding and a period of inactivity at mid-day, they sweep into the air by the hundreds or thousands, their brilliant red-and-black wings flashing in the sunlight as they fly to a deeper part of the lake. Here they group closely for the night, secure from land-based predators because of the depth of the water and their enormous numbers.

THE 'GHOSTS' OF MADAGASCAR

Without question, the most famous of the Madagascan mammals are the lemurs. The name 'lemur' comes from the Latin word for ghosts – possibly because of the unearthly howls that emanate from some of these forest dwellers.

LIVING ON ITS TAIL *The dwarf lemur stores food for the dry season in its furry tail, half its total length of 10–21 in. It feeds at night and spends the daylight hours in a twig nest built in a tree-top or hole in a trunk*

Lemurs may have inspired the legends of dog-headed men, for one large species, the indris, is tail-less and often stands erect.

The lemurs are primates and thus belong to the same group as monkeys, apes and men. Since monkeys and apes had not yet developed when Madagascar broke away from Africa, the lemurs on the island gradually evolved the way of life characteristic of monkeys elsewhere. Meanwhile, in Africa itself and in Europe and North America, the lemurs' ancestors were apparently driven into extinction by competition with the monkeys.

No one could mistake a lemur for a monkey in appearance; its sharply pointed face is totally unlike that of its more advanced cousins. Both lemurs and monkeys have succeeded in the world (although not in the same places at the same time), but the lemur never developed a brain as large or as complex as a monkey's.

Lemurs range from mouse-sized midgets to creatures the size of large dogs. Giant lemurs, as big as donkeys, flying lemurs and amphibious lemurs, all of which once lived on Madagascar, are now long extinct.

The so-called flying lemur of the Malayan and Philippine forests, the colugo, is not a lemur at all, but belongs to a different and separate group. The only lemur relatives outside Madagascar today are small and unobtrusive creatures of the deepest forests of Africa and Asia, such as the potto, loris and bush-baby. They are slow and retiring animals whose way of life does not bring them into competition with the more advanced monkeys.

On Madagascar, where there was no such competition, many of the 21 lemur species are highly energetic. They have diversified enormously, so that some lemurs have a life-pattern similar to monkeys while others are similar to small apes, mice, squirrels and other familiar mammals.

A lemur's hands and feet resemble those of a monkey and are well adapted to grasping limbs of trees and pieces of food. The eyes of most species are directed forwards rather than to the sides, a necessity for life in the tree-tops where wide gaps must be jumped and the distances between branches need to be accurately estimated.

They also possess some degree of colour vision, unlike most mammals, and this may help the species which are markedly different from one another in their range of colours – red, brown, gold, black, white and grey – to identify and court only mates of their own family.

Lemurs are divided into four groups: TYPICAL LEMURS, MONKEY-LIKE LEMURS, MOUSE and DWARF LEMURS, and the peculiar little AYE-AYES.

TYPICAL LEMURS Leaping from tree to tree, typical lemurs use their tails to maintain balance. They are busy animals, forever picking up food or grooming themselves. At rest, a typical lemur wraps its tail around its body and up over one shoulder; several animals sitting together may drape their tails around one another. A baby clings to its mother's belly until it is three weeks old, and then begins riding on her back. After another week or so, it leaves its mother for short exploratory trips, returning to eat and sleep. Most typical lemurs are vocal when disturbed, screeching in alarm to warn others in the troop. When they descend to visit water holes, adults frequently stand erect on their long hind legs to survey the surroundings.

The typical lemur usually seen in captivity is the handsome, friendly ring-tail that, unlike other lemurs, lives mostly among rocks. A vegetarian, it can adapt to whatever plant food is available seasonally. The ruffed lemur, also a member of the typical group, is equally attractive, with dramatic black-and-white patterns on its woolly body and limbs, and a bright-eyed head surrounded by a thick collar of fur. It is the only species to build a nest, and is more nocturnal than other typical lemurs. As the sun rises over the forest, the ruffed lemur ascends to the tree-tops and stretches arms and legs to face the warmth, giving rise to a legend that it is a sun-worshipper.

MONKEY-LIKE LEMURS Unlike the nocturnal typical lemurs, monkey-like lemurs are active mostly in the daytime. They are also considerably larger than typical lemurs. With tails streaming behind them, they make prodigious 30-ft leaps between trees, landing with audible thumps, their powerful hind legs acting as pincers to keep them in position. They climb upright hand-over-hand, and descend trees tail first. On the ground, they leap or hop upright, arms held to the front or up in the air.

The most remarkable of the monkey-like lemurs are the sifaka and the indris. The sifaka is a beautiful animal with golden eyes set in a black face, which in turn is surrounded by thick, silky, white fur. Its name is an imitation of the odd cry it utters when alarmed. Like the ruffed lemur, it raises its arms to the sun in the early morning. In

the heat of mid-day it spends several hours lying horizontally on a lower branch, arms and legs dangling below. Or it may find a comfortable fork in a tree and sit there, leaning back completely relaxed and comfortable.

Sifakas live in family groups but seem to lack the strong leaders that monkey tribes have. With no natural enemies, they spend their days in carefree eating, grooming and resting. The inquisitive youngsters scamper about, investigating the forest or wrestling. The infants can move about on their own from five weeks old, but they normally stay close to their mother for five months after that.

The tail-less indris is probably the noisiest animal in Madagascar. At mid-morning, an indris troop may 'sing' in concert, rending the air with deafening howls. It is not always easy to pinpoint the source of the noise because the troop is frequently on the move, and even a stationary indris, by forming its lips into a funnel, has the powers of a ventriloquist. Lucky is the observer who gets more than a glimpse of these handsome, black-and-white creatures. They can be seen only in the wild, for they do not survive even brief periods of captivity. The Government of Madagascar has wisely placed a ban upon the capture of these and nearly all other lemurs.

MOUSE AND DWARF LEMURS Very different from the big lemurs are the two other groups. The first of these, the little mouse and dwarf lemurs, are large-eyed, secretive and mostly nocturnal, hiding during the day in foliage or in hollow trees. The mouse lemur, not much bigger than the rodent for which it is named, has a simple brain and is among the most primitive of all primates. It eats fruit, leaves and possibly honey, but feeds primarily on insects; in turn, it is hunted by the island goshawk. The dwarf lemur, which assumes a state of torpor during the hot, dry season, has a more vegetarian diet, although it also consumes insects.

THE AYE-AYE With its big ears, the aye-aye is unlike any other lemur and is the sole member of its group. A strictly nocturnal forest-dweller, it secures its insect food in an unusual manner. Aided by acute senses of smell and hearing, it searches for wood-boring beetle larvae, tapping on tree trunks with its highly developed long middle finger. If a trunk sounds hollow, the aye-aye bites into it and the middle finger probes into the tree for the grub and extracts it. Fruits and bamboo shoots, chewed with sharp, powerful teeth, add variety to the aye-aye's insect diet.

ISLAND PREDATOR *Madagascar's largest flesh-eating animal, a fossa, stalks its prey in a tree. During the breeding season fossas form into bands and may even attack human beings. When annoyed they release an unpleasant, skunk-like odour*

CURLED UP FOR SAFETY *A prickly setifer tenrec, 6–7 in. long, curls up when it is disturbed. It lives in Madagascar's arid scrublands, fattening itself on grubs and other insects before hibernating – or being eaten by local farmers*

The ferocious fossa

No very large flesh-eating animals ever lurked in the rain forests of Madagascar or preyed upon the creatures of its grasslands. The largest carnivores on the island belong to the mongoose family.

One of these is the fossa, which grows to a length of 4 ft including the tail. It looks a little like a puma and is a ferocious predator. The soft, smooth fur is usually reddish-brown, but a very few black individuals have been found in rain forests.

Like other members of the mongoose family, fossas have scent glands beneath their tails and can discharge a noxious smell when they are angry or frightened. They walk in a somewhat flat-footed way, like bears, but they can move fast in pursuit of prey and follow lemurs into high trees.

Their favourite prey appears to be birds and small mammals, but they have been known to kill wild hogs and oxen – and there have been reports of attacks on human beings.

Primitive insect-eaters

Other mammals of Madagascar which are especially interesting to scientists include the tenrecs. Like the lemurs, they were isolated on the island long before more advanced forms of mammals evolved; and like the lemurs they have developed life patterns which are similar to the ways of life of totally unrelated mammals elsewhere.

Among the most primitive of mammals, the 20 or more species of tenrec differ widely in appearance, but they have a similar internal structure. Nearly all are nocturnal and secretive.

With a shield of stout, sharp spines on their backs and flanks, two tenrec species, 5–7 in. long and nearly tail-less, look like the European hedgehog (a more advanced insectivore). If they are disturbed, they hunch up into a partial ball.

Some of the burrowing tenrecs are better adapted to subterranean living than any other mammals except moles. The weak-eyed mole tenrec, with powerful front claws and fine, soft fur, is an efficient tunneller. Another burrowing tenrec, which is about the size of a rabbit, is believed to be the most prolific of all the world's mammals (it averages 25 young per litter).

The smallest tenrec weighs about one-third as much as a field mouse. It is a shy little creature and lives among fallen leaves or tangled grasses.

The marsh tenrec has evolved a way of life utterly remote from its insect-eating ancestors. Resembling a water shrew, it browses in streams and ponds, swimming by means of webbed feet and feeding primarily on water plants.

ISLAND FROM THE DEPTHS *Several of the coral islets in the Aldabra atoll group were pushed up from the bottom of the sea long after their formation. Their mushroom-like shape is caused by the eroding action of the waves over hundreds of years*

ISLANDS FOR POSTERITY

Scientists guard the refuge of the giant tortoise

Aldabra atoll and the group of similar small islands in the Indian Ocean between the Seychelles and Madagascar are now the scene of international conservation efforts. Once threatened with almost complete devastation – when it was planned to build an airfield on the island – Aldabra has now become the symbol of man's belated efforts to protect his planet.

RARE SURVIVORS *Indian Ocean giant tortoises survive in the wild only on Aldabra atoll. The huge reptiles, which weigh several hundred pounds and may live for more than 100 years, have been exterminated elsewhere by man*

Islands without mammals

Seventy million years ago when the islands of New Zealand broke away from Australia, the great invasion of the earth by mammals had not begun. So it was that this fertile island group developed its own animal kingdom almost entirely without the intrusion of the conquering mammals that were to become dominant throughout the lands that drifted slowly northwards beyond the Equator.

Until the arrival of the first human invaders – a group of Polynesian settlers, in the 8th century AD – the only mammals to inhabit New Zealand were two varieties of bats and the seals which basked upon the island shores.

Just as Australia provided the ideal environment for the evolution of advanced marsupials, so New Zealand, without mammals, was a natural paradise for birds. In the absence of predators on the ground, many birds became ground foragers. Gradually, they evolved heavier bodies and smaller wings, becoming flightless and relatively fearless. These two qualities were to make them pathetically vulnerable when mammals – in the form of human beings and the few animals they brought with them – arrived on the scene. The first settlers, adapting to their new environment without the pigs and poultry of their former home, fed instead on New Zealand's flightless birds.

THE VULNERABLE FLIGHTLESS BIRDS

As Madagascar had its great elephant birds, so New Zealand produced enormous flightless moas. One moa was about the size of a turkey, but the largest of the 19 species reached a height of over 11 ft. These vegetarian giants perhaps knew no predators of any kind until Polynesians arrived. From that day on, the moas were hunted for food and for bones and feathers used in personal adornment. Although accounts are confused, it is certain that not more than a few individuals, if any, of one species were left by the time Captain Cook arrived in 1769. The whole gigantic family is now almost certainly extinct.

Other New Zealand land birds have dwindled as their forest territories have been cleared for farming. They have also faced competition from about 25 species of foreign birds introduced by man – in addition to their other new and powerful enemies, the predatory mammals. About 12 of the 250 species have now vanished for ever.

Even domestic animals are a danger to the islands' rare birds; there are, for example, now only three species of chubby, nearly tail-less 'wrens', 3–4 in. in length. A fourth lived on Stephen Island, a tiny spot of land between New Zealand's two main islands. The only European to see it alive was the lighthouse keeper, who reported that it ran rapidly but 'did not fly at all'. The last remaining individual was killed by the lighthouse keeper's pet cat.

Sometimes, however, species that were thought

VARIED VEGETATION

New Zealand's vegetation includes nearly subtropical forests in the north, temperate forests in the south-west, and grasslands in the south-east. Central mountains, with 223 peaks more than 7500 ft high, force westerly winds to drop much of their moisture on the west coast. Most areas of New Zealand have a uniform rainfall from month to month and a seasonal temperature range of less than 17°C (30°F)

SAVED FROM EXTINCTION *For 50 years the flightless takahe was believed to be extinct. In 1948, however, some survivors were found in a mountain valley, and naturalists are trying to save the takahe by breeding birds in captivity*

to have died out come unexpectedly to light. For 50 years the brilliant blue-and-green takahe was believed to be extinct. Then in 1948 a colony of these iridescent flightless birds was discovered in a remote South Island mountain valley.

But the takahe's future is still uncertain. After millions of years of development in a world without mammals, the last survivors are threatened even in their mountain retreat by two creatures introduced by man: stoats which pursue them for prey, and deer which compete for their grassland food. New Zealand naturalists are now attempting to save the takahe from its otherwise inevitable extinction by breeding it in captivity.

Another flightless bird which appears to be in danger of extinction is the kakapo, a handsome green parrot which climbs trees. It uses its well-proportioned wings to make long glides to the ground, but its breast muscles are not strong enough to sustain true flight.

The national bird

The kiwi, New Zealand's national bird, is so shy and elusive that few observers see it in the wild. It does not look at all bird-like: the feathers are more like hairs; there are no visible wings or tail; and sensitive bristles at the base of the long bill serve the same function as a cat's whiskers. Three species are still found in New Zealand.

By day, the kiwi hides in dense vegetation, or perhaps in a burrow between tree roots or in a

GIANTS THAT VANISHED *Most moas were killed off by early Polynesian settlers before the Maoris reached New Zealand in 1350. According to tradition, the 500-lb. giants were very slow-moving, a trait which probably hastened their extinction*

hollow log. At night, it emerges to feed along the forest floor. Nostrils are at the tip of the bill; the kiwi is thought to have a keen sense of smell and the ability, unusual among birds, to scent out prey, although actual proof of this awaits further study. It stalks through thick undergrowth with quick, nervous movements, picking up fruit and insects in the leaf litter and probing the soil with its bill for earthworms.

At nesting time, the female kiwi produces one or occasionally two enormous eggs, glazed and ivory white or greenish, which may be a sizable fraction – even a quarter – of her own weight. Apparently this is quite enough for her, for the male, smaller than his mate, spends the next 70–80 days on incubation duty.

Scurrying scavenger

The island forms of land rails seem particularly prone to lose the power of flight. One New Zealand rail, the weka, is totally flightless and scurries about in a variety of environments, from dry scrub to wet forests and open beaches, seeking insects, lizards, mice, rats or young rabbits. Its wings are apparently used only as a means of balance when it makes its fast ground-level attack on prey. Its shrill, penetrating whistles may become a chorus as more wekas join in.

BIRD THAT HAS NO TAIL *The flightless kiwi has traces of wings hidden in its plumage but it lacks any vestige of a tail. The Mantell kiwi, seen here probing for food in forest litter, is one of the larger species and grows to 14 in. high*

SCURRYING FOR FOOD *A fast runner and good swimmer, a weka scurries across a beach in search of food. The birds live even in towns, where they raid rubbish bins*

PARROT THAT CAN STILL FLY *A kea parrot soars through the New Zealand skies. Keas breed all the year round and raise broods of one or two chicks. Female young continue to be fed by their father for some weeks after the males have become independent*

Flesh-eating parrots

The kea parrot is one of the few native New Zealand birds that has not suffered from European colonisation, for it has adapted to exploit the new source of food that the immigrants brought with them – sheep. The mountain-dwelling bird at first ate only dead sheep but gradually became able to kill weak or helpless sheep that had been trapped in snowdrifts. The keas are the only flesh-eating parrots in the world. Like other parrots they also eat fruit.

Despite concerted efforts by sheep-farmers the birds have never been in danger of extinction, possibly because they hatch and rear their young in virtually invulnerable eyries high in the Alps of New Zealand's South Island.

The shy songsters

Song is usually highly developed in honey-eating birds, and the forests of New Zealand resound to the liquid notes of two honey-eaters, the bell bird and the tui. The shy bell bird is seldom seen, but its song sounds like the chiming of silver bells.

Early New Zealand settlers called the tui the 'parson bird' because it has predominantly black plumage with white at the neck. Tuis are among the world's sweetest songsters; and they can also produce notes of such high frequency that a tui is sometimes seen to be singing vigorously although the human ear can detect only a few faint squeaks.

The vanished marvels

Not many years ago, the clear chuckling calls of the huia were heard in the forests, but huias are now believed extinct. These birds had developed woodpecker-like habits, as some of the Galapagos finches, Hawaiian honeycreepers and Madagascan vanga shrikes have done. Rather than each individual's acquisition of a similar method, however, each pair of huias may have formed a dependent partnership. The male, with a short, nearly straight bill, excavated holes in soft, dead wood, while the female used her longer, scimitar-shaped bill to probe harder wood. Unfortunately, observations of the huia were inconclusive; since a male was seen alone only once, some biologists have assumed that the male depended on the long reach of the female's bill for the succulent grubs of longhorn beetles, the birds' favourite food. But there is little direct evidence of such co-operation.

SEABIRDS' HAVEN

Many birds dependent upon the sea inhabit New Zealand's shores. One of the most common, the sooty shearwater, is among the most abundant bird species on earth. They are great wanderers and sometimes a parent bird flies 600 miles on fishing trips, leaving its solitary chick without food for as long as two weeks.

Many albatrosses, such as the royal albatross of New Zealand waters, breed only every other year. The fledgling period lasts nearly a year, however, so colonies are always occupied. A single colony on South Island accommodates only a few pairs; many more nest on small, remote islands far from the mainland. The royal albatross is ungainly on land, but a wingspan of nearly 10 ft makes it a spectacular bird in the air.

At least seven species of penguin breed on New Zealand's islands, and others visit there. At night, their courtship displays are accompanied by raucous trumpeting and braying. The penguins venture some hundreds of yards inland, not always choosing the shortest route, their trails wandering up and down rough and scrubby country. They scramble up steep coastal slopes, occasionally following little streams, where they slip and slide over water-worn stones.

BIRDS THAT NEED HEIGHT *Just before dawn, sooty shear-waters leave their burrows and assemble on lofty seaside rocks. Unless strong winds are blowing, the heavy-bodied birds have to launch themselves from a height*

MATES FOR LIFE *Crested penguins hop ashore on Snares Island, south of New Zealand, the only place where they have breeding rookeries. Year after year, the 29-in. birds choose the same mates and the same nesting sites. They lay two eggs in nests of sticks lined with grass and leaves*

SURVIVORS OF A 'LOST WORLD'

The New Zealand island group, isolated since the dawn of time, preserves life forms which have provided scientists with important clues about the evolution of life on earth.

One reptile, the tuatara, is the sole survivor of a group of reptiles which otherwise has been extinct for 135 million years; a group of primitive frogs, New Zealand's only amphibians, preserve fish-like characteristics which clearly show their proximity to the very first backboned creatures to emerge from the sea; and some New Zealand plants are now believed to be remnants of a 'lost world', the great super-continent of Gondwanaland.

Secrets of the slow dragon

New Zealand's most famous reptile, the tuatara, is a true relict, the only survivor of the 'beak-heads', a group of primitive reptiles which were wiped out elsewhere in competition with more advanced types during the Age of Reptiles. The ancient beak-heads – most of them larger than tuataras but like them in general structure – resembled the 'stem' reptiles, from which evolved dinosaurs, big marine reptiles, flying reptiles, modern reptiles, birds and mammals. The tuatara retains features possessed by the ancestor of all land animals.

A tuatara requires less heat than any other modern reptile. Although it suns itself in its burrow entrance, its body temperature never rises very high because of New Zealand's cool climate. Under similar conditions, most reptiles would be sluggish; lizards, for example, typically need temperatures of at least 24°C (75°F) in order to be active. But the tuatara is fully active with a body temperature of just 11°C (52°F); indeed, its highest recorded temperature is only 13°C (56°F).

The tuatara's metabolic rate is the lowest known for any vertebrate animal. A perfectly healthy tuatara with a body temperature of only 9°C (48°F), observed for an hour, did not breathe once. Two results of such a low metabolic rate are slow growth and considerable longevity. An adult

SPINY REPTILE *The Maoris named this primitive reptile the tuatara, which means spiny. Fewer than 10,000 still survive on some islands off the coast of New Zealand. Tuataras have a tiny third eye, covered by scales and apparently useless. They* *become able to breed only when they are about 20 years old. An entire year may elapse between mating and egg-laying and another year after laying before the tuatara hatches. They may live for at least 100 years*

tuatara may grow no more than $\frac{1}{2}$ in. in eight years, and life-spans of close to a century have been thoroughly authenticated. New Zealand zoologists have been inclined to regard Maori records of 300-year spans as quite possible.

Even a tuatara's start in life is unusual. In summer, the female tuatara excavates a shallow hole and lays yellowish, leathery eggs. The embryos begin to grow readily enough, but as winter approaches, their development almost ceases and they do not resume growth until the eggs are warmed by the spring sun. Eventually, 13 or 14 months after the eggs were laid, a 3-in. tuatara opens each shell with a horny 'egg tooth', a temporary device that has grown at the end of its snout and will fall off within a week. Scrambling to the surface through the loose soil covering the nest, the hatchling hides under a nearby stone or in a crevice. In the next four months it will double its length – probably its most rapid rate of growth.

One of this reptile's remarkable physical characteristics is a third eye, called the pineal eye. Only one-fiftieth of an inch in diameter and covered by scales, it is not at all obvious in the adult tuatara. A few other animals have such an organ, although less well developed; among them are the anole lizards of the Caribbean islands and the primitive jawless fish known as lampreys. Even frogs have a tiny brow spot. All higher creatures, man included, have deep within their heads a structure called the pineal body, which seems to have a glandular function. The tuatara's pineal eye is perhaps the remnant of a former pair of eyes that some distant ancestor had; it still retains a lens and a light-sensitive layer, but has no iris and only a degenerate connection to the brain. Although it does not work as an eye in the adult, it may be of some importance for very young tuataras, where the covering scales are transparent and not yet thickened and darkened by age.

Some of the structural features of a tuatara's skeleton are found in no other living reptiles. The presence in the skull of bony arches known only from fossils and crocodiles alerted zoologists to the tuatara's significance in the chain of evolution. Its backbone is more like that of a fish than that of a lizard, and so are its teeth, which are enamelled projections of the jaw instead of separate elements that fit into sockets. The front teeth are shaped into a beak like that of the tuatara's ancestors, the beak-heads. The tuatara further resembles some of the great reptiles of the past in having, besides ribs, a series of rib-like bones in its abdomen.

The handsome, 2 ft long tuatara is dark gold in colour, with beautiful gold-flecked eyes. The scaly, wrinkled skin often shows irregular reddish patches, which upon close inspection turn out to be masses of tiny orange mites. Large ticks lodged on the skin also fatten on the reptile's blood, but apparently cause no real harm to their host.

The adult tuatara either excavates its own burrow or takes up residence in the burrow of a 'muttonbird' (local name for shearwaters and petrels). It may also readily share its burrow with any muttonbird that chooses to use it during the few months the bird is raising its young. Although the reptile tolerates its burrow-mate and usually ignores the latter's eggs and chicks, it does prey on other birds, skinks and geckos, land snails, beetles and large flightless crickets.

How has the tuatara managed to survive when all its relatives and a host of more highly evolved reptiles became extinct millions of years ago? Part of the answer probably lies in New Zealand's lack of mammals. If these quick competitors and predators had reached the islands, the small, cold, slow dragon could hardly have continued its way of life for 135 million years.

The fish-like frogs

Stephen Island, the tiny rocky satellite island between New Zealand's two main islands, is still the home of some of the most primitive species of frog on earth. These frogs have played an important role in establishing New Zealand's antiquity. One is restricted to an area of not more than 500 sq. yds of rock-strewn slope towards the top of the island; others are found high in the mountains of the larger islands. All are exceedingly rare and are protected by appropriately strict laws.

These native New Zealand frogs are very small, less than 2 in. in length. Their various upland habitats have one thing in common: an absence of open water. This lack has led to a peculiar adaptation. The frogs lay eggs in damp crevices under boulders and rotting logs, where they can remain moist, each egg enclosed in a fluid-filled gelatinous capsule. The embryo becomes a froglet directly, not a tadpole. The young simply hatch out as miniature replicas of their parents.

Despite their specialised individual development, these frogs have several primitive features. Their backbones, like tuatara backbones, resemble those of fish, and they retain tail-wagging muscles although they absorb their tails at about a month. They lack external eardrums, and the vocal apparatus is so undeveloped that the best they can do is to chirp slightly. They have little or no

PRIMITIVE FROG *Hochstetter's frog, a primitive species found only on a few New Zealand islands, guards its hidden cluster of eggs. The young use long tails, that disappear after birth, to break open the tough egg capsules surrounding them*

webbing between the toes. They swim, but they kick their hind legs alternately rather than together as most frogs do. These species retain features characteristic of the ancestors of all frogs.

The primeval plants

Biologists studying New Zealand's vegetation have found that even more remarkable than its distinctiveness is its close similarity with South America. Although most of New Zealand's plants are found nowhere else in the world, many of them are closely related to South American types – as if they have been separated just long enough for differences to occur but not long enough for them to become truly distinct. The southern beech trees and the pines called poda-carps, together with the tough cushiony plants that carpet the mountain bogs and many shrubs of the coastal zone show this dramatic trans-Pacific relationship. In the same way, many of the insects of New Zealand have close relatives in southernmost South America – and nowhere else.

It is clear from fossils that some at least of these plants grew in Antarctica before the ice cap formed there, some 4 or 5 million years ago. Others grew on the isolated continental island of Kerguelen, in the mid-Indian Ocean, which now has a sub-Antarctic climate. It seems possible, therefore, that in New Zealand and southernmost South America – the only two large areas of land in the Southern Hemisphere that now have a cool-temperature climate – remnants of the ancient life of the great super-continent of Gondwanaland have lingered on, while evolution has thrown up new dominant forms on the mainlands elsewhere.

When Sir Arthur Conan Doyle wrote about a 'Lost World', he imagined a cliff-girdled plateau still inhabited by creatures from the age of dinosaurs. But islands such as New Zealand are 'lost worlds', sheltering fragments of vegetation and animal life of ancient continents – species that elsewhere vanished long ago.

THE ERODED WORLD *A weird, desolate landscape in Western Australia – known locally as The Tombstones – has been carved from limestone bedrock by millions of years of erosion. The formations average 10 ft in height*

LAND OF DROUGHTS AND FLOODS

The major fact about Australia's climate is its dryness, especially in the interior. Rain, wherever and whenever it falls, is unreliable, thus creating long droughts and sudden treacherous floods. But the island has not always been so dry. It is known that some 60 or 70 million years ago, a profusion of vegetation grew across its southern half, and primitive conifer and cycad trees were widespread. Rain forests occupied vast stretches and giant marsupials lived on the lush foliage or preyed on other animals.

When the last Ice Age ended the land dried up, deserts developed and advanced southwards, and the ancient rain forests gave way to a remarkable proliferation of that most Australian of trees, the eucalyptus. With the disappearance of forests and grasslands over much of the island continent, large marsupial grazers and beasts of prey also gradually vanished.

Curving along the east coast of the continent and extending off its south-east tip to the mountainous island of Tasmania, the mountains of the Great Dividing Range are today fertile and well watered, in some places receiving as much as 100 in. of rain a year. They prevent easterly trade winds from reaching the interior and are thus largely responsible for preserving the desert conditions of central and western Australia.

The arid plateau that occupies the western half of Australia is largely an expanse of hardened sand dunes. The only part of the territory that is humid enough for forests is the south-western corner. Large parts of the interior receive no more than 5 in. of rain a year, whereas the coasts get 10–20 in.

FERTILE COAST *The coastal strip of north-east Queensland, to the east of the Great Dividing Range, has high rainfall, good soil and steady warmth all the year round, making it a land of fertile valleys and lush, tropical rain forests*

In the central eastern lowlands, the salty remains of enormous lakes are indications of ancient invasions by the sea. Many of them still appear and disappear with the rare and scanty rainfall. The erratic streams that feed them are dry for most of the year, but even when they are filled with water, the ancient lakes are too salty for most life. Dinosaurs once lived there, but neither their numbers nor the nature of their environment can be deduced from the few fossils that have been discovered by scientists so far.

Off the continent's north-east coast lies the Great Barrier Reef, a product of corals and the largest single modification of the world's land area by living things. It runs parallel to the coast of Queensland for 1250 miles, enclosing a sea area of about 80,000 square miles. For millions of years the reef has served as a protective shield for the coast, absorbing the Pacific Ocean swells and preventing heavy surf from crashing on to the continental shore.

The northern third of Australia is in the tropics; the southern part experiences mid-winter cooling in June and July, although not enough to have created extensive plant and animal adaptations to a cold climate. Only in the high mountains of the south-east and Tasmania have alpine forms of animals and plants evolved and survived.

A HARSH WORLD FOR PLANTS

Most Australian vegetation is composed of plants able to survive in exceptionally poor soil. At its harshest, the interior desert consists mostly of gravel and stone. Gradually it shades into desert country dotted with sparse tussocks of hardy grass. As the desert becomes steppe, scattered shrubs and wide tracts of mulga and other acacias appear. Mallee scrub – dwarf eucalyptus growing on open grasslands – develops where rainfall is a little greater. Then, as high country and some parts of the coastline are approached, open woodland gives way to dense forests; wherever coastal areas are moist enough, there are rain forests, reminiscent of the continent's earlier lush vegetation.

Australia's three major zones of vegetation are the result of distinct climatic influences. Northern vegetation, much of which arrived long ago from the Indo-Malayan region, is tropical; mangroves, for example, commonly line the northern coastal shores and bays. The cooler southern zone is influenced by the Antarctic, and its vegetation is somewhat like that found in South America and South Africa, both also once parts of Gondwanaland. Nearly all the plants in this region are specialised as a result of long isolation. The third zone, in the centre of the continent and extending towards the west, has many plants found nowhere else, most of them highly adapted to resist drought and fire, which often ravages the tinder-dry scrub. A small fourth zone includes the high mountains of the south-east, where Australia's only alpine vegetation and animal life exist.

The simple, spore-propagating plants of Australia and the nearby islands are not nearly as unfamiliar as the more highly developed kinds, for the microscopic spores of fungi, mosses and ferns are easily carried by wind to all parts of the world. They settle and grow wherever they can. Seed plants, on the other hand, achieve long-distance dispersal, especially over vast areas of ocean, with much greater difficulty.

Ancient ferns

Australia has been a storehouse for the preservation of cycads, plants apparently derived from ancient seed-bearing ferns abundant when dinosaurs trod the earth. Cycads are squat plants, palm-like in appearance (although unrelated to

PREHISTORIC FERNS *The feathery crowns of ancient cycad trees resemble palm fronds, but they are in fact descendants of primitive ferns. Every tree bears male or female cones*

FLOWERING TREES *The widespread acacias, or wattles, derive their colour from clusters of stamens, the male reproductive part of a flower. They range from bushes to lofty trees of 100 ft*

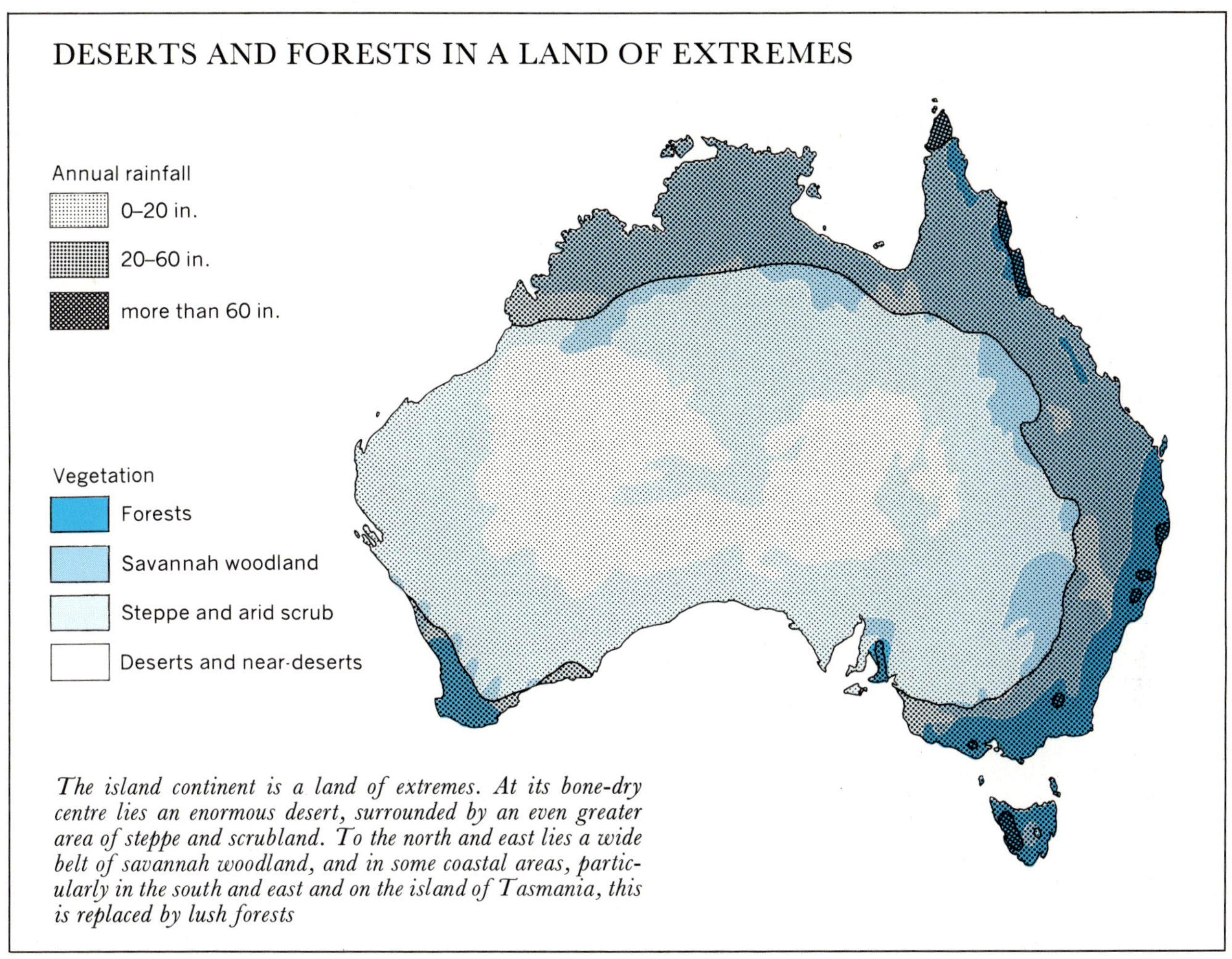

The island continent is a land of extremes. At its bone-dry centre lies an enormous desert, surrounded by an even greater area of steppe and scrubland. To the north and east lies a wide belt of savannah woodland, and in some coastal areas, particularly in the south and east and on the island of Tasmania, this is replaced by lush forests

palms) and little changed through millions of years; the first ones were established in Australia long before the separation from Gondwanaland. They are able to survive in environmental zones ranging from desert to rain forest. Because of this wide tolerance, they have changed little in form or function. Of the 100 species of cycad scattered throughout the Southern Hemisphere and the tropics, 25 grow only in Australia, mainly in the south-west, which is their last major home. Usually cycads are not impressive in size – most are no more than 10 ft tall, although one species grows to 60 ft. Many are believed to live an exceedingly long time; some, on Tambourine Mountain in southern Queensland, are estimated to be more than 10,000 years old.

Australia has only 36 kinds of native conifers, including the ancient primitive araucarias. One of the most curious of these conifers, which grow only in Australia, the kerosene bush – so named

because of its combustible foliage – is found at high altitudes as in Kosciusko State Park where, gnarled and twisted, it sprawls flat on rocky ledges. This small mountain plant, which lives for several hundred years, grows at maturity no more than 1 in. in diameter in a century.

One of the great tree groups especially identified with Australia is that of the acacias or wattles. Of the 700 species in the world, 600 grow in Australia and its related islands. Acacias can withstand arid conditions, even far into the desert, where their fluffy yellow or cream-coloured flowers give colour and scent to a withered landscape in the spring. They play an important role in providing food, shelter and organic material for the sparse animal communities occupying much of the continent. In the past 25 million years, acacias have spread into different ecological niches elsewhere in the world, but Australia is the greatest showcase of their achievements.

PARTNERSHIP WITH ANIMALS
Plants benefit from visitors

Some of Australia's flowering plants have evolved in unique partnership with birds and mammals that live partly or wholly on their nectar. The plants also benefit from their association with the visitors, which help them reproduce by transferring pollen from an anther of a flower, where it develops, to the stigma, where pollination occurs. Many flowers have specialised structures to pick up and deposit pollen as the visitors feed. Some lure insects, which attract birds and mammals, as does the showy red or yellow coloration.

PICKING UP POLLEN *Two rare dibblers – speckled marsupial mice – cling to a cluster of banksia. As the near-extinct animals drink the nectar their soft bushy fur picks up pollen which is transferred to the next flower they visit*

FAIR EXCHANGE *A yellow-fronted honeyeater thrusts its bill into a Christmas bell. To reach the nectar the bird must dip its whole head into the flower so that its head feathers brush against stigmas and anthers at the flower's lip*

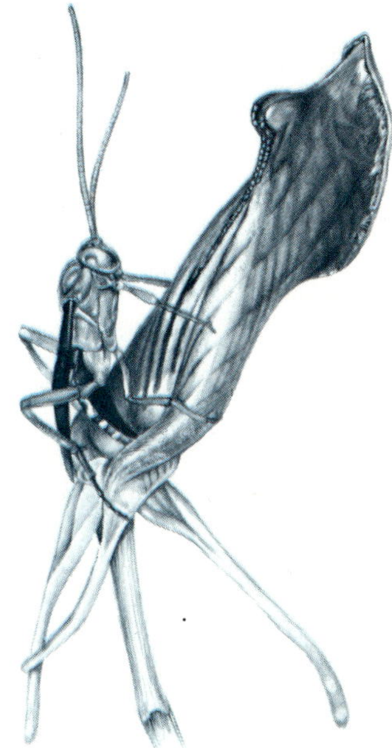

Australia's tongue orchids are remarkable for their close resemblance in shape and colour to certain female wasps. The imitation is so realistic that some male wasps mate only with the flowers. When they do this, the pollen is deposited on their abdomens, and when the wasps visit other orchids, they deliver the pollen to them

FAVOURITE FOOD *A tiny honey possum grasps the stem of a banksia, its favourite source of nectar. It eats by probing with its long tongue and sucking through its mouth and snout*

A SUBTLE FLOWER *As a spinebill honeyeater sucks nectar from a grevillea blossom, anthers press pollen on to the bird's head. Later in the flower's development, its stigmas become sticky and pick up the pollen from the honeyeaters*

GIANTS OF THE FOREST *Karri gum trees grow up to 250 ft high in south-west Australia. Their rapid growth – up to 35 ft in ten years – makes them valuable timber trees. Like many other eucalyptus trees, they shed their bark in patches*

UNHARMED BY FIRE *Many eucalyptus trees survive raging forest fires. They grow in an extremely dry climate and have adapted to the consequent hazards. Their outer bark may explode into flames, but the trees usually recover*

THE REMARKABLE EUCALYPTUS
Strange and varied trees from Australia's past

The best-known of Australia's plants are the several hundred varieties of eucalyptus trees – from huge gum trees and wizened scrub brush to the alpine snow gums that grow at 6000 ft in Kosciusko State Park. Absent from the completely arid deserts as well as from the dripping rain forest of the east coast, they flourish in all kinds of intermediate conditions. Botanists believe eucalyptus were much more diversified in ancient times, and that many were destroyed when the central region of the continent dried up.

Despite many differences in size and form, all eucalyptus trees are aromatic evergreens sharing several basic similarities of leaf shape and flower structure. The diversity of species is so great, however, that several kinds of gum may exist side by side, each specialised to succeed in its own way, without challenging the survival of the others.

Probably the most impressive eucalyptus trees are the gums in Tasmania and in the states of Western Australia and Victoria. In Western Australia's Valley of the Giants, their great, shining, silvery trunks rise to heights of over 200 ft. The gums of Victoria are among the world's tallest trees. One record reliably describes a 374 ft tall specimen found in Victoria; less reliable records claim heights of 500 ft.

Many eucalyptus trees have pointed leaves that droop downwards away from the sun and form hundreds of thousands of drip points during rain. Other members of the family have different leaf structures, the oddest leaf being that of the

HOW THE TREES VARY

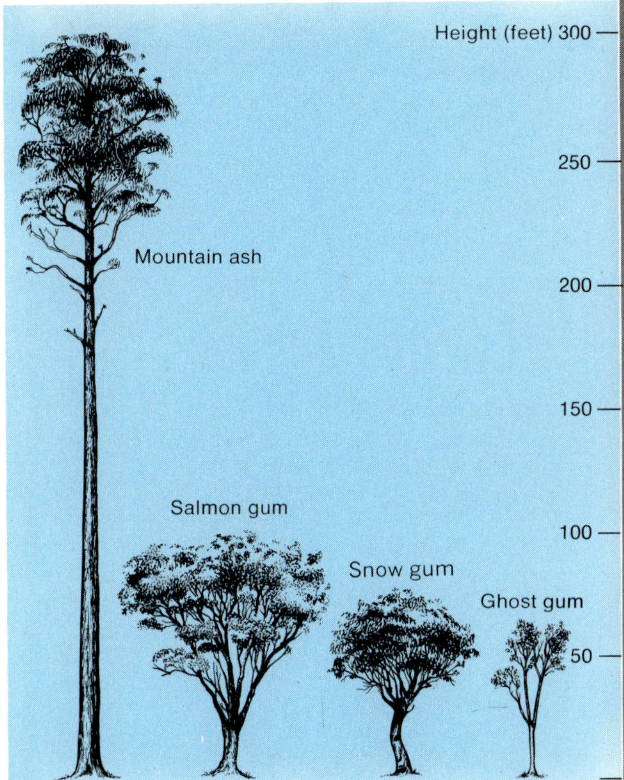

Eucalyptus trees grow widely in Australia and there are several hundred varieties. For example, in the tropical north there are ghost gums; the mountains harbour hardy snow gums; the dry south-west has drought-resistant salmon gums; and the temperate south-east is the homeland of the 300-ft mountain ash eucalyptus

spinning gum. An inhabitant of the cold, high altitudes of Victoria and Tasmania, the spinning gum has circular leaves growing at intervals around its twigs, each twig serving as an axle for the leaves. When the leaves die they remain in place, spinning like catherine-wheels in the wind and producing a loud whistling sound audible all over the mountain slopes.

Because eucalyptus trees do well in poor soil, much of Australia is dominated by the many-stemmed mallee scrub, only 10–20 ft tall. Its scant foliage provides only moderate cover, but enough to secure the existence of the remarkable mallee fowl. The scrub, like most eucalyptus trees, grows from an underground root-stock and is able to withstand the brush fires that may destroy the exposed part of the plant but not its buried portions.

EXTRAORDINARY CREATURES

The diversity of Australia's climate and vegetation is reflected in the wide variety of animals that have evolved only on the island continent. Even the creatures that crawl and wriggle over the face of the land are of unusual interest: giant worms; spiders that can kill human beings; frogs that can survive long desert droughts; and three extraordinary families of lizards – the skinks, geckos and goannas.

The world's largest worms

Australian forest soils harbour several impressive relatives of worms that go almost unnoticed in other parts of the world. Here the lowly flat-worm – represented elsewhere by tiny and insignificant species – takes a number of striking forms. One flat-worm, 3 in. long, is a deep Prussian blue with a white stripe running down its back. Another, much longer, with a shovel-shaped head and dark brown stripes on its yellow body, attacks and eats a wide variety of insects, earthworms and land snails.

Some of the earthworms in north-eastern Australia are the bulkiest worms in the world, and among the longest. Several species are 1 in. or more in diameter, and 11–12 ft long; when one of these giants contracts, it swells to the diameter of a man's wrist. Some species are dark; others are salmon-pink or banded with alternate cream and black stripes. Most of these big worms live amid tangled tree roots, emerging into the open when the forest litter is wet with rain. Their progress through the soil is often marked by loud gurgling and sucking noises. One species, the barking worm, emits a weird moan from within the soil when anyone walks overhead.

Even when they are not making audible sounds, worms may sometimes be detected by a strong, sickly odour, a little like creosote, possibly derived from plant oils in their diet of fallen leaves. If a squirter worm, which inhabits the brush forest near Sydney, is disturbed, it shoots jets of sticky fluid 4 ft into the air from pores along its body.

SINGING CRICKET *The 4 in. long king cricket 'sings' by rubbing its wings together to produce a rasping sound which attracts the female. She cannot answer the call because she lacks the resonators that are found in the male cricket*

INSECT-FANCIER'S PARADISE

Among Australia's 50,000 insect species are many which are unique to the continent. One grasshopper changes from blue-green to blue-black according to the temperature – the only known insect in the world that can change colour. In Western Australia, a small grey-green grasshopper so perfectly mimics the woolly, rounded leaves of certain desert plants that, although it is not microscopic, it is difficult to see even under a magnifying glass. In eastern Australia lives the giant king cricket, a 4 in. grasshopper with truly formidable jaws borne on a head one-third of the insect's total length. The world's largest moths and some of its largest butterflies also live there, as does a stick insect, 10 in. long, with tiny wings that serve only as glide planes when it appears at night.

Termite hills are found elsewhere, but those of the outback are constructed by the most primitive termites known. Yet their hill cities are marvels of construction and engineering. Each is built of particles of earth cemented by saliva and hardened to a concrete-like mixture. A large hill may measure 12 ft in circumference and rise more than 20 ft. The shapes of the mounds vary greatly.

Other hills built by meridian termites, up to 10 ft high and 8 ft wide but only 3 in. thick, are called 'magnetic' because their narrow edges point north and south, and their broad faces catch both the rising and setting sun. Inside the hill, a maze of rooms, tunnels and living galleries surrounds a central chamber packed with grass cut during the wet season for food. In early morning, the sun's rays warm the insects, sending them into a frenzy of communal activity. At noon, with the sun overhead, the narrow top of the mound receives few hot rays, but the workday is prolonged by the heat of the setting sun.

Australia's spiders can be dangerous; the bite of the Sydney funnel web spider has caused a number of human deaths. The large, hairy barking spider, which can kill newborn chicks, produces a barking sound by rubbing a stiff-bristled comb on its leg-like pedipalps against a similar comb at the base of its fangs.

Several spiders are remarkable hunters. Hanging head-down, the ogre-faced spider stretches between its front legs a small, thickly woven silk net, which it throws over a passing insect to enmesh it. Both the hairy imperial spider and the magnificent spider are aerial anglers, catching their prey with a single long line with a sticky droplet at the end. As a moth is lured to the glistening droplet, the spider swings the line directly at it, seldom missing.

264

HOW THREE AUSTRALIAN SPIDERS CATCH PREY

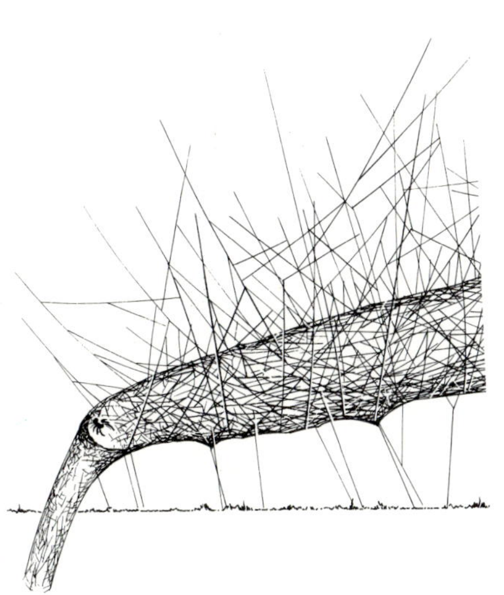

The platform spider forms a funnel to channel prey to its burrow

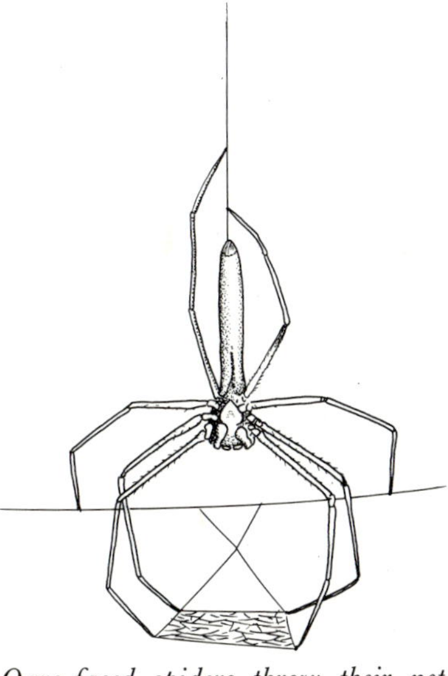

Ogre-faced spiders throw their nets over insects venturing near by

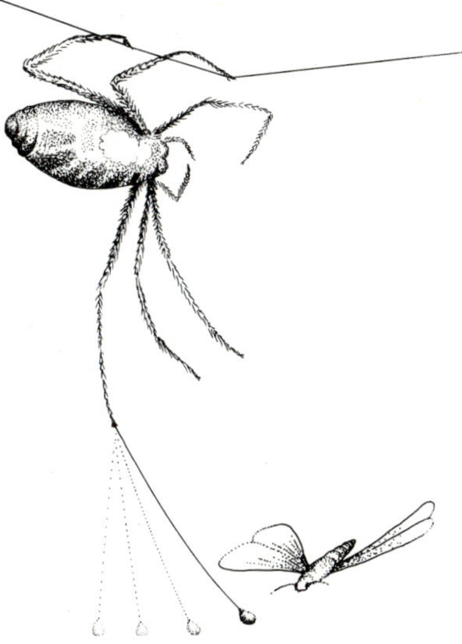

Hairy imperial spiders trap prey by spinning threads with sticky ends

A LIVING TWIG *Giant stick insects mimic the eucalyptus twigs they live on, and their eggs resemble the seeds of certain shrubs. Like their European equivalents, they remain motionless during the day and feed at night*

TERMITE COLONY *Two million termites, including a queen 15 years old, may inhabit a single colony. The queens, which can produce an egg every two seconds at certain periods, are groomed by worker termites who carry away the eggs*

LINKS WITH PRE-HISTORY

Although Australia is remarkably dry, its sparse and unreliable inland waters support their own forms of life. Some of the rivers and billabongs (deeper parts of rivers that retain water in the dry season) contain ancient and fascinating animals.

The major river system in the continent is the Murray and Darling in the south-east, totalling over 2000 miles in length but with a low volume considering the area it drains. The wide annual fluctuation from torrent to trickle is characteristic of many Australian rivers. Native aquatic animals are adapted to contend with either prolonged drought or massive flooding, which for a brief time may cover vast flatland areas. Such floods are important to the distribution of all forms of aquatic life on the continent.

Like their counterparts elsewhere, Australia's freshwater clams and snails live on river bottoms. In times of drought, the molluscs are able to burrow 4–5 ft down into the mud, where they have been known to survive in a state of torpor for as long as three years before returning to active life in the water. In structure, they show strong relationships to the molluscs of South America and Africa, and lesser affinities to those on the ancient islands of Madagascar and Ceylon – all former parts of Gondwanaland. Distinguishable from these old species are newer forms of molluscs that arrived more recently from Asia by raft or were attached to visiting birds.

Primitive shrimps

The smallest crustaceans living in the Australian rivers and lakes differ little from others around the world, for the durable eggs of such creatures are often carried great distances by wind or birds. There are a few notable exceptions, however. In the cold mountain waters of Tasmania and southern Victoria live some of the most primitive crustaceans in the world, the syncarid shrimps. These elongated little animals have few modern modifications. Their many legs are all of the same type, unspecialised for different functions; since there are no appendages for carrying eggs, a female must deposit her eggs immediately after fertilisation – a rare necessity in the crustacean world. Syncarids closely resemble the fossils of crustaceans that lived nearly 300 million years

LIFE IN A BILLABONG *During droughts, many Australian rivers dry up, except for isolated pools – called billabongs – which were their deepest parts. Life continues to flourish in the stagnant water basins for months or even years*

BURROWING SHRIMPS *Ghost nippers dig burrows in the muddy bottoms of Australian estuaries. These pale shrimps spend their lives in burrows sifting through the sand and mud for food. They use their enlarged claws for defence*

ago in the waters of Europe and North America.

Lake Surprise in Tasmania harbours a parasitic crustacean and its host fish, which are interesting to scientists. Small crustaceans of this sort were known previously only from Africa and South America. But this species was found on a host fish that could not possibly have migrated from either of those locations. The only explanation is that the ancestors of both host and parasite accompanied Australia when it was separated from Gondwanaland.

Although Australia has several other such crustaceans from ancient times, most of its larger freshwater species, while themselves found only in the continent, have cousins now distributed throughout the world. Australia has both the largest and the smallest crayfish: the Murray River 'lobster' reaches 6 lb., and a similar Tasmanian form may reach 8 lb.; at the other extreme, a tiny Queensland crayfish is scarcely 1 in. long at maturity. In abundance and diversity of crayfish, Australia is exceeded only by North America.

Several Australian crustaceans have taken to life in damp habitats. Amphipods or scuds, isopods or slaters, and crayfish can be found under fallen leaves and branches in the forest litter. Closer to the sea, in salt marshes and estuaries, more recent marine invaders live in abundance. Some beaches glisten with thousands of pink-and-blue soldier crabs, which disappear immediately beneath the sand when anyone approaches.

Ancient fish

Fewer than 200 species of freshwater fish inhabit Australia's rivers – a meagre number compared with that of other continents. Moreover, only two of them – the barramundi and the lungfish – are ancient freshwater forms; the rest came directly from the sea. Because of prolonged conditions of extreme drought and flood, not only must the freshwater fish be extraordinarily tough, but their populations must be able to recover quickly after suffering a devastatingly high mortality.

The 3-ft barramundi belongs to an ancient group found elsewhere in Africa, the Americas and the Far East.

The Australian lungfish is considerably more ancient than the barramundi, dating back at least 325 million years. It is much more primitive than the lungfish of South America and Africa (to which it has no direct relationship) and far more interesting to scientists. Australia's lungfish, covered with enormous scales – a primitive charac-

teristic – may reach a length of 6 ft, weigh 100 lb., and live for 100 years. In the muddy rivers in Queensland, it grubs slowly across the bottom, its small mouth searching for soft foods, mostly aquatic plants and fallen leaves, together with small crustaceans and molluscs. Unlike its African and American counterparts, however, the Australian species cannot survive very long droughts. It has not adapted as they did by building a mud burrow to live in. If a river dries completely, it dies.

The greenish eggs of the lungfish are unlike other fish eggs, somewhat resembling those of modern amphibians. They are deposited in large clusters, each egg surrounded by a thick envelope of gelatinous substance. The young hatch out in the form of tadpoles, to develop gradually into adults.

The lungfish has a dual breathing apparatus: gills and a capacious airbladder-lung. When gills alone cannot take enough oxygen from the muddy water it inhabits, the fish rises to the surface to breathe through its lung. In this feature, and in its flipper-like fins, the lungfish strongly resembles the distant ancestors of all backboned animals that first emerged from water to take up an amphibian existence. Without a knowledge of such an archaic, relatively unchanged animal, it would be hard to conceive of the first slow invasion of the land.

LINK WITH THE PAST *The lungfish can breathe air to survive in stagnant water. Although it represents a blind alley from which nothing evolved, the lungfish resembles fish which left the water – the ancestors of all backboned land animals*

FROGS OF THE DESERT

Australia has no frogs as primitive as those of New Zealand, but some of its amphibian residents are remarkably well adapted to life in the desert, where streams run only occasionally. The frogs living in the centre of the continent are likely to be almost spherical during wet periods. They have absorbed enough water to tide them over the long months of the dry season.

Many ground-dwelling frogs live in swamps and dig burrows in which to lay eggs. When the rains come and the water rises, the eggs are washed out and hatch into tadpoles, which soon turn into miniature frogs. Other ground frogs whip up masses of a sticky froth with which they are able to protect their eggs.

Tree frogs are small in most parts of the world, but one of the largest frogs in Australia is a tree frog. Its green body, not including the legs, is more than 6 in. long.

The history of reptiles in Australia is obscure. Fossils from dinosaur times are scarce – the most impressive remains are a few claw marks, some giant marine reptiles found far inland, giant lizards and a few bones. Nearly 400 modern species inhabit the continent, and in the deserts they may be as numerous as birds.

One of the most common reptilian groups is that of the skinks, those quick, slender, often brightly coloured lizards. The more than 100 species of Australian skinks have diversified greatly in form and habits. When the blue-tongued lizard, a heavily built skink, is frightened or irritated, it inflates its body with air, hisses slightly and, head held high, faces the attacker with its mouth wide open, displaying a broad, bright blue tongue. If it manages to bite an opponent, it is difficult to dislodge, but it feeds only on insects, snails and plants.

STORING UP WATER *The holy cross frog – named after the black pattern on its back – looks bloated in the rainy season when it stores water for the estivation period. Australia's water-storing frogs which remain underground during the dry season are hunted by Aborigines for food and water*

COMPARING THE GROWTH

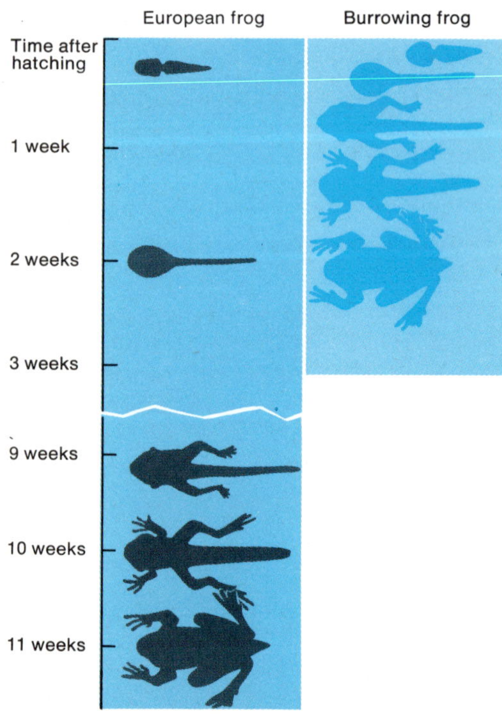

Frogs mature faster in the Australian desert than in milder climates. Burrowing frogs breed in pools created by floods or storms. Their eggs hatch very quickly and tadpoles become frogs within two weeks, compared with 11 weeks in Europe

TAIL THAT CONFUSES *The desert skink, like other lizard species, can shed its tail when it is attacked by a predator. Because the tail continues to thrash about violently after it has separated, the predator is confused and the skink escapes*

REPTILE THAT DOES NOT LAY EGGS *The blue-tongued skink does not lay eggs like most other reptiles: instead its 10–15 young are born alive. The skink, which can grow to 2 ft long, lives on small insects and fruit*

THE MOST VOCAL OF REPTILES

Geckos, which are nocturnal lizards common around the Pacific and Indian oceans, are highly diversified in Australia; one, with a bizarre bony head, is regarded as the most primitive member of the group. There are more than 40 Australian geckos, some 12 in. long but most considerably smaller. In several species, the broad tail serves as a reservoir of fatty food; some can go without food (but not water) for a year.

Most geckos are rather flabby lizards with fine, bumpy scales that do not overlap. The skin on their undersides is often so thin and translucent that internal organs may be seen as they climb

SPECIALISED EYES AND FEET

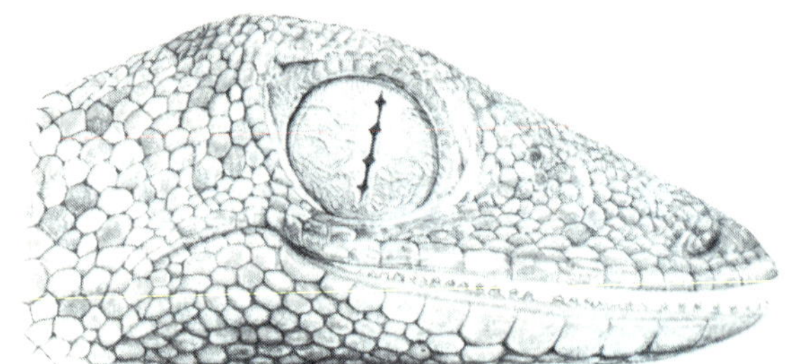

The pupils of some geckos' eyes contract to four small pin-holes that transmit merging images to the retina

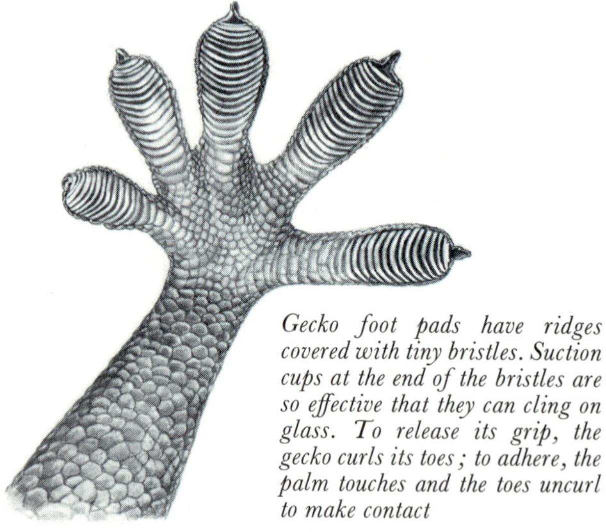

Gecko foot pads have ridges covered with tiny bristles. Suction cups at the end of the bristles are so effective that they can cling on glass. To release its grip, the gecko curls its toes; to adhere, the palm touches and the toes uncurl to make contact

HEAD OR TAIL? *The resemblance between the spade-shaped tail (top) and diamond-shaped head of the leaf-tailed gecko probably confuses predators. It sheds the tail when attacked*

across a window-pane. A gecko's feet enable it to climb any surface at all, no matter how smooth. On the bottom of the splayed toes are adhesive pads, consisting of parallel vanes in the skin that can be raised or lowered, creating many microscopic vacuums. Its small claws help in running over wood or stone. When a gecko stalks prey (insects, spiders and centipedes), it rushes to within a short distance, then slowly creeps closer. With a sudden final lunge, it seizes the victim in its jaws, and swallows it whole without chewing. The gecko's large, protruding eyes lack lids, but are covered by transparent scales. It keeps its eyes clean by wiping them with its tongue.

Geckos are among the most vocal reptiles; the name suggests the kind of cry made by a few of the smaller ones. Others emit sharp clicks, barking sounds, or shrill, almost human wails. Their defence behaviour is impressive: they puff up, open their mouths, rise high on their legs, and leap forwards – but they are harmless to man.

The 'devil' dragon

Other kinds of lizards abound, some of which Australians call dragons. The strangest is a creature known as the mountain devil or moloch. Superficially resembling the horned toad of the south-west United States, the mountain devil actually has no close relatives anywhere. A thorny armour of stout spines covers its body and head, and a spiked hump stands on its neck. If it is alarmed, the lizard pulls its head towards the hump, possibly for protection, although the true function of the hump has not yet been discovered. This grotesque, slow-moving animal waddles along in search of its main food, black ants. It eats up to 2000 in a meal but it does so without haste, one at a time.

The mountain devil, which lives in arid, rocky areas where rainfall is scarce, has a remarkable skin that helps it to acquire water. If the moloch merely settles one foot in a tiny puddle, its entire skin is wet within seconds. A network of microscopic canals running between the scales provides a capillary, blotting-paper action that instantly soaks up moisture and channels it to the mouth. The skin also changes colour according to temperature, helping the animal to absorb or reflect heat while at the same time affording protective camouflage against predators.

GROTESQUE BUT PLACID *The prickly spines of the mountain devil belie its generally placid nature. It moves slowly looking for its main food – black ants. The 6 in. long female produces as many as ten 1-in. eggs two months after mating*

FEARSOME BLUFF *Threatening an intruder, a frilled lizard displays a wide neck ruff, which makes it appear that the head has increased several times in size. In fact it is mainly an act of bluff, for the lizard runs off as soon as possible*

Frilled lizards

Another spectacular dragon is the frilled lizard, which, as it walks on all fours through the underbrush after insect food, does not look unusual except for its length of 2 ft or more. A membrane extends back from the head and covers the animal's neck. Usually this membrane remains neatly folded, but when one of its natural enemies, such as a snake, or a man approaches, and escape is impossible, dramatic events follow.

First, rising high on its legs, the lizard opens its mouth wide to display a bright yellow lining. As the jaws open, a bony apparatus at the base of the tongue automatically erects a wide, umbrella-like frill around the head. This stands out at almost right angles to the body, making the creature suddenly appear many times its former size. It puts on such a show that no predator or uninformed person would want to get closer. At the first chance for escape, however, it runs off, soon attaining sufficient speed to rise on its hind legs, dashing and staggering along at a furious pace, body almost erect, mouth still open. Although it

is mainly bluff, few sights in the world can equal this bizarre exhibition.

Some biologists believe the umbrella frill may serve another use besides bluff, however, for it is plentifully supplied with blood vessels. This wide, thin membrane may function as a radiator, allowing the escape of excess heat built up through the absorption of sunlight. Such an adaptation would not be unique; erect membranes on the backs of several dinosaurs very likely served the same purpose. The frilled lizard does spend some time trying to regulate body temperature. It lies facing the sun at noon, and at right angles to the sun in early morning and in the evening to catch as much heat as possible from the weakened rays; on the other hand when the day becomes too hot the lizard searches for shade.

Even at rest, the frilled lizard behaves oddly. Now and then its head bobs up and down; no one knows why, but perhaps a frequent change of head-position may help it to see more clearly by providing some measure of depth perception. Or the dragon may wave its arms about, slowly if it has been resting for a time but more actively after one of its mad dashes. Again, the significance of this action is not known.

Australia teems with small and medium-sized lizards, many found nowhere else. Even better than the frilled lizard at running on its hind legs is the bicycle lizard, which seems to be forever hurrying about in a frenzy of activity.

Lizards without legs

A group of burrowing lizards found only in Australia varies oddly from the lizard body plan: they are entirely legless. Unlike snakes, they have external ears, which are simply holes leading into the hearing organ. The small, scaly flaps that replaced the hind legs can be held at right angles to the body to give better traction in a burrow.

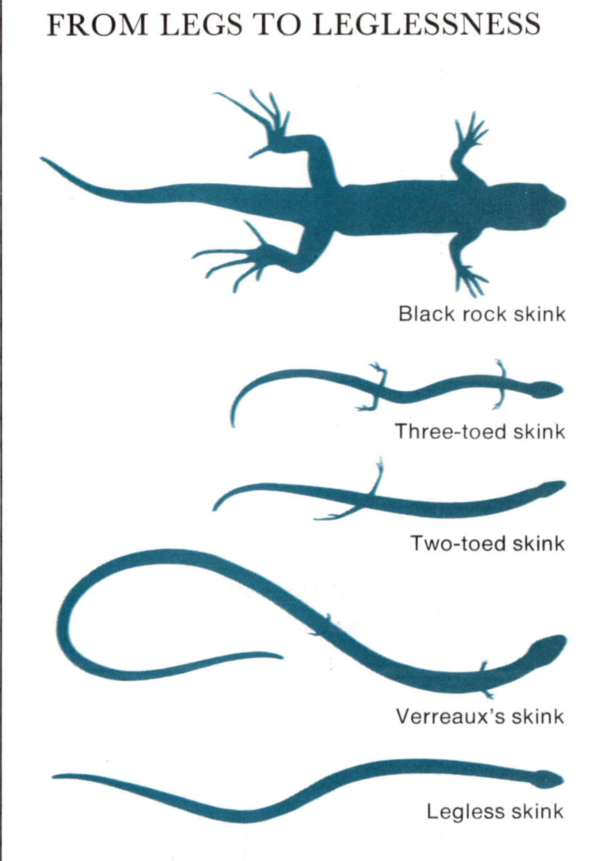

Living skinks show the stages in the evolution of legless reptiles. The black rock skink has four limbs; the three-toed skink has reduced limbs. The two-toed desert skink has no front legs; and the four tiny limbs of Verreaux's skink are useless. The final step is the completely legless skink

Like other members of the lizard family, but unlike snakes, their tail is longer than their head and body and they can shed it if attacked. When this happens, they grow a replacement within a short time.

Legless lizards apparently evolved from legged forms, as has occurred elsewhere. Since some lizards normally live under stones and wriggle through tight places, muscular thrusts of the body may be more important than the efforts of feeble legs. Thus deterioration of lizard legs seems to be an instance, like flightlessness in birds, of losing an organ through chance mutation without penalty to security or food-gathering.

Giant goannas

The most impressive lizards of the continent are the monitors or, as Australians call them, goannas (probably a corruption of 'iguanas', although there is no relationship). These are usually larger than the dragons. The largest lives not in Australia but in Indonesia on the small island of Komodo, and other monitors are found as far away as Africa. But Australia contains the greatest variety. Australia was also the habitat of the largest monitor the world has known. About 3 million years ago, long after the dinosaurs, a dinosaur-sized monitor about 15 ft long roamed the plains. Even today, the perenty goanna, at 7–8 ft, and a larger goanna from New Guinea, are among the giants of the world of reptiles.

Australia's goannas come in all sizes and occupy a wide range of habitats. The sand goanna lives far out in the desert. Others, such as the lace monitor, live high in the trees of dense forests. The perenty, despite its size, is primarily a burrower. Another goanna wedges itself into rock crevices by means of a spiny tail which it also uses to store fat for the dry season. There is even an aquatic monitor to be found swimming in some of the continent's rivers and lakes.

Goannas attack their prey with speed and agility, sometimes using their powerful muscular tails to knock over and stun a victim. They may swallow their food whole, or first shred it with long, curved claws. A goanna's teeth, like those of most snakes, curve backwards; food bitten, therefore, is not easily dislodged and must be swallowed. The lizards give the impression of being nervous as they walk along, flicking a long, forked tongue in and out; chemical particles picked up from the soil convey information about food to a special head organ. At times, a goanna may run short distances on its hind legs.

A BLIND SNAKE *Smooth scales and strong skulls equip Australia's blind snakes for their lifetime underground, burrowing for termites and ants. They are not venomous*

SNAKES OF THE SEA *Well adapted to their marine environment, sea snakes have a flattened tail paddle, a flattened body and valves that seal the nostrils under water*

LAND OF STRANGE SNAKES

Sharing the underground habitat of legless lizards are more than 20 species of blind, non-venomous snakes. Blind snakes use their projecting snouts as scoops or wedges to take advantage of the least crevice or soft spot in the soil.

The other snakes of Australia fall into peculiar pattern types. Some of the pythons arrived well after the last Ice Age, probably from Malaysia. Some Australian species reach considerable size, one attaining a length of 20 ft or more – the largest snake on the continent.

Everywhere in the world except in Australia the most numerous snakes belong to a group called colubrids – either non-venomous or with venom harmful only to small animals. In Australia, the colubrids are outnumbered by the venomous elapids: 60 per cent of the 130 species of snakes. Elapids living elsewhere include African and Asian cobras, African mambas, Asian kraits and American coral snakes. Fortunately, the Australian species are not especially aggressive.

The taipan, 10–11 ft long, is a contender for the title of the world's most dangerous snake. The venom from a single taipan could kill 200 sheep; Asia's king cobra could kill only 34.

The amount of venom injected by many Australian elapids is harmless to man, but quickly kills small animals – birds, frogs, lizards, or other snakes. Although Australia has the highest proportion of venomous snakes of all countries, the non-urban human population is so sparse that snakebite fatalities are much less common than in Asian countries.

Sea snakes related to the elapids inhabit Australia's shores. Although venomous, they are shy and easily frightened; only fishermen handling crowded nets need fear their bite. Nostrils, located high on the head, are equipped with valves that close during submergence, and the tails are flattened vertically into efficient sculling paddles. Some sea snakes spend much time on coral reefs or in mangroves along the tropical northern shore; a smaller number is found in the temperate southern regions.

SEARCHING FOR FOOD *A long-necked turtle plods across a dried-up stream bed in the Murray River in search of food, its neck extended to about the same length as its body. It cannot* *retract the neck straight back, as other turtles and tortoises do, but when it is threatened it tucks both head and neck sideways just under the rim of its strong protective shell*

Turtles and crocodiles

Only a few freshwater turtles live in the area. Close relatives of one freshwater species, the pitted-shell turtle of New Guinea, lived in Eurasia and North America over 60 million years ago. The present distribution of freshwater turtles in Australia is at first puzzling, for a few live in the most arid regions where river flow is undependable and the rate of evaporation high. Their range of habitats is thus reduced to certain spots, seemingly unconnected. But the explanation is clear. During times of flood, the turtles swim over what was formerly desert; when the waters disappear, many turtles undoubtedly die, but some are near enough to the remaining streams to survive until the next floods cover the arid plains.

From the standpoint of geological time, these freshwater turtles – long-necked and short-necked – are apparently newcomers. They may have arrived from Asia, but many are related to South American turtles. Certainly they have no close connections with the ancient land tortoises that lived in Australia when it was a wetter, more heavily forested land. Sea turtles come ashore on deserted beaches around the continent to lay eggs as they do on many other islands and atolls throughout the tropical Pacific and Indian oceans.

The largest Australian reptiles are two species of crocodiles. The estuarine or sea-going crocodile, a monster reported to grow to 30 ft in length (but averaging only 15 ft), is known throughout the Oriental tropics. It frequents north coastal river mouths, but in times of flood may swim 50 miles or more upriver. It sometimes ventures as far as 40 miles out to sea. Although the estuarine crocodile feeds mostly on fish, it also preys on other reptiles, water birds and mammals. Like other crocodiles, it seizes its terrestrial victims – including man – at the water's edge, dragging them below the surface to drown before it devours them. Johnston's crocodile is an 8-ft freshwater animal found only in the rivers and billabongs of northern Australia, where it eats waterfowl and shoreline animals.

Australia's living things never lose their fascination; no other island can boast so many unique species. Superlatives become commonplace – the largest coral reef in the world, the biggest worms in the world, the most kinds of poisonous snakes in the world. But even these hardly prepare a visitor for the bower birds and egg-laying mammals.

A living laboratory of evolution

Australia's birds are unequalled in their diversity of form, colour and behaviour. The egg-laying platypus and the various pouched animals provide clues to the island continent's age-long isolation

Australia, a continent and the largest island in the world, has more than 650 species of birds. Parrots and cockatoos flashing brilliant colours, tailor-birds stitching leaves together to form nests, crested bell birds with their ringing cries, brolga cranes in group courtship dances, magnificent black swans with crimson beaks, muttonbirds taking flight by the thousands for the yearly journey to the northern Pacific – all these birds are exciting enough, but Australia harbours others which can be ranked among the most remarkable in the world.

The Australian emu is the world's second largest bird, ranking just behind the African ostrich, with which it is often confused. An emu may weigh as much as 120 lb. and stand 6 ft tall. Like some kangaroos, this flightless bird is a grazer, and thus occupies an ecological niche filled elsewhere by deer, antelope and wild cattle. The emu, camouflaged by its shaggy brown plumage, inhabits Australia's plains and forests, where it dashes along as swiftly as a galloping horse, swaying from side to side, and swims across creeks and rivers. Scarcely discriminating about what it eats, it picks up grass, flowers, berries, fruits, leaves and insects. One emu's stomach was found to contain 3000 caterpillars.

A characteristic of all running animals is a reduced number of toes, with a consequent lessening of frictional contact with the ground. Horses, for example, run on only one toe; deer and ostriches have two toes on each foot. The three-toed emu runs with great bursts of speed, although not quite so swiftly as the ostrich with its longer legs and only two toes.

At mating time, the male emu calls out in deep, guttural tones as he approaches a female, driving off rival males with powerful forward kicks. The female replies in low-pitched booming sounds that resonate in an air sac in her neck. After courtship and mating, she builds a nest of grass, bark and dry leaves, in which she lays 9–15 dark green eggs weighing up to 2 lb. each. The eggs are tough enough to withstand the attacks of most predators, but black-breasted buzzards drop stones to shatter them.

Egg-laying is the female emu's sole task, and she then leaves with no further maternal concern. The male broods the clutch for two months until the downy, yellow-and-brown striped chicks hatch out. But his responsibility has not ended. The chicks follow him everywhere, running back and forth between his legs, keeping in touch by means of intermittent piping calls. At night, when the male settles down, the chicks cluster beneath

A BIRD PARADISE *No group of Australian creatures is more varied than its native bird population. The 12-in. mulga parrot (left) is only one of more than 50 colourful parrots that have spread into every part of the country. In all, Australia has 58 bird families, ranging from enormous emus to exotic birds of paradise*

PROTECTIVE HELMET *The cassowary's horny crest provides protection from thorns as the bird runs swiftly through undergrowth. These 5 ft tall flightless birds are glossy black and well camouflaged in dense tropical brush*

EXPECTANT FATHER *Emus form pairs in December – the early Australian summer – and the female lays her eggs in winter. The male, who is left to guard and incubate them, lives on stored-up fat and rarely leaves the nest until they hatch*

his plumage, there to sleep hidden and protected. The paternal guardianship and guidance go on for at least a year as the chicks slowly learn to get along by themselves. In the meantime, the female has mated again with another male, and has laid a new clutch of eggs. Males, on the other hand, are free to mate only every other year, spending nearly all their time bringing up their chicks.

Pugnacious heavyweight

In tropical northern Australia and New Guinea lives another large, flightless bird, the cassowary. It is 5 ft tall and it stalks alone through dense tropical brush, a tough horned helmet keeping branches and leaves from its face. Although primarily a vegetarian, the cassowary catches crabs and fish by wading up to its neck in streams. It is a pugnacious creature, capable of delivering furious kicks with its powerful, heavily clawed feet. Despite its great weight – sometimes well over 100 lb. – the cassowary is a prodigious jumper, able to clear 7–8 ft in one bound. In contrast to the drab emu, it has glossy black feathers (which are used as currency by New Guinea natives), a blue neck and red wattles, yet the two birds belong to related families. The eggs are similar and, like its emu counterpart, the male cassowary has full responsibility for brooding and shepherding the young until they can care for themselves.

Bluff and camouflage

Not all of the incredible variety of Australia's feathered inhabitants are pugnacious, however: some defend themselves in other ways. For example, the perfectly camouflaged tawny frogmouth, a member of the nocturnal whip poor-will (or nightjar) family, often spends hours perched on a branch stub. If it is disturbed, it may fly away or it may present a horrifying spectacle of great opened beak and staring golden eyes – pure bluff. It hunts nothing more than insects, centipedes, scorpions or possibly an occasional small mouse, frog or lizard.

Apparently the first Australian bird seen by European explorers was a black swan, which confirmed their impression of the Southern Hemisphere as an upside-down place. In most regions of Australia, flocks of black swans are seen at dusk or occasionally on moonlit nights, flying in V-formation from one feeding ground to another.

Two groups of songbirds, mostly 5–6 in. long, with several species each, bear names belonging to unrelated European birds, obviously because they reminded English settlers of their homeland. The robins have crimson, flame-coloured, pink or yellow breasts; some use cobwebs to bind together their carefully constructed nests. The fairy wrens have long, erect tails, and most have some blue in their plumage. Their nests are usually domed.

Penguin parade

One of the best-known bird attractions in Australia is the nightly parade, during the nesting season, of the little fairy penguins on Phillip Island south-east of Melbourne. Each evening, just before dark, the sea suddenly reveals amid its breaking waves hordes of small penguins skimming in towards shore. As each wave breaks and recedes, hundreds of the little birds pop upright and walk up the beach, flapping their fin-like wings and conversing loudly.

Determinedly they march uphill to their burrows, where vociferous exchanges occur between the returning bird and the mate that stayed behind to incubate their two, sometimes three, eggs during the preceding few days. Duties are exchanged the following day, and the other goes forth, after noisy farewells, to fish in the sea. This procedure continues for three months while the eggs hatch and the chicks grow, although towards the end of the period the young are left alone a good deal, existing on their accumulated fat and growing sea-going plumage before they finally leave for the water.

DEFENSIVE BLUFF *A tawny frogmouth gapes, showing its pale mouth lining, to scare off an intruder. Like the owl, the frogmouth flies silently on downy feathers. It does not hunt from the air, but preys on insects and spiders on the ground*

UPSIDE-DOWN WORLD *The earliest European explorers were convinced that Australia was a world upside-down when they discovered that the native swans were black. The young cygnets, usually in broods of four to six, are dark grey*

BRILLIANT ORNAMENTS FROM THE SKY

Australia harbours about 60 species of parrots and their cousins – lorikeets, parakeets, cockatoos and budgerigars. The colourful budgerigars, popular in pet shops all over the world, darken the inland Australian sky in enormous numbers and create a sound like thunder with fluttering wings. A flock looks like a dazzling, wheeling storm until suddenly it settles in open forest, decorating every tree with brilliant, living ornaments.

Few birds anywhere can excel the bold colours of Western Australia's red-capped parrot. Below the red crown of its head are yellow cheeks; its body is purple, with green wings and a yellow rump; the leg plumage and underside of its tail are red, while the upper tail surface is blue with a green overlay. The parrot looks like a child's attempt to produce the most colourful bird possible with an elementary painting set containing only the gaudiest pigments.

Of the several species of cockatoos in the world, nearly all are native to Australia. They range in form and colour from the sulphur-crested cockatoo to the black palm cockatoo, which is able to crack large, hard nuts that no other bird can break. Cockatoos feed in flocks; when any bird becomes alarmed it screeches, and the whole flock explodes into the air with ear-splitting shrieks that would confuse any predator.

BIRDS OF MANY COLOURS *Rainbow lorikeets perch in a thicket where, with their brush-like tongues, they take nectar from the blossoms of flowering trees. Flocks of the raucous birds migrate across Australia, following the seasonal flowering of trees and shrubs*

ACROBAT OF THE PLAINS *A sulphur-crested cockatoo dangles from a branch. Flocks of these handsome birds roam across the plains of northern and eastern Australia, feeding on beetle larvae and seeds – especially wheat*

COLOURFUL TARGET *A red-capped parrot bursts into flight in south-west Australia, disturbed in its meal by shots from an angry farmer. The bird uses its hooked beak not only to crack open eucalyptus capsules but also to raid orchards*

CREST OF FLAME *A pink cockatoo displays its highly coloured crest. With its strong bill, the bird digs insects and roots from the hard ground of the arid Australian interior. It is known to nest in the same tree hollow year after year*

281

PREPARING THE MOUND *A male mallee fowl (right) scrapes debris with its large feet to build the incubating mound, while his mate waits to lay an egg. Mounds are built in April, and the winter rains ferment the plant material to create heat for the ½-lb. eggs*

BIRDS THAT BUILD INCUBATORS

The most extraordinary Australian birds are the mound builders, bower birds and lyrebirds. Many mound builders (or megapodes) shape heaps of forest litter into incubating chambers for their eggs; some mounds that have been examined measure 30 ft in diameter and 15 ft in height, although most are much smaller.

The mallee fowl of the hot, dry, inland regions is probably the best known megapode, famous for regulating the environment of its eggs with an ability and energy that defy explanation. The male constructs a massive, intricate nest while the female stays near by, doing little to help. He first excavates a hole 2–3 ft deep in the soil. Then he scratches up leaves and twigs over a 50-yd radius to fill the hole and raise the mound 12 in. or more above ground level. Finally, he digs a chamber in the mound, fills it with sand and more vegetable debris, and smooths over the surface.

When the rotting of the vegetation has raised the mound's internal temperature to 33°C (92°F), the male exposes the egg chamber, and the female, now allowed to enter the nest, digs a niche in the side wall of the chamber and lays a single egg. She retires, and the male spends the next two hours building the mound back to its original height. At intervals of several days, the female lays 5–35 more eggs in the mound. The male attends the nest daily, probing into the mound with his bill and determining the temperature by his tongue – with a degree of sensitivity that scientific research has not yet been able to explain satisfactorily.

At first, there is danger that the mound's temperature may rise excessively; later on, as the rate of decay diminishes, it may fall below the optimum. Since egg-laying and nest-caring activities may go on for six to eight months, changing seasonal temperatures can further complicate the problem of environmental control. But

MATURE AT BIRTH *Just hatched, a brush turkey chick emerges from the nesting mound. The chicks develop to an advanced stage in the egg and, when they hatch, they can dig themselves to the surface. They fend for themselves and often fly within a day*

it is the male's responsibility to maintain the mound at 33°C, and he does so by a variety of corrective activities. If internal heat raises the temperature too high, he uncovers the egg chamber enough to allow cooling; if the internal temperature falls, he adds more vegetable debris, which accelerates the rate of decay and increases the amount of insulation. In summer, he digs up the egg-chamber material very early in the morning, spreading it on the ground and allowing it to cool before returning it to serve as insulation against the heat of mid-day. In autumn, he waits until the sun is overhead before digging up the chamber; the leaves and debris are then warmed in the noon sun before being returned in early afternoon. Each adjustment takes him hours to perform, making him surely one of the busiest animals and most solicitous fathers alive.

Because egg-laying extends over several months, the first chicks begin hatching before the last egg is laid. After seven weeks of incubation, they emerge from their shells as much as 3 ft below the surface of the mound and struggle upwards, at about 12 in. an hour, purely by their own efforts.

With all that the parents have had to do, it is perhaps understandable that they now take no interest in the chicks. The offspring are at once independent and go about their solitary business. In due course they too will mate and become slaves to the inherited ritual that provides for the survival of the species.

Two other types of mound builders are found in Australia, but their behaviour patterns differ slightly. Whereas the mallee fowl usually builds a new nest each spring, the brush turkey of the eastern coastal area is content to use the same mound year after year. By adding to it each spring, brush turkeys often create structures some 50 ft in diameter and 8–10 ft high. The third variety – the jungle fowl – is found on Pacific islands as well as in northern Australia. On some islands its incubating problems are easily solved, as it makes its nest on the slope of dormant volcanoes and, because the heat source is constant, the nest can be left unattended. All the bird has to do is pick a nesting site with the proper temperature. In Australia, where there are no volcanoes, the jungle fowl builds mounds of organic debris.

THE ELABORATE DISPLAY ARTISTS

Another feathered wonder of both Australia and New Guinea is the male bowerbird. With much artistry, it builds a structure with the sole known function of attracting a female. Bowerbirds are rather drab, and the behaviour of the male has suggested to some observers that the bower, decorated and even painted, may serve as a substitute for male finery. Bowers built by different species fall into several major categories, ranging from a simple platform decorated with leaves and stones to an elaborate structure that early European explorers assumed must be the highly skilled work of native people.

The arbour of avenue builders, such as the satin bowerbirds, consists of a small platform surmounted by two walls of fine sticks about 18 in. high, and decorated with blue feathers, flowers, scraps of paper, shells and cicada cases. The

VARIETY OF BOWERS

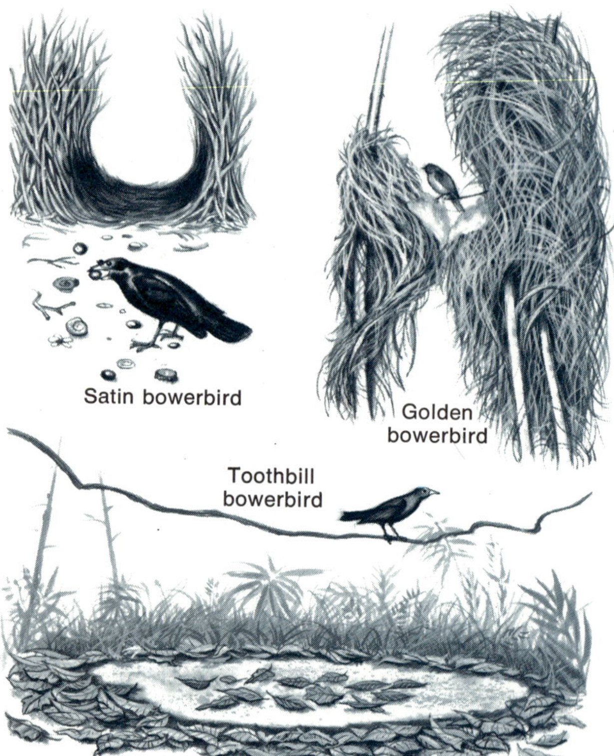

The bird-built bowers range from the simple leaf-decorated clearing of the toothbill to the walled avenue prepared and painted by the satin bowerbird and the golden bowerbird's maypole mound – sometimes 9 ft high

DECORATING THE BOWER *A fawn-breasted bowerbird makes up for its own drabness by decorating a walled avenue with chewed-up plants. The dark green juice dries a red-brown. 'Unprocessed' plant material lies at the bottom of the picture*

bower of the golden bowerbird, one of the may-pole builders, is constructed around two thin tree trunks spanned by a vine, upon which flowers and lichens are placed as decorations. The walls of this bower may be as much as 9 ft high.

Even after erecting a bower, the male spends nearly all of his time improving and rearranging his collection of ornaments. Dried berries and withered flowers are discarded and replaced with fresh ones. Hundreds of shiny pebbles, animal vertebrae and knucklebones, snail shells, colourful feathers, iridescent beetle wing-covers – the kind of collection depending on the species – are constantly cleaned and arranged. Some species of bowerbird are even more artistic: the male chews colourful berries, mixes them with saliva, strips off a piece of frayed bark to use as a brush, and paints the walls of his bower.

When a female begins to show interest, the male hops up and down in increasing excitement, displaying his coloured neck feathers. When mating is completed, the female builds a conventional, purely functional nest in a tree. In some species, the male helps to feed the female while he continues his own elegant housekeeping activities in the original bower, putting things in readiness for the next female passer-by.

The gifted mimic
Other mounds, used only for display, are built by lyrebirds, which exhibit magnificent feathered tails, and possess an unequalled gift of mimicry. A species known as the superb lyrebird is one of the largest perching birds in the world; it spends much time on the ground, however, where it appears pheasant-like when its tail is unopened. As a male embarks upon his courtship display, the curved bronze side feathers of the great tail swing far out to each side, and a fine filigree of shining white feathers spreads over the wide space between. Each morning the lyrebird stands upon his display mound and sings his own natural song, as sweet and rich as that of any bird known.

But this is only the beginning, for he then embarks on a vocal performance of truly astonishing virtuosity. He utters a fantastic variety of sounds, often with baffling ventriloquism. In his repertoire is the song of every bird in the forest, as well as a composite of these sounds. A single lyrebird can imitate the noisy flight of a flock of parrots, complete with dozens of simultaneous calls and the rush of wind through their feathers or the rustle of leaves. It sings, shrieks, trills, squawks, warbles and caws in a continuous

exhibition of versatility. One of its favourite squeaking calls is that of the pilot bird, a creature usually near by when the lyrebird ceases its morning song and begins scratching about for insects and worms. Dancer, architect, singer, mimic – the lyrebird is in many ways the most remarkable of Australia's birds.

Laughing snake-killers
But the favourite of many Australians is another, more common bird, not remarkable except for its raucous, hilarious, infectious laughter – the kookaburra. Its cries are among the best known in the world, for hardly a jungle film has been made without the kookaburra's rollicking laughter. This large kingfisher does not feed on fish, preferring instead to hunt snakes, lizards and rodents. The Australians have so much regard for the prowess of kookaburras as snake-killers that they have introduced the birds into new regions. Kookaburras often hunt in groups, and they can attack prey larger than themselves.

LAUGHING JACKASS *The kookaburra is famous for its weird calls of mocking laughter, and it is popularly known as the 'laughing jackass'. Kookaburras often feed on poisonous snakes and, surprisingly, are members of the kingfisher family*

THE SPECTACULAR BIRDS OF PARADISE
Their courtship ritual ensures that only the strongest and most beautiful survive

The glorious and ostentatious birds of paradise that inhabit the rain forests of New Guinea and Australia are the unlikely relatives of the smart but unhandsome common crow. Until a government ban in 1921 they were for centuries hunted by natives and traders to make tribal headdresses and ornate hats for European ladies. They received their English name when the first explorers were presented with skins, from which the natives had removed the legs. Unable to believe that they were normal birds the Europeans christened them 'legless wanderers from paradise'. Only the males have the gay colouring and long, elaborate feathers. They use their fine plumage to attract mates – often many times in a single breeding season. When one mating is complete, the male leaves his drab companion to hatch and rear the young chick while he moves on to his next mate. Ornithologists have found that although the many species have different courtship displays, there are some points of similarity. All birds of paradise conduct their courtship by elaborate ritual dances. All the males in any group use the same 'arena' for these dances, but each has his own site within it; only the finer, stronger males can command the most favoured central positions, and so they are luckier in attracting mates. Some poorer specimens may not mate at all, so the characteristics of only the best birds are passed on to succeeding generations.

LEGENDARY LEADER *The 6-in. king bird of paradise, with its long, wiry tail feathers tightly coiled at the ends (bottom left), is the smallest of the species; but according to legend it is the monarch of all the birds*

ALL SET FOR COURTSHIP *A male magnificent bird of paradise makes sure that he is seen to best advantage during his courtship display by first removing leaves and branches from the tree to let the sun shine through*

NOBLE AND ROYAL NAMES *The broad-tailed Princess Stephanie bird is one of many that were given the names of European royalty and nobility by explorers who hoped to find favour with a patron at home*

THE ONLY EGG-LAYING MAMMALS

PROTECTING HER YOUNG *A female platypus sprawls across her burrow, shielding her young. Blind for 11 weeks, the baby remains dependent on the mother for five months*

Looking on as comparatively insignificant spectators during the age of the giant reptiles were the first mammals. Although their remains are rare, enough fossils have been found to indicate that they were delicate, rat-sized creatures that fed upon insects. Details of the anatomy and habits of these ancient animals may never be known, but scientists believe that the monotremes of Australia and New Guinea, the world's only egg-laying mammals, provide some clues to the probable nature of the earliest mammals.

The most revealing features of the two monotremes – the duckbilled platypus and three species of spiny anteaters (echidnas) – are certain skeletal and other anatomical parts and their means of reproduction. All else, although fascinating, is of much less significance.

Both the platypus and the echidna have chest bones that support forelimbs, as well as bones jutting out from the pelvic girdle that are clearly of reptilian origin. The brains of these monotremes, although not much inferior to those of

SHAPED FOR LAND AND WATER *Propelled by broad, webbed forelegs, a platypus swims under water, leaving a bubble trail. Insulated by fur and fat, the platypus is well shaped for swimming and burrowing. Skin folds protect its eyes and ears while it probes with its leathery bill. The flat tail acts as a rudder in water and is used to tamp down earth when the platypus is burrowing. The hind legs of male platypuses have spurs containing venom*

marsupials, the pouched mammals such as kangaroos, have certain primitive features that have been greatly improved upon by placentals – the more advanced mammals which nourish growing embryos for a comparatively long time in the womb. A monotreme's abdominal organs terminate in a single opening through which the excretory, digestive and reproductive systems empty; such a single orifice is characteristic of amphibians, reptiles and birds, but not of most of the modern mammals.

Other primitive features are the relatively poor regulation of body temperatures (which are often quite low) and milk glands that exude their fluid not through nipples but through pores in two areas of the abdomen.

The single most important characteristic of both animals is the laying of small, rubbery, compressible eggs. The platypus usually produces two or three light-coloured eggs; the echidna rarely lays more than a single egg.

The platypus is a highly efficient aquatic animal that grubs along the bottom of cool upland streams, its eyes and ear apertures tightly closed by skin folds. It gobbles up worms, molluscs, crayfish, aquatic insects and anything else that comes into contact with its soft, sensitive bill. The food is packed into expandable cheek pouches, together with pebbles, which assist in crushing whole organisms to a manageable size. The platypus swims by sculling its forelimbs vigorously. Since the fan-shaped webbing on its front feet spreads under and well past the stout, sharp claws, the animal folds the webbing out of the way under its palms when it walks on dry ground.

A platypus burrow extends deep into the bank of a stream, perhaps as much as 50 ft, with the entrance opening 6–12 in. above the waterline. The separate resting chambers of the male and female give off a musky odour that is the result of pungent secretions from the animals' throat glands.

At mating time, the female makes a more elaborate tunnel with a spacious chamber at the end, which she pads with moist leaves and reeds brought in tucked under her broad tail. She then plugs the tunnel to the nest at several places with thick, firmly tamped earthen barriers. Finally she lays two or three small eggs, which usually adhere to one another, and she incubates them for 10–14 days by curling her body around them. After the young hatch, naked and sightless, they lap milk from her fur. The mother holds the tiny creatures in the curve of her body, where their feeble, spasmodic movements stimulate the milk

SPiny anteater *A short-beaked echidna probes for ants and termites with its bird-like bill. If threatened it digs straight down with its foreclaws; its splayed hind claws actually point backwards and are very effective at digging*

flow, but she has to leave them periodically to feed herself, wash and carry out waste. After each trip, she carefully plugs the tunnel again.

Toothless anteater

The echidna is more widely distributed than the platypus; it is found in Tasmania and New Guinea, as well as on the Australian continent. In deserts, savannahs and rain forests, it digs rapidly in search of ants and termites, capturing them with its darting, sticky tongue. Its thick covering of spines parallels the defensive coats of the tenrecs, hedgehogs and porcupines found on other continents. Although toothless, the echidna uses horny structures on the roof of its mouth and back area of its tongue to crush insect prey.

The female echidna incubates her single egg for about a week inside a temporary pouch; here the defenceless, squirming infant remains during its early growth, stimulating the flow of thick, yellowish milk. It emerges after about ten weeks when it has become too big and spiny for its mother's comfort.

STRANGE ANIMALS WITH POUCHES

As the Age of Reptiles drew to a close and the super-continents began to break up, the pouched mammals (marsupials) and the more advanced placental mammals co-existed. However, on the larger land masses, the superiority of the placentals (animals whose young develop to an advanced stage in the womb before birth) gradually resulted in the elimination of marsupials, except for the opossums of the Americas. But in the isolated island continent of Australia there were no placental mammals and, in the absence of competition, Australia developed its own unique range of animals with pouches.

With little competition and an enormously wide range of diverse land to occupy, Australia's marsupials evolved into a great diversity of specialised forms, some of which no longer exist. Two or three million years ago the continent supported kangaroos that were 10 ft tall, swamp-dwelling animals the size of rhinoceroses, and marsupial 'lions' that were probably the largest pouched carnivores the world has ever known. Changes in climate and vegetation eliminated their feeding grounds, however, and these creatures eventually disappeared.

The present marsupials that most resemble Australia's early inhabitants are all carnivorous, preying upon insects, lizards and other small ground creatures. Their primitive traits include five distinctly separated fingers and toes, a rudimentary pouch and, usually, many small front teeth. Tiny marsupial 'mice' feed on a steady diet of insects, snails, worms and anything else they find on or beneath the soil. Silky-furred, blind marsupial 'moles' tunnel a short distance under ground and consume insects.

The alert and sprightly bandicoots are larger than the 'mice' and 'moles', but their diet, too, consists of insects, lizards and other small animals. The numbat (or banded anteater) is an insectivore with 52 teeth – more than any other land animal. It licks up termites with a long, worm-like tongue. The female has four nipples but no pouch, and the young cling to the teats and grip her fur with their forefeet as she walks. The rearward-facing pouch of the female rabbit-eared bandicoot, like that of the marsupial mole, is especially convenient when the animal burrows in the sand.

HOW MARSUPIALS FILL THE NORMAL

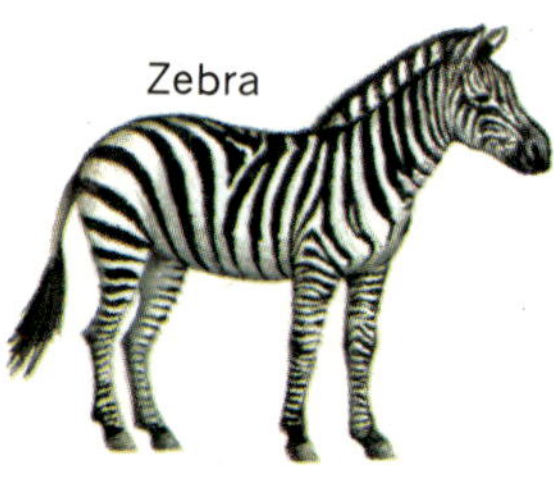

Zebra

Open grasslands offer food to large grazing animals like Australia's kangaroos and the zebras of Africa. Grazers eat enormous quantities of grass to make the maximum use of its low nutritional value. Lack of concealment in open country forces both the zebra and kangaroo to be wary and swift

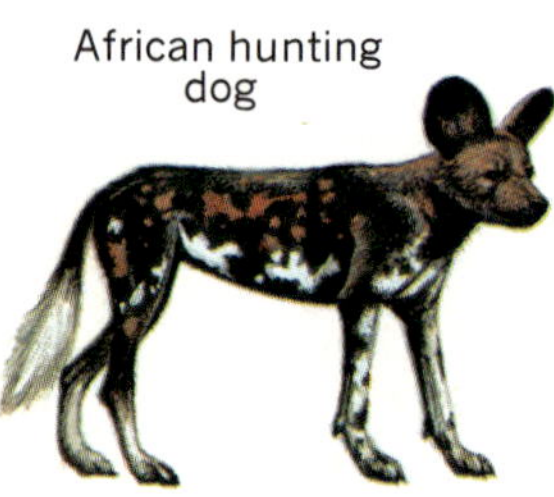

African hunting dog

Large carnivorous animals feed mainly on the large grazers. Like the African hunting dogs, the Tasmanian wolf (which may now be extinct) relied on endurance rather than speed when running down its prey. With its strong jaws, the wolf – called a thylacine – was able to feed on the native kangaroos

Grasshopper mouse

Like the grasshopper mice and other small predators of the plains, the marsupial mice of Australia feed greedily on insects. Because they provide easy prey for larger hunters, these species live in hiding and make quick dashes to find their own prey. The flattened head allows it to probe in crevices

Marten

Tree-hunting animals are solitary, agile and quick. They prey – mainly at night – on small animals and birds' eggs. Both the marten of Europe and America and the Australian tiger cat have long bodies and short legs with tree-gripping feet. Adapted for tree-climbing, they can also hunt on the ground

Anteater

Mammals that feed on ants and termites have long snouts and sticky tongues to thrust into crannies and lick up prey. Powerful clawed forelimbs rip open termite mounds and rotten logs. Teeth are weak and degenerate, as in the numbat, or disappear altogether as in some South American anteaters

Mole

Burrowing mammals, such as the European mole and its marsupial equivalent, rarely appear above ground. With strong forelegs and enlarged nails they dig below the surface in their unending search for worms and small insects. Their eyes have degenerated, but they have acutely sensitive organs of touch

ECOLOGICAL ROLES

Australian marsupials (below) have developed in isolation to fill roles occupied elsewhere by placentals (left)

SCAVENGING FOR FOOD *Fallen logs and leaf litter provide the delicate, unaggressive numbat with its main source of food – termites. The species, which is now very rare, shelters in logs hollowed by termites to escape dogs and foxes*

A LOST 'WOLF' AND THE DEVIL 'CATS'

The so-called native cats and tiger cats are another group of relatively primitive and unspecialised marsupials. They are usually nocturnal predators living principally on the ground. Tiger cats, or quolls, climb well in pursuit of reptiles, birds and small mammals. The inaccurate 'cat' designation probably derives from the way these creatures attack their prey. They approach stealthily, usually from above or behind, then with a springing charge land on a victim's back and quickly bite through the spinal cord.

Mystery surrounds the fate of the thylacine, or Tasmanian wolf, which is also called the Tasmanian tiger because of the narrow, dark stripes across its back. No thylacines have been seen in several decades, although paw prints are reportedly found from time to time. This largest marsupial predator, if it still exists, stands 2–3 ft high and measures about 6 ft long. At first glance, the thylacine appears very dog-like until the proportions of its long hind legs and thick-based tail are seen. The rigid tail is unresponsive and does not wag. When a thylacine yawns, all dog illusions vanish, for its mouth opens at what seems an impossibly wide angle – among the greatest gapes in the mammal world.

A solitary hunter, the thylacine trots determinedly after its prey, putting on a burst of speed only at the end of the chase. Unfortunately the thylacine was not studied scientifically or even carefully observed in the wild and little is known of its habits. The Tasmanian wolf's overall effect upon sheep in the early part of this century is debatable, but since Tasmanian sheep raisers

292

NOCTURNAL PREDATORS *Two Tasmanian devils feed on a large lizard. These marsupials, which hunt mainly at night, eat any animal, living or dead. They share a den for two weeks before mating, and the young are born about six months later*

were convinced that the thylacine was harmful and economically undesirable, it was ruthlessly hunted into near or total extinction. Its coughing barks are no longer heard in the forest, and the world has perhaps lost one of the most interesting archaic predators.

The Tasmanian devil, a stocky, barrel-shaped animal, weighing 13–20 lb., with a fur-less pink nose, is the second largest marsupial predator. Although when living wild it appears to be an intractable bundle of fury, the devil can be tamed. It is a clean animal that washes its face much as the domestic cat does. It likes water and swims and dives with ease, but hunts and scavenges mostly on land. Its heavily constructed head bears powerful jaws able to crunch through muscle and bone without difficulty. A fair number of devils still exist in Tasmania, but they are nocturnal and seldom seen; they spend their daylight hours hiding in rock crevices or hollow logs in the island's forests and underbrush.

EXTINCT BREED? *The Tasmanian wolves, or thylacines, were last seen in 1933. Because of their secretive habits, however, zoologists cannot be certain that the species has disappeared altogether. Their main prey was kangaroos and wallabies*

GLIDERS AND BURROWERS

The marsupial plant-eaters are much more evident than the secretive, nocturnal carnivores. Kangaroos, among the best known animals in the world and almost synonymous with Australia, are only one of the forms that the vegetarian marsupials have assumed.

Several species of possum (quite unlike the American marsupial opossum, except for their prehensile tails) live in Australia's forests, where they eat foliage and insects, sip nectar, and are important as flower pollinators. The brush-tailed possum well illustrates the typical marsupial means of reproduction. Only $17\frac{1}{2}$ days after mating, the female gives birth to a single infant measuring about $\frac{1}{2}$ in. in length – so tiny that its

mother outweighs it 10,000 times. (A human mother outweighs her baby only 15–20 times.) Like other marsupial young, the baby brush-tail finds its own way into the mother's pouch, attaches to a nipple, and does not leave the pouch until it is four or five months old.

Some possums have taken up a way of life like that of flying squirrels and Malaysia's flying lemurs. One of them, the greater gliding possum, is more than 3 ft long; floating downwards from tree to tree or from a tree to the ground, it may span distances of 300 ft or more. Unlike a flying squirrel, it does not extend its front legs outwards to stretch the membrane joining the front and rear limbs, but bends its arms at the elbows so

OUTSTRETCHED FOR FLIGHT *A feather-tailed glider sails towards a eucalyptus twig where another sips nectar. The 4–6 in. long marsupials also eat buds and insects. They use their tails to grasp branches and steer through the air*

that the forearms point inwards and the paws meet beneath the head. This posture allows the membrane to billow up and form a concave surface. Since smaller marsupial gliders extend all four legs when gliding, stretching the membrane surface into a flat plane, most of them are less likely to cover as much distance, although yellow-bellied gliders can float even further.

Sugar gliders, among the most numerous of all eastern Australian mammals, are often seen climbing trees rapidly with nesting material clutched tightly in their coiled tails, or sailing down from the trees for 100 ft or more, using their tails for steering. Quite unlike these agile gliders, but closely related, are two species of cuscus, a beautifully furred tree-dweller which is found only in Queensland. Its moon face, with inconspicuous ears, has a perpetually wide-eyed expression, which appears both quizzical and placid.

Living on the ground are two kinds of wombat, tubby, almost tail-less creatures. Despite their size (3–4 ft in length) they are remarkably fine burrowers, able to dig wide, deep tunnels.

A wombat's tunnel may have a diameter of 2 ft and run underground for dozens of feet. A powerful digger and weighing up to 70 lb., a wombat uses the claws of all four feet to excavate vast quantities of earth in a short time. Even in its tunnel the animal defends itself effectively, lashing out violently and crushing an attacker's head or limbs between its body and the sides of the burrow. Wombats lead quiet and solitary lives.

As might be expected from its tunnelling habits, the wombat's well-developed pouch opens backwards so that the young wombat is protected as its mother digs and burrows. The vestigial muscles of the stumpy tail, and other features, suggest to scientists that the wombat's ancestors lived in trees.

CLINGING BY ITS TAIL *Coiling its prehensile tail around a branch a cuscus clings to its perch. About 3½ ft long from head to tail-tip, the cuscus forages at night for fruits, leaves, insects and occasionally birds and eggs*

TEETH THAT NEVER STOP GROWING *The Australian wombat munches grass and roots with teeth, like those of the guinea pig, that grow continuously throughout its life. To excavate its burrow it digs with its front feet and clears dirt with the hind feet*

THE SPECIALIST PLANT-EATER

Unlikely as it seems, the wombat's closest relative is the koala, one of Australia's most beloved tree-dwelling animals. Zoologists believe the koala may have returned to arboreal life from a land-dwelling, wombat-like ancestor. Many basic koala characteristics resemble those of a wombat, including the absence of a tail and a rear-facing pouch that may seem a hazardous arrangement in an animal that sits and climbs upright. Actually, it is a good design that prevents snagging on branches, and the pouch muscles are so strong that the infant cannot fall out. When the young koala is mature enough to leave the pouch, however, it climbs on its mother's back to ride around, hugging her tightly, for nearly a year before taking up life on its own.

The koala is an outstanding example of specialisation carried to a dangerous degree. Although Australia has hundreds of species of eucalyptus, the animal eats little but the leaves of about 12 of them, prefers five, and has one favourite. Even then there are hazards. The koala likes tender leaves when they are available, but must select them with care. In the Australian winter, young

HOW DIET RESTRICTS THE KOALA

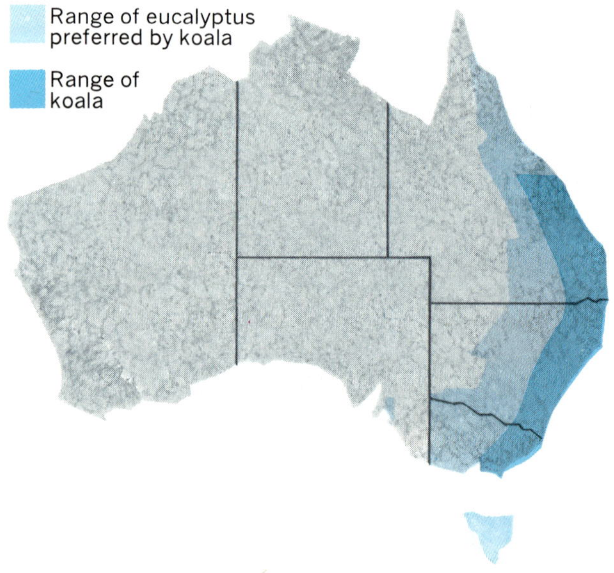

Koalas eat only the more mature leaves of certain favourite eucalyptus trees. They are thus confined to eastern Australia, the only part of the continent where their food trees grow

gum leaves and shoots produce excessive quantities of hydrocyanic acid – 4 oz. of such leaves is enough to kill a sheep. Moreover, the older leaves in the koala's diet contain tough fibres and strong-smelling oils that impair their digestibility. To most other animals the leaves of the eucalyptus are distasteful and poisonous, and indeed even the koala finds the new shoots at the beginning of the season to be too oily. When this happens it turns its attention from one preferred tree to another that has older sprouts. Perhaps it is to cope with such fare that the koala has a large, coiled appendix – measuring 6–8 ft long if unwound – in which digestion takes place very slowly. Swallowing such quantities of pungent oils, however, does seem to bring the animals one benefit; they are completely free from lice and other vermin.

Koalas are found only in the eucalyptus belt of eastern Australia – from Queensland to Victoria – but their diet differs in north and south, reflecting in the main the availability of different types of eucalyptus tree. It is the animals' total dependence upon one highly specialised type of food that gives rise to conservationists' fears that they may eventually become extinct. If, for example, a plant disease should attack their food trees, the koalas, in the wild state at least, would very quickly vanish.

Perhaps the specialised diet of koalas would be easier to understand if eucalyptus leaves were not so oily and tough. Young koalas, for example, cannot go immediately from milk to an adult diet: they must subsist for a time upon a regurgitated vegetable mush predigested by the mother. After weaning they rarely drink; the name 'koala' means 'no drink' in one aboriginal dialect.

Once successfully on its own, however, a koala may live out a long and quiet life for up to 20 years. Sluggish much of the time, it can move quickly from limb to limb. If hurt, koalas cry like babies; if they are caught in one of the explosive fires that occasionally rage through the gum forests, they have been heard to scream piercingly. Despite the dangers and uncertainties of life, these gentle, appealing little creatures with their woolly ears and sleepy eyes are still to be seen in parklands in Queensland and Victoria, under the protection of the Australian government, whose conservation policy has now successfully removed the additional threat of the pest-hunter from the koala's already precarious existence.

APPEALING LEAF-EATER *Grasping a eucalyptus branch with one paw, a koala devours eucalyptus leaves, its only food. An adult eats about 2½ lb. daily. Moisture in the leaves enables it to survive without water*

FIGHTING STYLE *Sparring male kangaroos clutch with their paws or kick sideways in a stylised trial of strength. Unlike the powerful downward clawing movements used for defence, these kicks are aimed to throw an opponent off balance, rather than to kill or injure him*

COMFORTABLE HOME *A large kangaroo baby, called a joey, peers from its mother's pouch. At birth, the baby kangaroo weighs less than 1/25 oz.; it stays in the pouch for almost a year*

KANGAROOS OF THE OUTBACK
Unique animals that are in danger of vanishing from the earth

The largest living marsupial mammals in the world are the kangaroos of Australia. The graceful, long-tailed wallabies inhabit the brushlands; the grey kangaroos are found in the open grassy forests; the richly coloured red kangaroos live in the continent's vast inland plains; and the powerful walleroos keep to the coastal mountains and parched interior. Male red and grey kangaroos are known to grow over 8 ft tall and to weigh considerably more than 200 lb. Kangaroos can render almost any enemy powerless with their huge hind legs, and they are more than a match for their natural animal enemies, such as the dingo. Yet they are in danger – from man. Several of the smaller wallaby species are considered by world wildlife conservation authorities to have no more than a doubtful future; and some are now confined to very small areas of the Australian mainland and Tasmania. Although the larger reds and greys are still too numerous to qualify for inclusion in any list of endangered species there is cause for concern that the combined effect of the continent's prolonged droughts and killing for the dog-food and shoe-leather trades has reduced numbers in some areas by more than half. In 1970, for example, it is known that more than 2 million of the animals were killed by man; if killing on this scale were to continue, combined with the effects of long droughts, it is thought by the official conservation research body in Australia that the red kangaroo could become a rare animal. Up to now, no grassland reserves of sufficient size have been set aside for the kangaroo, but the Australian Conservation Foundation has called for a proper conservation policy. State wildlife authorities, aware of the danger, are now imposing new restrictions on the size and number of kangaroos that can be harvested for their meat and valuable skins. Particular safeguards are being introduced for the younger breeding animals.

SIESTA TIME *Whiptail wallabies, like other members of the kangaroo family, travel in small groups. They rest during the heat of the day and feed at night on the grasses available. Bacteria in their stomachs help to digest the tough fibres*

SEARCHING FOR WATER *Water holes are vital to kangaroos in the dry interior regions. The animals dig holes with their sharp claws. To conserve vital nitrates, they do not overdrink and so do not excrete much nitrogen from the body*

SPEEDY JUMPER *A kangaroo can bound along at 25 mph for some distance, and can reach 30 mph in shorter spurts. It can leap as far as 27 ft and jump as high as 10 ft, depending to some extent on the condition of the terrain*

ROCK INHABITANT *A rock wallaby shelters by day in crevices. It emerges at dusk to bound over rocky slopes and climb trees in search of plant food. Rough footpads give the nimble marsupial a better grip when it jumps chasms as wide as 13 ft*

SYMBOL OF AUSTRALIA

Without antelope herds to graze on its dry grasslands, Australia had a vast, unoccupied area available for marsupial exploitation. When one considers the characteristics of antelope, deer, bison and other hoofed placental mammals, the marsupial response, although entirely successful, is unexpected. It is the creature that in the rest of the world *means* Australia: the kangaroo. When Europeans first discovered it, they had a hard time describing or even drawing it, so unfamiliar were the proportions. Yet a large kangaroo is a magnificent animal, admirably suited to its way of life. Few sights are more thrilling than a group of big kangaroos bounding across a plain at full speed, tails waving up and down and serving as counterbalances as they cover 20–30 ft with every graceful leap. Surely a kangaroo is one of the greatest achievements of evolutionary design.

The kangaroo pattern takes a variety of forms. There are arboreal kangaroos, wallaroos, wallabies and little rat kangaroos, in addition to the impressively large red and great grey ones. Rat kangaroos are the least specialised of all, while the tree kangaroo is a reversion to the arboreal life that kangaroo ancestors probably led before becoming ground-dwellers. Wallaroos and wallabies vary only slightly from the main members of the kangaroo family: the principal difference is in size more than anything else.

The larger kangaroos, although not predators, are formidable and often cantankerous antagonists to animal and man. When a fully grown male stands erect to a height of over 7 ft, balancing its 150 lb. on a powerful tail, it is an imposing sight. Only the males fight; the females are much smaller and less muscular. In a fight, the animal tears furiously with its forearms – the biceps are considerably larger than those of a strong man. Its characteristic method of attack is lashing out and downwards with its huge hind legs in one of the most powerful and lethal blows delivered by any living land animal.

A newborn kangaroo is ridiculously small, yet very much on its own. Guided only by instinct, the 1-in. creature must crawl from the birth orifice up the abdomen of its mother and into her pouch, a distance of a foot or more. Once inside, it descends to seek one of the nipples, which soon swells with milk inside the baby's mouth, making a secure anchoring seal between the almost embryonic infant and its mother. The tiny kangaroo has no real features yet: its eyes and ears are unformed, as are its hind legs and tail. However, its stubby front legs have minute claws to help it to climb, and since its nostrils and the olfactory portion of its brain are well developed, zoologists believe that it finds its way by scent. The mother, 30,000 times the weight of her baby, does nothing to help; there is no truth to the story that she licks a path in her fur to guide the young one. She does, however, lick whatever blood and mucus is left in the baby's wake, for at birth the infant is still connected to its rudimentary placenta by an umbilical cord. Only rarely are twins born, but if they are, the mother looks after both – a feat that seems incredible if one considers the cramped quarters of the pouch. Most of the young are born in winter; the breeding season of some kangaroos is tied to rain and the growth of plants.

Assembly-line production

The baby, known as a joey, spends much of its time in its mother's pouch for almost a year, until it weighs 8–10 lb., or occasionally even more. At the end of the year, the mother has had enough and no longer lets her joey enter; from then on it lives outside the pouch, although she still feeds and defends it if it is threatened.

A few days after a joey is born, the female mates again, but the five-week development of the fertilised egg is suspended by hormonal control until the pouch is finally vacated – many months later. With such assembly-line production, a female kangaroo is seldom without at least one joey in some stage of growth.

In the more arid regions of Australia, kangaroos have an important effect on the populations of other animals. With powerful limbs, they excavate water holes where no surface water is visible. These are then visited regularly by emus, native cats and cockatoos, as well as many other creatures. The margins of the water holes may also support short-rooted plants that would otherwise be unable to tap sub-surface waters. In their turn, these plants support other animals.

WILD AND WOLF-LIKE *The swift-footed dingo, about the size of a small alsatian, is probably descended from the Indian plains wolf. Introduced by the Aborigines about 30,000 BC, dingoes are wild and wolf-like; they howl but do not bark*

THE LATECOMERS

Reliable evidence indicates that no placentals except bats and rodents – rats and mice – lived in Australia until early aboriginal settlers brought an Asian dog – the dingo – about 30,000 BC.

The wild dingo is now widely distributed over much of Australia, in areas ranging from forest to arid country. Travelling in small packs, it prowls in search of kangaroos and other marsupials. It is also a predator of Australia's present huge population of rabbits, whose progenitors were first brought in by European settlers nearly 200 years ago.

Fruit bats (or flying foxes) have been in Australia since early times. Seals, not excluded by ocean barriers, come ashore regularly to bask or rear their young: the fur seal (once hunted nearly to extinction) and the sea lion are now protected by the government.

Despite the advent of Western man, the island continent of Australia remains the greatest living and largely unspoilt laboratory of evolution. No other island in the world supports so many truly distinctive and isolated plants and animals.

Islands born of fire

Volcanoes are one of the greatest natural hazards to life on this planet; yet they are at the same time a source of new life, creating new islands in our ever-changing world

Shortly after 7.30 on the morning of November 14, 1963, the skipper of a fishing vessel off the southern coast of Iceland trained his binoculars on a column of dark smoke rising from the sea about a mile to the south-east. As he studied the billowing black cloud, his questions gave way to suspicion, and finally to astonishment. He ordered his crew to sail closer and then called the nearest radio station. He wanted to report, he said, a volcanic eruption.

To scientists, the rising black eruption column – about 200 ft high at the time of the captain's report – clearly suggested that molten material from a fissure on the sea floor had built a volcanic mountain, and that the summit cone was already near the surface. Within three hours of the first report, scientists and journalists began arriving at the scene in ships and planes. By this time the column of smoke had attained a height of 12,000 ft. Explosions almost every half-minute were sending volcanic ash, dust and rock 'bombs' 500 ft into the air. Waves were breaking on something new just under the surface. All signs pointed to a single outcome. An island was about to be born.

That same night, the volcano's smoking cone pushed up through the waves, and by the following morning had grown to a height of 33 ft. As the explosions roared on, the flanks and summit continued to build. Five days later, the island was 200 ft high and 2000 ft long.

The Icelandic government named the island Surtsey, after Surtur, a fire demon in Norse mythology. Surtur was a powerful giant, and Surtsey has proved a worthy heir to the reputation of its namesake. When the eruptions quietened down in August 1965, after almost two years of intermittent volcanic activity, accompanied by spectacular displays of lightning, the island was 550 ft high, 6900 ft long, and a square mile in area; it continued to grow for two more years.

Born of volcanism, Surtsey was the first new known island to appear in the North Atlantic in more than 1000 years. The growing island greatly excited scientists, two of whom made a landing during a lull in the eruptions in December 1963. Now volcanologists could observe at close hand what the volcano was doing. Biologists saw the island as a totally new environment; there they could study in detail the arrival and development of life on isolated, barren land. For scientists, then, Surtsey was an unexpected bonus, an exciting new laboratory. For the rest of the world, it was a dramatic reminder that the earth is not a mass of dead rock, but a dynamic structure continuously undergoing change.

THE BIRTH OF AN ISLAND *A towering column of steam and smoke heralds the dramatic birth of the island of Surtsey, which emerged from the North Atlantic off the coast of Iceland in 1963. Within two years the island was a square mile in area, and the first hardy plants and animals had become established on its surface*

WHEN LAVA MEETS THE SEA *Clouds of steam hiss where the red-hot lava from Surtsey's volcano comes into contact with the ocean. Lava solidifies at the shore, protecting the cinders from sea erosion and binding loose material together*

LIVING ON A BALL OF FIRE

In contrast to the moving winds about our heads and the tumult of the seas, the earth beneath our feet may seem stable and solid enough. But in reality we are standing on a thin crust stretched over a seething mass of boiling metal, and even this crust is in a state of constant change. The dense core beneath it, composed largely of iron and nickel, is still close to the temperature that prevailed when the planet was formed – about 2964°C (5400°F). The core, 4400 miles in diameter, is surrounded by a zone called the mantle, nearly 1800 miles thick; together they account for most of the earth's volume.

Early in the history of the earth, heat from the interior was radiated into Space, until the final layer of lighter-weight rock, the crust, solidified as a crystalline skin floating on the mantle. The earth is a sphere 8000 miles in diameter, and its crust is remarkably thin – thinner proportionately than an eggshell, since under the oceans it is only 5 miles thick and elsewhere about 30 miles. Stretched, wrinkled, torn and pushed about ever since its formation, the crust has developed all kinds of irregularities and weaknesses.

Beneath the crust, mantle rock (or magma) is still hot and under great pressure; many geologists believe it is plastic enough for slow convection currents to pass through it, carrying superheated material up towards the surface, where it cools slightly and sinks again. In some places, the magma becomes trapped in a reservoir or magma chamber pressing up into the crust, where the weight of overlying rocks is less confining. As the magma rests in its chamber, some of the minerals in it crystallise, and the freed gases expand. When the pressure on the surrounding rock becomes intolerable, the crust above the chamber breaks to form a conduit or passage, and material from the chamber is ejected in a volcanic eruption.

Nearly every land mass has remnants of ancient volcanism. Of the volcanoes in the world today, about 500 are known to have been active within the span of recorded history. Sixty or seventy million years ago, a stupendous amount of volcanic activity throughout Europe covered thousands of square miles with lava and ash. There is no reason to doubt that volcanic eruptions will continue into the future, with the rate of activity fluctuating over time. The fires of the earth burn on, and the moulding of its crust is far from finished.

HOW VOLCANOES ERUPT

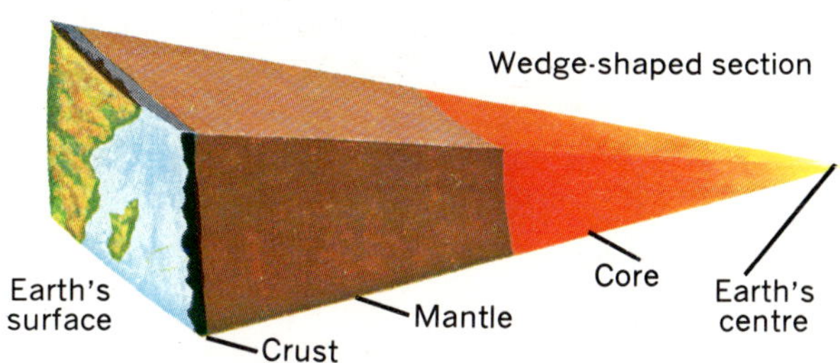

The earth's interior comprises two main regions: the central core, 4400 miles in diameter, and the mantle, 1800 miles thick. The crust is about 5 miles deep under the ocean floor and no more than 30 miles elsewhere

BIRTH AND DEATH OF A VOLCANIC ISLAND

A volcano on the sea floor begins where liquid rock from the mantle rises through a weak spot in the crust

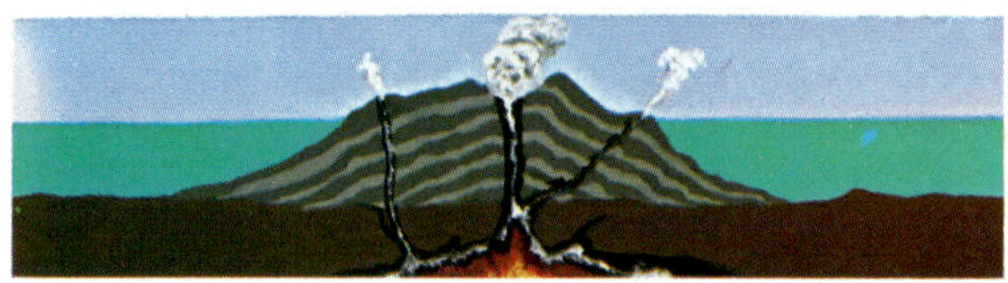

A volcano that continues to erupt and build until it reaches the surface of the ocean becomes a volcanic island

A high island with a peak several thousand feet above sea level may have a life span of millions of years

In time, the weight of the island pushes the sea floor down, and winds and waves erode the peak to a plain

When the island disappears below the sea, it is called a guyot; shallow-water marine life continues around it

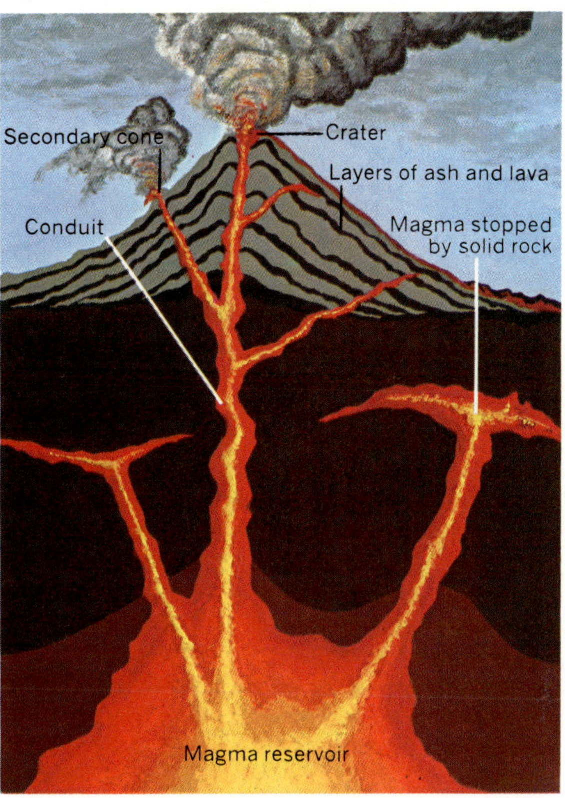

Convection currents carry molten magma up through the earth's mantle. Where the crust is weak, the pressure of expanding gases in the magma causes a volcanic eruption. The magma may not escape to the surface through every conduit it makes: if it is trapped by underground solid rock layers, it forms granite or basalt

TYPES OF VOLCANIC ERUPTION

HAWAII *Lava with little gas flows quietly without explosive eruptions*

STROMBOLI *Gaseous lava, hurled into the air, freezes into bombs*

MOUNT PELÉE *Lava with more gas than solid material becomes froth*

VESUVIUS *Lava with high gas content erupts in clouds of pumice and ash*

SEETHING VOLCANIC CONE *Lava bubbled in Surtsey's volcanic cone soon after it emerged from the sea in 1963. For almost four years, rivers of molten rock continued to flow intermittently from cracks below the rim of the crater*

CREATION OF A CRATER *In some eruptions, as the showers of lava fall back from the sky to build the volcanic cone, repeated explosions from the conduit keep its opening comparatively clear. The result, when the lava solidifies, is a sunken crater*

THE RISE AND FALL OF A VOLCANIC CRATER

From a bowl of lava to a plant-clogged lake

In the fantastic time-scale of this planet's evolution, all volcanoes and volcanic islands are probably transitory. The birth and growth of a volcanic island may be spectacularly rapid; its decline through the long processes of erosion may be so slow as to be imperceptible. The eruption usually leaves in its wake a bowl-shaped depression, the crater, at the top of the conduit. Later, other eruptions may form small cones in the old crater. But gradually all volcanic activity ceases. Rain may fill a crater lake; vegetation, followed by an animal population, may fill the crater. In the end, slowly but surely, the crater is eroded.

RAIN-FILLED CRATER *Long dead, the crater of Rano-Raraku volcano on Easter Island in the Pacific holds a shallow lake of trapped rainwater. The lake, which has neither inlet nor outlet, is being invaded gradually by vegetation spreading in from the rim*

ERODING VOLCANO *Lehua volcano, probably an offshoot of the nearby larger Hawaiian island of Niihau, has been worn away into a crescent shape by the action of the sea. Lehua is a cone of ash and volcanic glass fragments, which erode easily*

THE PACIFIC'S 'RING OF FIRE'

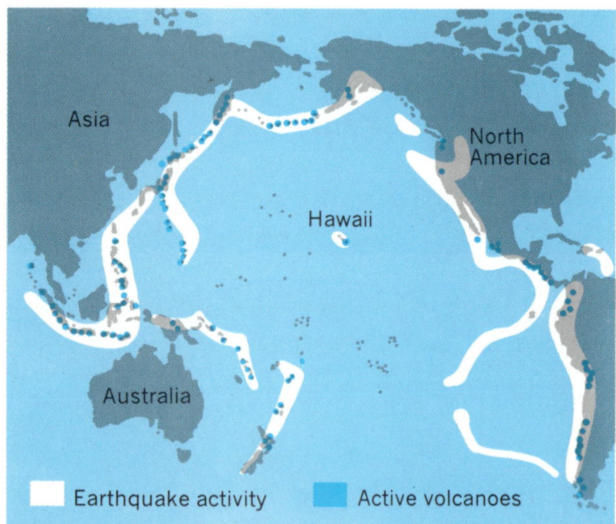

GIRDLING THE OCEAN *A circle of volcanoes – active and dormant – skirts the edge of the Pacific Ocean, along the major fault lines in the earth's crust. This 'Ring of Fire' zone is the world's main area for earthquake and volcanic activity and it includes scores of volcanic islands*

The Pacific Ocean is ringed by fire – a great chain of volcanoes. Most of the large islands along the western rim are probably continental rather than volcanic, but many contain volcanoes; and nearly all the smaller islands further from Asia are of volcanic origin. Some are exposed summits of volcanoes; others are atolls, coral reefs built on the flanks of volcanoes that long ago sank back into the sea or were eroded to below sea level. A ring of coral islets may surround a volcanic cone.

The small Pacific islands appear in greatest profusion to the south and west. Clustered in groups, strung out like beads in gently curving chains, or aligned in straight rows, they form patterns that correspond to areas and lines of

MENACING FOUNTAIN OF FIRE *The village of Kapoho (foreground) was eventually destroyed by the 1960 eruption of Kilauea on Hawaii. The quick-flowing lava added 500 new acres of land to the island's coast*

structural weakness in the floor of the Pacific.

Most of the 10,000 volcanoes in the Pacific are hidden beneath the sea and no longer active. The undersea volcanoes are distributed throughout the entire ocean. A considerable group lies under the Gulf of Alaska. One recently discovered submarine volcano known to be active today is 750 miles south-east of Tahiti in a straight line with Neilson Reef, Rapa and Morotiri. Ten thousand feet above the sea floor, its summit still lies 1500 ft below the surface of the ocean, and it continues to build upwards. Such volcanic sea-mounts – huge underwater mountains that have never broken the surface – are characteristic of this submarine world. Other undersea mountains of volcanic origin, with flat tops, are called guyots. Once visible islands, they were planed level by water erosion and later became submerged. In a sense, both sea-mounts and guyots are 'islands'. If their summits are close to the surface, they offer isolated, relatively shallow, dimly illuminated environments for marine life, and each one is the centre for a small marine community.

Wrath of the gods

Since ancient times, man's terror and awe before the spectacle of an erupting volcano has been reflected in myths and folklore. Between the 5th and 8th centuries BC, Mount Etna in eastern Sicily figured prominently in Greek mythology as a site of the gods' activities. In all ages and places, a volcanic eruption has been taken as a sign of anger among the gods. After an eruption in AD 1104, Icelanders believed their great volcano, Hekla, was the entrance to a hell populated by furious demons.

No two volcanoes are alike. Each is an individual, differing from others in structure, eruptive activity, composition of ejected materials and life cycle. Despite such variations, however, all volcanoes obviously share one common feature: the passage and ejection of mantle material from the interior of the earth through a conduit and out of a vent.

When ash, rock and gas spew forth from a vent, differences in volcanic activity become apparent. Some eruptions are quiet, massive and long-lasting. Others are sudden and so cataclysmic that they dwarf all other known earthly events. Mauna Loa and Kilauea, active volcanoes on the island of Hawaii, grow slowly and predictably, but Bandai-san, dormant in Japan for more than 1000 years, gave only two minutes' warning before it blew apart in 1888. Most eruptions fall some-

where between the two extremes, with days or weeks of ominous preparations, followed by dramatic ejections of incandescent gas and dark ash, showers of rock 'bombs' and, most important of all, great flows of lava.

A few volcanic islands, such as Bogosloff Island north of the Aleutians and Ilha Nova in the Azores, do not always remain above water, but appear and disappear – then bafflingly appear again. If a cone composed of loose cinders is formed, the sea may erode it away before lava has had a chance to consolidate it. The permanence of Surtsey was much in doubt until a flow of lava began four-and-a-half months after the island first appeared. Syrtlingur, a nearby islet that had emerged as Surtsey quietened down, failed to produce lava and disappeared after only a few months of existence.

SHORT-LIVED ISLET *Volcanic material spouted hundreds of feet into the air in 1965 to form Syrtlingur, less than a mile from Surtsey. The islet was 37 acres in area when its ash and cinders were swept away by a storm*

HARNESSING THE EARTH'S ENERGY *A geothermal power plant has been built near Wairakei in New Zealand to tap the underground reservoirs of volcanic steam, and so produce nearly 10 per cent of the country's electric power*

NATURAL LAUNDRY *Hot springs, boiling mudholes and geysers – like Pohutu, which spurts steam 60–100 ft high almost continuously – abound in New Zealand's Whakarewarewa Thermal Region. Maoris wash clothes in hot springs*

LIVING UNDER AN OMINOUS SHADOW

Volcanic processes are often disastrous, but they may sometimes benefit life. For thousands of years, the magma reservoir beneath a volcano continues to heat water that seeps from the surface to the chamber zone. The mixture of steam and other gases with heated water may then issue at the surface as hot springs and steam sprays, called fumaroles, and sometimes as geysers. The warmth from such springs and fumaroles provides opportunities for plant and animal life to exist in otherwise unfavourable latitudes. In New Zealand, Iceland and Italy, subterranean steam and hot water are major sources of heat and electrical power. Several other nations are planning to tap the earth's heat energy.

Although volcanic soils look sterile and hostile to life, they are rich in basic plant nutrients and these eventually encourage the growth of lush vegetation. Unfortunately, the richness of volcanic soils is indirectly tied to an extensive loss of human life. From ancient times, men have cultivated the slopes of volcanoes and the valleys below, since they are among the most fertile areas on earth. Sizable populations on Stromboli and on Mount Etna in the Mediterranean illustrate the willingness of men to settle close to active volcanoes despite awareness of the ever-present danger.

The same kind of clustering has resulted in devastating human disasters throughout Indonesia and in the area around Mount Taal in the Philippines. Taal has had at least 18 major eruptions since 1572 and it is believed to have killed two out of every three men, women and children who have made their homes in nearby

EXPLOSIVE SHADOW *The menace of a volcano looms over the island of Stromboli off Italy. For more than 2500 years, ash and lava have flowed from the volcano. Every few years pressure builds up to an explosive level*

villages. Yet the area around it is regularly recolonised after each eruption has run its course. Even on the flanks of Mount Etna, Europe's most continuously active volcano, farmers and vine-growers returned to salvage what they could after the eruption in April 1971.

Volcanic eruptions can affect established life on islands in two fundamental ways. First, they may add to or subtract from an island's mass and land area. Second, they may destroy part or all of an island's plant and animal life. It is not only direct contact with volcanic material and gases that causes destruction of life. Hurricane-force winds resulting from the violent explosion in 1815 of Tomboro, east of Java, pulled whole trees, large animals and houses straight up into the air. A huge sea wave (tsunami) produced by an eruption may be even more destructive.

Two of the most dramatic geological events of the last hundred years were the terrible eruption of Mount Pelée on Martinique in the Antilles, and one of the greatest volcanic explosions known, that of Krakatoa in a strait of the Malay Archipelago. When the side of Mount Pelée burst in 1902, it set free an enormous, dense cloud of mud, ash and incandescent solid matter. The fiery cloud swept down a broad valley at estimated speeds of up to 180 mph, within minutes destroying everything in its path, including the city of Saint-Pierre. Only bare land was left in its wake. Of the 30,000 people in the city, two survived – one at the very edge of the catastrophe, the other in a completely enclosed dungeon where he was awaiting execution. Despite the vast destruction, plant and animal life from the margins of the area began to invade the barren soil in the months that followed, and in only a few years the whole region was again clothed in vegetation and populated by animals.

EXPLOSION THAT SHOOK THE WORLD

The world has seldom experienced so terrifying an explosion as the one that destroyed Krakatoa in 1883. Possibly the most tremendous eruption in recorded history, it occurred in a region where a great rift in the ocean floor crosses the curve of the Malay Archipelago at the Sunda Strait, between the two large islands of Sumatra and Java. Here, truly, *X* marked the spot. In the strait lie a number of volcanic islands, one of which, Krakatoa, had been formed by the slow coalescence of several smaller islands and craters.

In May 1883, the residents of the large islands in the chain heard explosions and saw 7 mile high columns of pumice and ash coming from Krakatoa. The eruptions continued all summer, growing increasingly violent. By late August they had become a series of continuous roars, audible hundreds of miles away.

Finally, on the morning of August 27, came four explosions that no one near by lived to tell about. In the third, most of the island, including the two northern cones, one 1400 ft high, vanished leaving a 1000 ft hole beneath the surface of the sea. As the summit cone collapsed, the sea engulfed 2 cubic miles of rock. Giant tidal waves – one reaching points on coastal hills as high as 135 ft above sea level – thundered out across the strait, destroying all seaside villages on neighbouring islands and carrying ships with their dead crews miles inland. The waves travelled far enough to raise the water level in the English Channel.

Ash and gases rose many miles into the sky, spreading total darkness for 150 miles in every direction. Nearly $4\frac{1}{2}$ cubic miles of volcanic material had been hurled into the air. Sounds of the explosion were heard clearly over one-seventh of the earth's surface; as far away as Rodriguez Island, 3000 miles distant in the Indian Ocean, residents thought great naval guns were firing near by. For two years, dust in high-altitude jet streams created the most brilliant sunsets, sunrises and twilight glows ever seen.

The disaster brought death to more than 37,000 human beings on islands throughout the region,

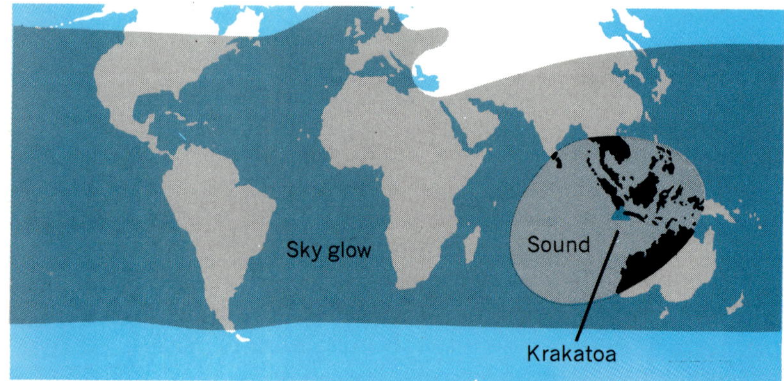

Effects around the world

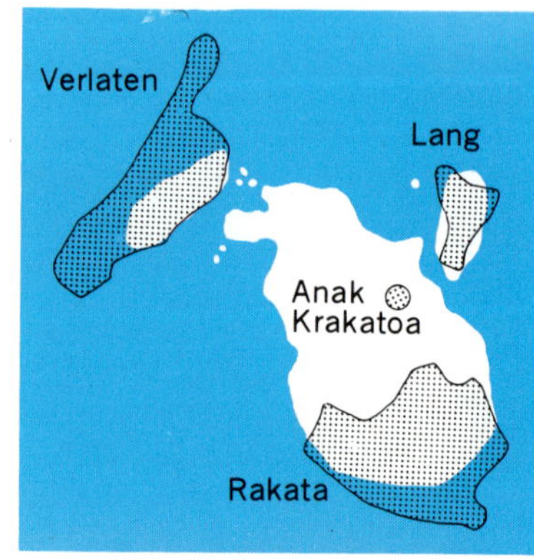

How Krakatoa changed

When the island of Krakatoa in Indonesia exploded in 1883 – with the force of a million H-bombs – the sound travelled 3000 miles in all directions; the seismic wave raised the sea level more than 12 in. in South Africa, 4550 miles away, travelled round Africa and was still a noticeable 2 in. surge in the English Channel. Airborne dust particles, reported to have reached a height of 17 miles, gradually spread to cover more than 135 million square miles, and unusually vivid sunsets, caused by the fine dust, were seen throughout most of the world during the following two years

Prehistoric Krakatoa is believed to have been 6000 ft high. The 1883 eruption shattered most of the island but created some new land (dotted areas). New volcanic activity in 1927 gave rise to Anak Krakatoa – the child of Krakatoa – in the centre of the three present-day islands

some of them hundreds of miles away. The loss of plant and animal life was incalculable, but certainly, that morning, every living thing on the fragments of Krakatoa and on the small neighbouring islands, including worms and roots deep in the soil, ceased to exist. For a while Krakatoa, or what was left of it – the islands of Rakata (also called Krakatoa), Verlaten and Lang, as we know them today – was as lifeless as the earth had been 4000 million years before.

The rebirth of life

The first scientific visitor to Krakatoa after the eruption was a botanist who carefully went over the island in the autumn of 1883 but was unable to find a single living thing. A few months later, another investigator found the first recorded animal: a tiny red spiderling just arrived on its silken-thread balloon. Instinct had already led it to spin a web on the barren ground, despite the absence of prey that it could catch.

The first summer after the eruption, a few shoots of grass appeared. In 1886, botanists found 34 species of plants, including blue-green algae, mosses, ferns and different seed plants. After this, colonisation by plants progressed steadily. In 1897, the total number of species had risen to 61, and in 1906 to 108. By 1928, some 276 species of plants of all kinds were present on the islands.

All the plants found by the first collectors belonged to varieties dispersed by winds or ocean currents. None, of course, could have grown from the rootstocks that existed before the eruption and were now all dead – buried as deep as 260 ft beneath the new surface. The early plants were grasses, herbs and flowering shrubs; low-growing forest plants could not get a footing until trees had taken hold and established the necessary dense thickets. Not until 14 years after the eruption did plants grow from seeds carried by bats or birds, for until then the island had not attracted animals.

Some of the plants that arrived soon after the eruption flourished in surprising ways. *Cyrtandra*, normally rare in the Javanese forests, formed a dense scrub covering the upper slopes. As this occurred, mosses, ferns and orchids began growing abundantly in the branches. Then trees invaded and took over, reducing later generations of *Cyrtandra* to its usual subordinate role of small shrub. As a result, the mosses were again limited to their normal, rather sparse populations.

Well established today, less than a century after the great eruption, the plant communities of Verlaten, Lang and especially Rakata show signs

BARREN NEW ISLAND *Nearly 50 years after its creation, the new island of Anak Krakatoa, in the centre of the old crater, is still almost completely barren. The remains of Krakatoa, called Rakata (bottom) has made the best plant recovery*

of becoming the climax forests that grow throughout the Malay Archipelago.

Like the pioneering plants, the first animals to resettle Krakatoa were those carried by wind and sea. Hordes of butterflies, as well as the first spider, starved to death. But 14 years after the eruption, 132 species of birds and insects were found on the island. Fifty years after the event, the total number of animal inhabitants was still only about two-thirds of that normal for islands in the region.

Two weeks after the new island of Surtsey rose in 1963 from the sea south of Iceland, seabirds and seals came to rest on its still-warm shores. Undoubtedly, many kinds of birds stopped also at Krakatoa within days after its eruptions subsided; unless they found food in the surrounding waters, they must have soon departed for the islands on either side of the Sunda Strait. Although their brief visits may have brought occasional seeds and spores, the birds probably contributed little to establishing an animal population. Eggs of snails, worms or insects may have hatched after arriving on the island attached to their feet or feathers, but without plant food they could have had little future on the barren land. The animal life of Krakatoa is unlikely to regain its former richness until long after the vegetation has been restored to its former lushness and variety.

The sequence of animal colonisation of Krakatoa is not well recorded. There was the spiderling found among the cinders, and scientists soon afterwards observed the arrival of flying insects, which left or died. In 1908, no earthworms were found but 20 years later, a population of earthworms had become well established on Rakata; their forebears probably floated across the straits on a 'raft' of vegetation.

Land snails had been numerous on Krakatoa before the eruption, but six years later the only snails were marine species living along the shore. Today, land snails of many types again flourish on Krakatoa, as do many species of ground insects, not all of them strong fliers.

First animals

The first backboned animals to resettle Krakatoa seem to have been a monitor lizard and a gecko. Monitors swim well, and the distances through the straits are short enough to be negotiated without difficulty. Gecko eggs could have come on driftwood.

Forty years after the extinction of all life on Krakatoa, almost 600 animal species had become established, including a python (also a good swimmer), a crocodile, lizards, more than two dozen resident species of birds, three bats and a rat. Eighty per cent of the animal inhabitants on the fragmented island had wings, and a further 10 per cent were species that had arrived by air, even without wings. Those animals that reached one of the islands and survived usually increased prolifically before settling down as part of a more balanced community. A species of lizard that arrived in the early 1920's was extraordinarily abundant by 1924; in 1928, one species of rat swarmed on Lang Island.

Whether an immigrant animal took hold or not depended largely on the availability of food. Scavengers able to survive on dead plant and animal material, scarce as it may have been, probably were the first to become established. Second were the grazers, which fed on the increasing vegetation. Last came predators and parasites.

Krakatoa illustrates how an island denuded of life is repopulated by a succession of accidental arrivals. Pioneers establish themselves and set the stage for more and more complex communities, until a stable climax is reached. The earliest arrivals may be unseen or scarcely noticeable: bacteria, tiny spiders, insects, fungi, lichens. But the pace of development quickens.

TWO KINDS OF LAVA *Weathered and dun-coloured, blocks of chunky aa lava (foreground) contrast with the billowy black smoothness of a pahoehoe flow behind. Aa is more crumbly and can support plant life more quickly*

FIERCE HEAT AFTER ERUPTION *The walls of a Hawaiian lava tube glow at about $815°C$ ($1500°F$). Lava tubes form when the surface of a lava flow solidifies and the liquid lava drains away. Lava caves, open at one end, have the same origin*

HAWAII – HISTORY WRITTEN IN LAVA

Despite shortcomings of early investigations, the pattern of Krakatoa's resettlement is fairly clear. Far greater mystery surrounds the original arrival of life on the Hawaiian Archipelago. This great chain of volcanic islands is thousands of miles from the nearest continent and hundreds of miles from other islands. Because the older islands of Hawaii appeared over 10 million years ago, no accurate reconstruction of the sequence of plant and animal colonisation is possible today. Fossils of land forms are extremely rare, and since men did not settle on the islands until 1000 years ago, not even an oral history of the early years is available.

Nevertheless, by observing what occurs on a new lava flow at high and low altitudes, and under water, scientists have learnt much about the way colonisation takes place. Such new flows are continually available on the large island of Hawaii, which is less than a million years old and still growing to the south-east. For example, the destructive 1960 eruption near the shore at Kapoho, a seaside village on the flanks of Kilauea volcano, added nearly 500 acres to Hawaii's land mass.

Most magma – the mantle substance lying deep within the earth, composed of plastic rock and gases under great pressure – emerges as lava, which may flow in either of two distinct forms. The forms have been given Hawaiian names: *aa* lava is rough, chunky and very dense; *pahoehoe* lava is smooth and billowing. The kind of lava depends on the quantity and condition of gases in the rock as it begins to cool. Aa, with its multitude of crevices, encourages the relatively early establishment of life; pahoehoe, until it weakens and cracks, provides few places where roots can penetrate or insects hide.

The 1960 Kapoho eruption produced both kinds of lava, but especially vast fields of aa rubble. When the river of crumbling aa reached the sea, it spread out into a fan and poured down the submerged slope over old rocky shores and coral reefs. The lava fan created a new shoreline of rough and initially lifeless rocks and low cliffs that cooled quickly in the water.

315

BEAUTY BORN OF FIRE AND WATER
Hawaii's strange volcanic heritage

Many millions of years ago, immeasurable volcanic forces first began to erupt and pour out molten rock along a rift in the Pacific floor to start the formation of what is now the 1000 mile long chain of the Hawaiian Islands. The volcanic activity ran from west to east: the western islands, their volcanoes long extinct, have been eroded by the sea; Hawaii – the easternmost, newest and biggest island – still has active volcanoes. If the Pacific were drained of water, the twin volcanic peaks of Hawaii, Mauna Kea and Mauna Loa, would stand revealed as earth's highest mountains – rising to more than 30,000 ft from the sea-bed. Over the ages, dense vegetation has sprung up and living things have colonised these islands born of fire. At the same time, rain has worn deep ridges and rivers have cut channels. Together, fire and water can be seen to have shaped some of the most beautiful of all the island landscapes in the world.

CRATER FORMED BY EROSION *The crater of Haleakala on Maui Island, near Hawaii, was carved by erosion, not volcanic activity. The volcano was active until 1790*

LUSH VEGETATION *Scarlet gingerlilies bloom near Rainbow Falls on Hawaii. The rich volcanic soil of the Hawaiian Islands supports a profusion of plants, and the heavy rainfall fills great rivers which plunge down many waterfalls on their way to the sea*

RAIN-GOUGED GROOVES *The steep sides of Napali Valley on Kauai island have been scarred over the centuries by rain. On the gentler slopes, the soil supports plants like the red ohia lehua*

BLACK SAND *Surf and wind have reduced black lava to sand on beaches such as this one, at Kalapana, Hawaii*

FORCES OF NATURE *The steep cliffs and rugged depths of Waimea Canyon on Kauai were created by several weathering forces: rainfall, diverted streams, water seeping into rocks*

THE FIRST SIGNS OF LIFE'S RETURN

The mountain above Kapoho on Hawaii is one of the most active volcanic zones in the world. Near the centre of the island, Mauna Loa rises from a base 18,000 ft below sea level to a peak more than 13,000 ft above the surface. The large oval caldera (crater) at its summit is 3 miles long. Some 10,000 ft down from the summit, on the south-east slope, is Kilauea, currently more active than Mauna Loa itself, with a caldera $2\frac{1}{2}$ miles long. Its magma chamber supplies not only Kilauea but also several smaller vents, including that of Kapoho on the shore. Most lava flows from this reservoir originate at or near Kilauea rather than at Kapoho.

In 1959, a few months before the Kapoho eruption, volcanologists, who had detected a swelling of Kilauea, were expecting an eruption from its fire pit, Halemaumau. Instead, Kilauea Iki, a smaller adjacent crater, began one of the most dramatic displays in Hawaiian history. Incandescent fountains of ash and cinders rose nearly 2000 ft into the air, where trade winds caught and blew them as far as 2 miles away. When the eruption subsided, the floor of Kilauea Iki hardened into a massive crust of pahoehoe.

Immediately afterwards, scientists began studying the lava, the enormous cinder cone just down-wind of the crater, and all the regions of diminished fall-out within a 2-square-mile area, which ended where ash was only 1 in. deep. These investigations are among the most thorough ever undertaken. They concentrated on the development of soils in which life gradually appears, and on the nature and interaction of that life, both plant and animal.

Plant-supporting lava

On bare continental granite and other ancient surfaces, hardy lichens are always the first pioneers. This is not necessarily the case on ejected volcanic materials, which may contain many basic plant nutrients and thus be able to support less tolerant plants. The first plants to become established on the floor of Kilauea Iki crater were certain algae (mostly blue-green algae). Mosses followed, then a native fern and, finally, hardy native Hawaiian seed plants. Although lichens did appear in the early stages, they did not seem to play a substantial role in helping other plants to become established. The ferns, especially sword

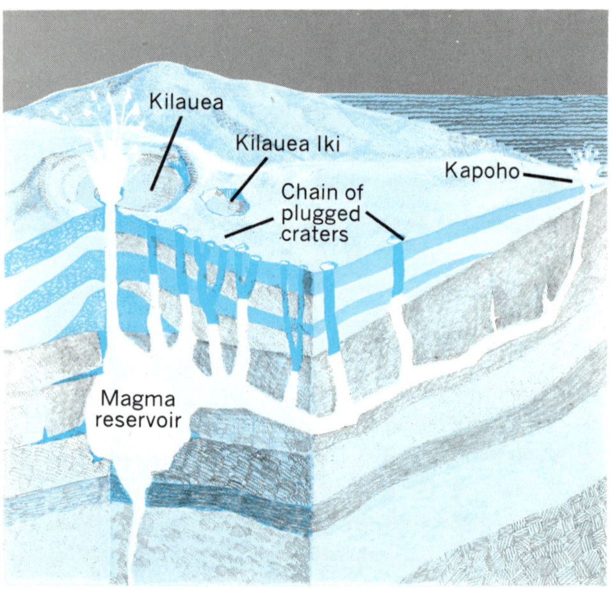

CASCADES OF LIQUID FIRE *The 1959 Kilauea Iki eruption set a 1900-ft record for Hawaiian lava fountains. The eruption lasted for about five weeks and poured nearly 160 million cubic yards of lava over the island's surface*

MOST ACTIVE VOLCANO IN THE WORLD *Kilauea, on Hawaii, has erupted frequently in recent years and is now the subject of much of the world's volcano research. Its complicated structure includes several craters connected by underground conduits*

WEATHER-RESISTANT LAVA *The pahoehoe lava floor of Kilauea Iki crater appears lifeless ten years after an eruption, contrasting with the lush vegetation round the rim. Pahoehoe is slow to become soil in which plants can grow*

FIRST PLANT LIFE *As the pahoehoe lava gradually cracks, small ferns take root in the mineral-rich lava. When the pioneer plants die, their remains add to the organic debris collected in the cracks and help other types of plants to colonise*

fern and two or three others, were able to begin growing soon after lava or cinders cooled; in one spot, tiny forms of some of these ferns appeared only six weeks after the eruption.

A visitor to Kilauea Iki today sees light yellow and white lines and spots on the massive pahoehoe lava crust where hot steam rising from cracks has altered the basic composition of the basalt, principally through oxidation. Here new plant life is more prolific, and blue-green algae are able to invade such cracks, called fumaroles. As the crater cools, many kinds of more highly developed pioneer plants appear. The invasion of vegetation from the margins is slow, for the surface at the centre – at 77°C (170°F) – is still too warm for any but the simplest life. Not only does the black lava absorb the sun's heat, but temperatures only 80 ft beneath the crust are still approximately 1093°C (2000°F).

The same succession of pioneer plants occurs on the cinder cone. Steam escapes through the cone's structure of loose material, especially from the great cracks caused by settling. As the emerging gas becomes mostly hot water, chemical alteration of the loose rocks and cinders proceeds more rapidly, permitting the growth of ferns, mosses and lichens. A little further away, well down the slope, seed plants have begun to take hold.

Other craters near by that have not erupted for a century or more show small trees spaced out across their floors, almost as in a well-planned orchard. In cooling, the floor surfaces cracked into a network of polygons, each crevice becoming a place for lava to alter into more usable soils and for organic matter to collect. With the trees spaced out as they are, there is little competition between individual plants. Indeed, in the dry lava and cinder deposits of the Kilauea region, there seems to be little or no competition among the uncrowded plant and animal species.

Trees which survive in the path of molten lava

Because the Hawaiian Islands have been volcanic from their very beginning, many of the first plants to invade new surfaces are hardy native species already adapted to volcanic soils. Non-native plants often follow soon after and may flourish for a while, but in the long run some of them seem to lose their vigour and are replaced by native forms.

The native trees of the Hawaiian Islands are extremely durable. After an eruption, the ohia lehua tree, unless it was completely buried by a shower of burning cinders, usually makes a remarkable recovery. If a tree's conducting tissues on the side towards the volcano are destroyed, new shoots often appear on the sheltered side and take over the production of food for the entire tree. Another whole root system grows just beneath the new surface, which may be several feet above the old, and often this is enough to ensure survival. The injured tree grows at a phenomenal rate, soon covering over the burnt side and sometimes in as little as six years nearly doubling its circumference. Trees or shrubs which remain stunted produce heavy crops of fruits and seeds. Plant hormones undoubtedly induce both the active growth and the seed production, perhaps stimulated by the holocaust which the tree has experienced.

Trees killed in an eruption may contribute in several ways to the renewal of life. New lava and cinders contain nearly all basic plant nutrients, but they lack one nutrient necessary for life: nitrates, which normally take some time to accumulate from dead organic material and bird wastes. Where lava has streamed over a forest, however, the burning of trees may have introduced an appreciable amount of nitrogen into the volcanic materials, enabling plants to invade more quickly than would otherwise be possible.

If a burnt tree remains standing, it serves as a

A TREE THAT WITHSTANDS FIRE *An ohia lehua tree catches fire as lava flows past. The tree has remarkable powers of recovery. If one side burns, shoots from the other soon cover the damaged area; if it dies, it contributes nitrogen to the lava*

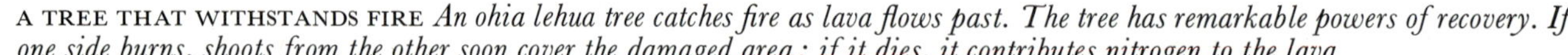

channel for rain and condensed fog, and invading plants cluster thickly at the foot of the stark skeleton. Ferns, mosses and other plants may grow under the length of a fallen tree, especially at points where water drips on rainy or damp days. If a tree is completely incinerated, a tubular lava mould may be left standing, the inner cavity – with its shade, moisture and collected nutrients – often sprouting a luxuriant growth of ferns and other similar pioneer plants.

In the short time that has elapsed since the 1959 eruption, only small forms of animal life have re-colonised Kilauea Iki. Among the earliest settlers were minute one-celled organisms, which appeared in the water-holding cavities of porous lava and cinders along with their food plants, bacteria and blue-green algae. As at Krakatoa (and Surtsey), the first visible land animals were tiny spiders. Ballooned to the region as airborne spiderlings, they immediately constructed webs, which caught newly arrived flies and organic matter blown across the wasteland by trade winds.

Land snails from unaffected neighbouring terrain are now found on the lava around the periphery of Kilauea Iki. Large wolf spiders live in the loose cinder deposits, where they hunt on the surface at night. During the day, they seek moisture and protection inches down in the loose material. Here they may come across food in the form of mites, small insects and other creatures. Many of these smaller animals feed upon the algae and fungi that grow an inch or two beneath the rough cinders on the surface.

Islands of the future

The rewards of observing plant and animal re-colonisation of volcanic terrain do not end at simply seeing what appears first. The emergence of materials from deep within the earth, their slow chemical alteration into soils capable of supporting life, and the appearance of bacteria and blue-green algae are events that resemble what happened on the primordial earth. To study the ways highly developed plants and animals invade and settle a totally new and barren environment is to be struck anew by the awesome persistence of life.

Scientists have only inexact estimates of the number of volcanic islands that have existed in the past, and an incomplete understanding of how they were populated and used as stepping-stones for the dispersal of life to other islands. Our sense of history and time is as limited as our knowledge in this difficult field. We tend to think of ourselves as at the end of a series of events, when it is

LIFE RETURNS *A scarlet-flowered ohia lehua seedling sprouts at the base of a dead tree. Water seeps from a 'reservoir' in the hollow trunk, which also gives shade; the burnt tree's ashes give nourishment for plant growth*

more likely that we are somewhere in the middle.

It seems evident that the earth is far from complete. The continents continue to drift and to create new stresses and weaknesses. Volcanologists see persisting signs of opening rifts, underwater volcanism and vast alterations of the sea floor. Stretching into the future are millennia of continuing volcanic events that will bring new islands rising from the oceans. As they emerge and cool, these islands of the future will be populated by plants and animals according to the same means of dispersal, colonisation and succession that have existed through past ages.

Standing at this moment of time, we may look in either direction, wondering what lived on ancient guyots now drowned 1000 ft under the sea, and what will take hold and grow on some garden island 1 million or 10 million years from now. We may wonder, too, whether men will still be here to see the dramatic events of the future. Will our descendants still be able to watch in awe as new islands thunder up from the deeps?

The growing islands

*Not all tropical islands are the direct products of geological forces.
Billions of tiny sea organisms build islands of coral, and a single seedling
in shallow waters may mark the beginnings of a mangrove islet*

Most islands either separate from continents or roar up out of the depths as volcanoes, but some are created slowly and quietly by living island builders: corals, coralline algae and mangroves. Thousands of years ago in the western Pacific, Micronesian and Polynesian voyagers, possibly impelled both by population pressures and by their questing spirit of exploration, sought out such islands. In recent centuries, island folklore, Admiralty reports, literary narratives and films have intrigued Western man, luring him to the South Pacific with images of high, cloud-wreathed volcanic islands and coral islets lying low in the dark blue tropical ocean.

Although some individual species of the coral family can be found in many of the world's seas and oceans – as far north as the cold waters of Norwegian fjords and off the coasts of Devon and Cornwall – it is only in the warm equatorial waters of the Pacific that the reef-building species survive and die. The resulting underwater cities of spectacularly coloured coral rock, inhabited by myriad shoals of fish, sponges and underwater plants, are among nature's most amazing sights.

Most of the world's coral structures are under water, but occasionally a reef grows to such an extent that at least part of it emerges from the sea as an islet. By far the most common of all such coral growths is the atoll – a chain of islets, partly or wholly encircling a shallow sheltered lagoon.

Dissimilar though they seem, the atoll and the high volcanic island, also typical of the Pacific, both owe their origins to the same violent geological forces pushing up from the earth's hot, molten centre. For lying beneath every atoll and coral reef is an ancient hidden volcano. Because coral cannot live and grow in water deeper than about 150 ft, its development is restricted to shallow water – or at least to areas where the water was once that depth. Hundreds of thousands of years ago, the first coral colonies established themselves on the sides of small volcanic islands and started to grow upwards and outwards. Gradually through the centuries as the earth settled and changed, the original volcanic base subsided slowly downwards while the coral kept growing, an inch or two a year, up towards the surface of the sea.

The Pacific reefs and atolls are highly complex assemblages of organisms. Each is formed from the skeletons of innumerable tiny animals, living and dead, called stony corals or polyps. Dozens or even hundreds of different species may have added their skeletons over the centuries to the building of any one reef or coral island.

AN ISLAND GROWS *Living coral grows on the shoreline of a rocky island in mid-ocean and slowly over the centuries builds wide fringing reefs. Winds and waves erode the corals to form sand, which accumulates as beaches or even as new sandy islands. Seeds and seedlings will soon arrive, carried by wind, sea and birds*

KEEPING PACE WITH SEA-BED CHANGES

No reef or atoll is ever complete. The life, growth and death cycle of the coral animals is never still, and the coral community's plant life continues to develop year by year through the centuries. Although these islands seem to change little in size and character, they grow constantly below and above the surface of warm tropical water.

The creation of an atoll begins soon after a volcanic island rises above warm tropical seas. Coral larvae swimming in the plankton settle on the flanks of the island, where they form colonies. Upon the limestone skeletons of old corals, a fringing reef is built over the years. Hugging the island closely, the reef grows outwards from the shoreline, its leading edge harbouring the densest and most varied populations of coral, as sand and silt fill the old reef area left behind. By making waves break far out from shore, the developing

reef helps to preserve the island from erosion. Since the reef also grows upwards, rising near enough to the surface to be exposed at low tide, it effectively increases the area of the island. It makes a direct contribution to the shore when storm waves tear off chunks of coral and cast them upon the beach, or when the sea level drops over long periods of time and wide limestone benches become exposed.

If the sea floor on which the huge conical mountain rests begins to subside, the peak that forms the island, together with all the living things on it, sinks also. This process continues for hundreds of thousands of years, and if the descent is gradual enough, coral growing upwards at about an inch a year can keep pace. The distance between the shore and the reef slowly widens. Meanwhile, weather and waves work

HOW A CORAL ATOLL MARKS THE SITE OF A VANISHED ISLAND

Living coral grows in warm, relatively shallow waters, yet some reefs and atolls extend many hundreds of feet down into the ocean. Charles Darwin discovered the reason: coral grows around the fringe of volcanic islands which later sink

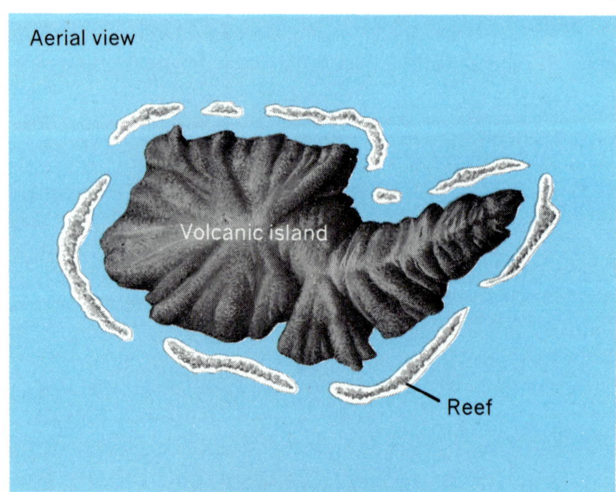

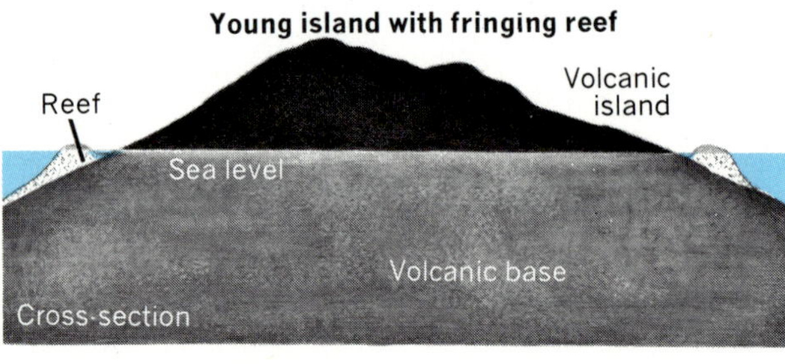

A fringing reef, hugging the shoreline of a volcanic island, is the first step in the slow formation of any coral atoll

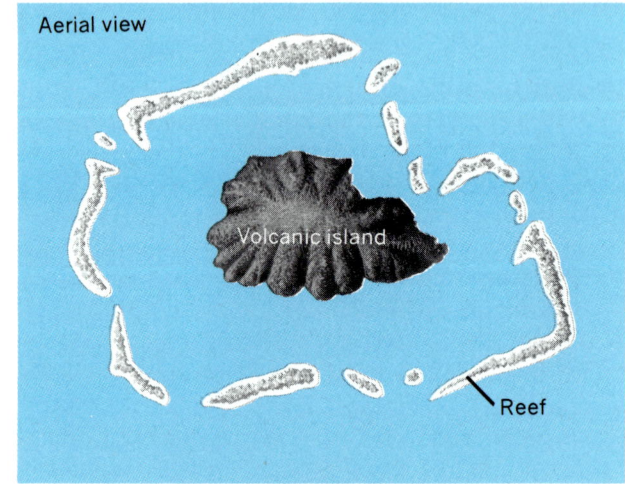

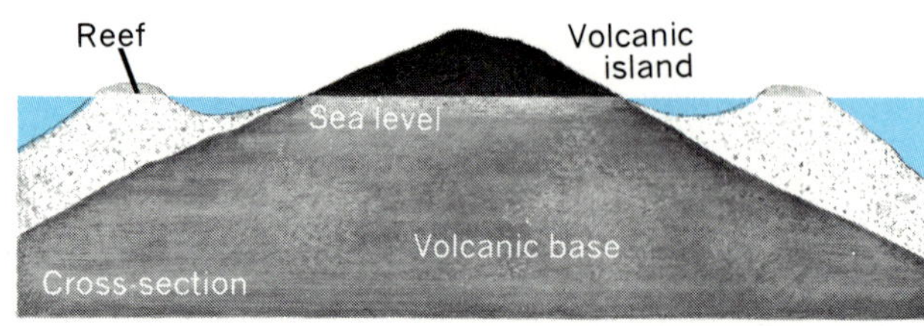

As the sea floor drops, the island sinks. Provided that this happens slowly, the coral continues to grow near the surface

their destruction upon the sinking peak of the basaltic mountain. Finally, as its last vestiges disappear from the centre of the ring-shaped reef, coral and coralline algae soon begin to grow in the vacated area. The reef is an atoll, and the water-filled basin inside it is called a lagoon.

Volcanoes are rarely perfectly conical. Since those that rise from the sea are often irregular, the fringing reefs, which follow the outline of the shores, do not form perfect circles in the sea. As an island sinks and the reef grows on, becoming a barrier reef far out from shore, the developing atoll may take on any shape. Some are roughly circular, some elongated, oval-shaped chains around their lagoon; others take the form of a crescent or horseshoe, indicating that the reef did not completely encircle the original volcano.

Other influences, however, affect the appearance of these coral atolls above water. Because of differences in winds and waves at different points in a reef, the conditions for the growth and development of the coral animals and algae may be very different within the same island structure. The coral may have grown more quickly in one area and have had a stunted development in others. The result is that the circle of an atoll may be completed only under water: above the surface it may appear as a series of islands.

Unfavourable conditions

Conditions for the growth of coral are not always as favourable as those indicated by the emergence of a massive reef or atoll chain. Across the whole breadth of the Pacific Ocean bed, divers and submariners have found flat-topped sea mountains, each with a crown of coral. These mountains, which are called guyots, once rose to the surface, where the action of the waves flattened their peaks; coral colonised their sides and continued to develop for thousands of years until the rate of the mountains' subsidence became so fast that the growth of the coral could not keep pace and so fell beyond the depth of water at which it could continue to survive and multiply.

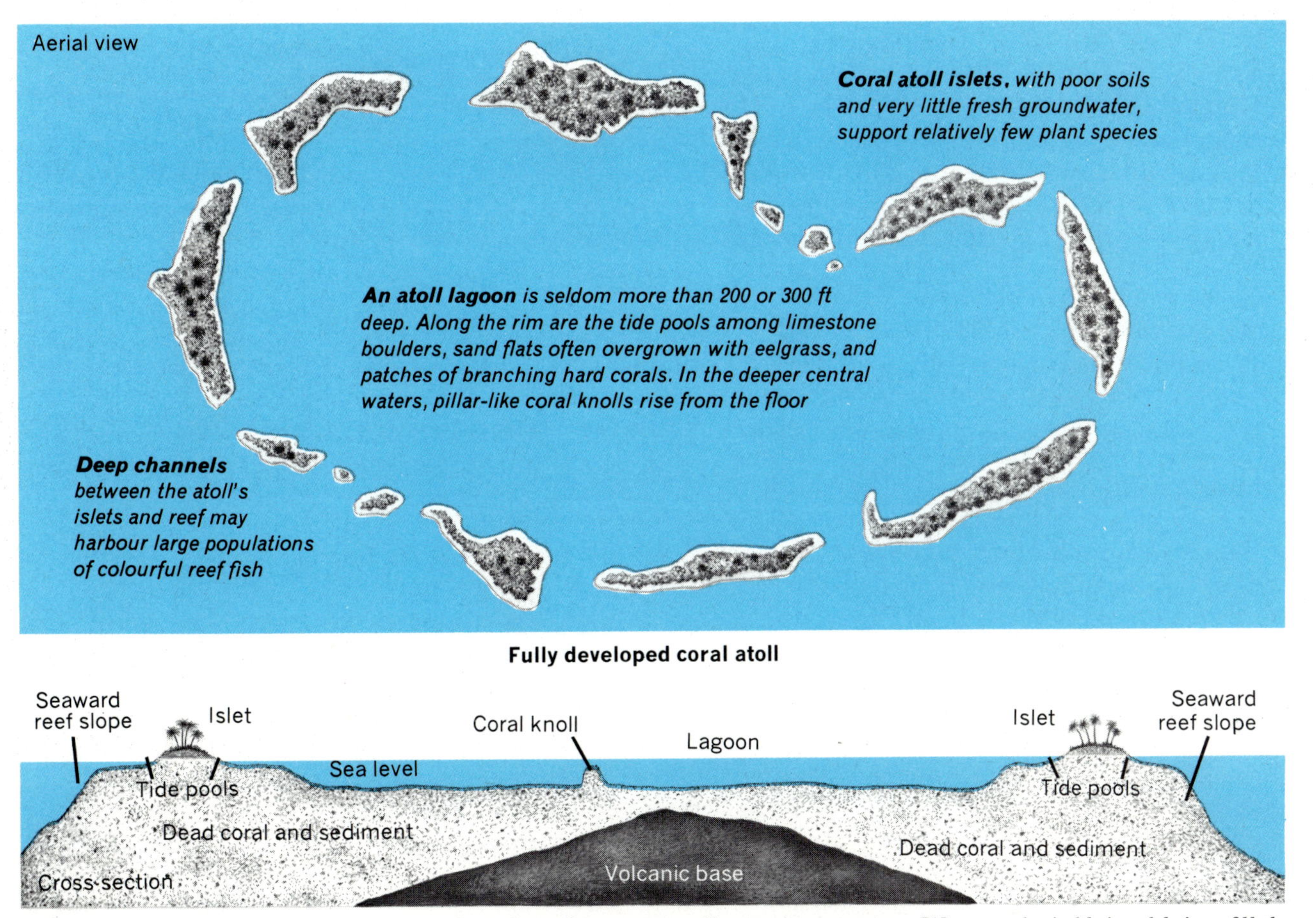

When the volcanic island drops below the surface, the reef – now an atoll – continues to grow. Waves and wind bring debris to fill the lagoon and erode the coral. It stops growing only if the sea floor drops too quickly to keep the coral in shallow water

A HARSH WORLD FOR PLANTS AND ANIMALS

The soil of newly emerged volcanic islands is rich in a variety of inorganic food for plant growth, but the basic limestone of a coral islet limits atoll vegetation to a comparatively few species. Since the phosphates and nitrates important to plant growth are in short supply and are slow to accumulate, an islet must be colonised by hardy pioneer plants such as those able to extract nitrogen from the air through the help of specialised bacteria inhabiting their tissues.

Coral sand and rubble, moreover, cannot support surface water. In a rainstorm, water immediately filters downwards, coming to rest at a level only a little above that of the sea. A plant must have long roots to reach this water, which remains fresh only if the islet is broad enough to prevent contamination by salt water.

Vegetation on atolls may at times show signs of food deficiencies: lessened fruit production, stunted growth and pale leaves. Handicapped by an unstable environment, atolls are generally incapable of supporting a wide variety of vegetation. Depending on their size and annual rainfall, islets in the South Pacific may harbour no more than 100 plant species, a far cry from the thousands found on typical volcanic islands.

Plants are carried to an atoll, as to any island, by prevailing winds, cyclones and typhoons, ocean currents, birds or man. Some of the plants that float in by sea – especially the low-lying strand plants that carpet the upper beach of an atoll – have seeds so remarkably resistant to salt water that they can drift along in a current for four or five months without very significant damage.

Plants usually first touch land on the windward side of an islet, but this is the most difficult place for them to become established. Only those able to withstand continuous salt spray can survive: the best examples are *Morinda* (a member of the coffee family), *Scaevola*, tree heliotropes, beach morning glory, and climbing *Wedelia* shrubs.

Tree heliotropes characteristically fringe atolls. These giant plants, occasionally up to 60 ft tall, can be found growing sometimes in a forest, sometimes individually. On low, dry islets, *Cordia*, which is usually no more than 20 ft high, can form an extraordinarily dense forest thicket, successfully preventing the growth of other plants. At one time, *Ochrosia* forests may have been dominant on many atolls in moist climates;

today, however, few of these slender trees with their spoke-like branches grow on the islets that man has cleared and settled.

Of course, many atolls lack these large trees. Their vegetation consists of small plants such as mallow scrub, which produces attractive yellow flowers that temporarily decorate the arid islets with bright patches of colour.

The hardy pioneers

Because of their ancient, mid-oceanic origins, most atolls are far from continental sources of supply for animals as well as for plants. They harbour bats, especially the fruit bats, introduced land mammals, and possibly introduced amphibians. The variety of all animals is severely restricted. Insects, scorpions, centipedes and crabs are usually present, as are a few kinds of skinks and geckos – all animals with relatively impervious skins that withstand the effects of salt spray during voyages on rafts of vegetation. Great marine turtles – green, loggerhead or hawksbill – may surface offshore, the females lumbering on to coral beaches to lay eggs during their breeding season in the summer.

The parent volcanic island of an atoll may once have been heavily populated, but as the island sank and the surrounding reef grew upwards from its crown, few if any of its animals could have made even a slow transition; most of them must have starved or been unable to breed. Early volcanic islands probably supported skinks and geckos, but most species of these lizards now present on atolls no doubt arrived much later on natural rafts or aboard the canoes of early travellers. That they are almost identical over vast reaches of the Pacific suggests a recent arrival.

The same supposition may be true for many kinds of insects, for rarely is an insect species peculiar to one atoll; rather, it is usually distributed widely over a number of atolls. Insects, despite their large number of species – more than those of the rest of the world's animals combined – may be represented on Pacific atolls by fewer than 2000 species altogether, and by only a fraction of that number on any one atoll. Because the small variety of insects meant fewer competitors, the species that did establish themselves often multiplied prodigiously. Similar circumstances explain the small number of varieties, but large populations, of land snails, lizards and other forms of native atoll animal life.

Other than birds, the most conspicuous animals on atolls are crabs. On many atolls, the dominant crustaceans are hermit crabs and land crabs, with far fewer of the large coconut crabs. Along temperate shores, hermit crabs live in the water and cannot stand exposure, but in the tropics some of them spend much of their time either on the beach or far inland in scrub or forest. The abandoned snail shells that hermit crabs inhabit protect them against both predators and the searing heat of mid-day, when they often rest quietly in the shade of shrubs or limestone rocks. In the evening, they migrate to the lagoon, where they immerse themselves briefly to renew the small supply of sea-water held in their shells. As the water bathes their gills during the day, oxygen from the air diffuses through it into the bloodstream. A female hermit crab seeks the lagoon for another reason. She must release her eggs in the water, where the young can pass through their normal larval stages.

Stoutly constructed land crabs, often almost as plentiful as hermit crabs, burrow into the coral sand and rubble. These crabs are so numerous that their excavations have a major effect on the surface structure of an atoll. They are omnivorous scavengers, but because they cannot climb trees they are restricted to food at ground level.

A formidable crab

The least numerous but most interesting of atoll crustaceans is the big, colourful coconut crab. Not only does it attain great size – reaching a length of up to 1 ft and a leg span of nearly 2 ft – but its powerful claws are truly formidable. Except for periodic trips to the sea to moisten its gills or to release eggs, it is completely divorced from its marine origins. It eats the fruit of screw pines and other atoll plants, as well as animal carcasses. It readily climbs coconut palms, but some biologists doubt that it can break off the tough nuts or even open an intact one that has fallen to the ground.

Through the dispersal of their larvae, coconut crabs are widely distributed in the tropical Pacific, although apparently they do not become established on arid atolls where there is little food. Nor have they reached the remote Hawaiian Islands. Since a crab of this size provides a good meal for several people, the species has suffered extensive hunting by island residents.

Giantism is often a characteristic of island life, especially on large islands, as exemplified by the

UPSIDE-DOWN MEAL *Fruit bats, or flying foxes, rest in a tree as they discard the seeds and pulp of its fruit. They digest only the fruit juices – and occasionally flowers. The small flying mammals roost in clusters during daylight*

giant sunflower trees and iguanas of the Galapagos and the emus and cassowaries of Australia. But on small, arid islets, some animal inhabitants exhibit precisely the opposite condition: dwarfism. Tiny lizards live on small islands near Madagascar, related to bigger lizards on the larger island. Other minute lizards live on small islands in the Gulf of California, and uncommonly small snakes occur on islands in the Gulf of California, in the Mediterranean and in the Florida Keys. Even domestic mammals such as pigs and introduced pests such as rats are smaller on many atolls. The reasons are not clear, but insufficient food may be one important cause.

RESTING-PLACE FOR MIGRANTS

Many birds that frequent atolls are small: the Micronesian pigeon, the Micronesian starling, a reed warbler, among others. The shore birds and seabirds that visit the atolls only to nest or feed naturally interbreed with their relatives ranging over the wide Pacific, and so are normal-sized because their genetic heritage is kept open.

With the coming of autumn in the Northern Hemisphere, thousands of birds leave Canada, Alaska, Siberia and other North Pacific lands to fly to milder climates. One of the migratory birds commonly seen on atolls is the Pacific golden plover; but other plovers, curlews, sandpipers and phalaropes also visit to rest briefly on their way to more luxuriant islands or to remain until it is time to return north. Even the familiar mallard ducks and North American pintails sometimes land on the lagoons and atolls of the Pacific.

The seabirds of coral islets range from the smallest terns, such as the white-capped noddy, to the three large boobies of the Pacific: red-footed, blue-faced and brown. Frigate birds and tropic birds are also present, although white-tailed tropic birds prefer to nest on the cliffs of mountainous islands. Besides a variety of noddies, several other terns frequent atolls, including the crested tern and the beautifully distinctive fairy tern. Most of the seabirds nest on the sand amid scrub vegetation, which offers a little shade, but the fairy tern precariously balances its single egg on a tree limb or on top of a rock; the young chick's claws can cling securely to the perch from the very moment it hatches.

Enriching the islets

The water birds that nest on atolls are among the most important animals in an islet's ecology. They consume both plant and animal life – especially fish, crabs and molluscs – and their waste products are sometimes abundant enough to alter the chemistry of the limestone soil, which in turn determines the kinds of plants that can live there. Bird wastes dropped in lagoons and pools enrich the waters and encourage the growth of algae.

If large numbers of shore birds live on an islet, the high mortality rate ensures potent enrichment; the carcass of a dead bird on land or in shallow water provides an intense concentration of fertilising organic matter. It is principally through fish-eating birds that nutritious food from the ocean comes to be deposited upon the islets of an atoll; accumulations of phosphate-rich guano (bird waste) are especially noticeable on drier atolls such as Baker and Howland islands. Members of the tern family and boobies are the principal producers of guano, although shearwaters, petrels, frigate birds and tropic birds also contribute to its accumulation. Guano deposits have, however, led to the exploitation and ruin of many small islands, as men have arrived to mine the valuable substance. Nauru, an elevated 'pancake' atoll, has been mined for its phosphates since the late 19th century.

Like birds everywhere, atoll birds show distinct preferences in feeding and nesting grounds as well

PATHS OF MIGRATION *Birds migrate over the Pacific on three main routes. Most spend April to August in the Northern Hemisphere, then fly south for the Southern Hemisphere's summer. Some species breed in the north; others in the south*

NOISY BREEDERS *Sooty terns nest among the sparse grasses of Johnston Atoll in the Pacific. They are called 'wideawakes' because they shriek day and night during the breeding season. Sooty terns rear one chick each year*

as different ranges in flight. The terns, for example, may be divided into those species that habitually feed close to shore and those that fish several miles out to sea. Birds of the open sea, such as tropic birds and albatrosses, are sighted hundreds of miles from the nearest land.

Just as birds living on the major continents use regular routes for their annual migrations, the birds of the Pacific follow established paths, using particular groups of islands as resting places. One route is far west in the ocean and includes the Palau Islands, 600 miles north of New Guinea; another connects Japan with the Marianas Islands, 1000 miles to the south; a third passes through the Hawaiian chain to the northern breeding grounds of Alaska and the Aleutians. On all the routes, small islands play an important role.

At nesting season, many atolls are alive with bird activity. Noddies and other terns, especially, fuss and cry constantly night and day. These little birds fly busily back and forth across the shoreline, hovering and repeatedly diving after the multitudes of small fish in the shallow water. A noddy often feeds only a few feet from shore, in enclosed pools or in shallow parts of the lagoon. It hovers skilfully, at times remaining for minutes

above a huge school of minnows before selecting an individual, plunging down, and plucking it from the horde of nearly identical fish.

Far above may loom the ominous, dark shape of a frigate bird patrolling the air currents with unmoving, sharply raked wings and forked tail. They sometimes fish for themselves, but more often hover high in the air, on the watch for a booby fishing down below. When the frigate bird sights a successful fisher, it plunges in a hawk-like and often accurate dive. It pummels the booby, perhaps even turning it upside-down, until the booby releases or disgorges its catch. The frigate bird then dives after the prize, snatching it in mid-air. On land, the frigate bird is still a predator, eating eggs and young birds of any kind that it finds unguarded.

The extraordinarily high mortality rate among birds nesting on atolls approaches or even exceeds 50 per cent in some species. Storm waves and winds can devastate a nesting colony, and if a small island is swept by particularly high waves, its entire bird population may be almost obliterated. If rats, dogs, cats or pigs are allowed by man to invade an islet, nearly all its nesting birds are inevitably doomed.

329

NAKED AT BIRTH *A brown booby huddles over her three-week-old chick on a crowded nesting colony on Johnston Atoll in the Pacific. The young booby is naked when it is hatched from the egg, but it soon becomes covered with a soft white down. It takes five months for the chicks to become fully feathered, and shortly afterwards they make their first flight*

WITHOUT FEAR *A red-tailed tropic bird broods its egg on Takutea Atoll in the South Pacific. Many birds breeding in isolated places have no fear of man; islanders can approach and remove tail feathers for ornaments*

FINDING SAFETY ON THE REMOTE ATOLLS

Breeding grounds near rich food supplies

Many bird species use the Pacific atolls as nesting grounds. Some – for example, boobies and frigate birds – rarely fly far from sight of land and therefore nest only on islands close to other land. Other, stronger fliers – such as the world-travelling albatross – navigate, perhaps by instinct, perhaps by the sun and stars, to find small isolated islets hundreds of miles from other land. In all cases, however, the nesting islands are near strong ocean currents, rich with life-supporting plankton – the first link in the food chain of all the oceans. Fish eat the plankton and they in turn provide the long-distance wanderers of the skies with food and even the water that is often missing from their lonely and barren island resting places.

BIRD WITH NO NEST *A fairy tern guards its egg, balanced on a tree limb without a nest. The birds, which never build nests, also lay their single egg on bare rocks. When the chicks are hatched they are immediately able to cling to their perch*

WAITING FOR FOOD *Young great frigate birds await the return of their food-gathering parents. Frigate birds prey on the young of other species nesting near by and even feed on each other's unguarded eggs and newly hatched chicks*

FOAM-FLECKED AFTER A STORM *A black-footed albatross sits out a storm in its nest of mud and grass on Midway Island. After hatching, the young chick often has to withstand intense sunlight until it grows its feathers and can fly away*

CREATING AN ISLAND *Hundreds of arching roots – called prop roots – trap sediment to create shallow-water islets. Barnacles, oysters and other animals settle in mangrove roots*

LUSH NESTING PLACE *White egrets speckle the foliage of an*

ISLANDS CREATED BY TREE SEEDLINGS

Mangroves are the trees familiar to cinema audiences as the landmarks of continental and island shorelines in the tropics. Their ability to extend the land mass of these areas is surpassed only by their role as island pioneers. Many Pacific islands have been established by these trees alone.

Their seeds probably germinate best in brackish water, but the trees also thrive in pure sea-water.

In the Caribbean, the comparatively small red mangrove is the first of the family to appear in shallow water. Its fruit, still on the trees, grows a dart-like root; on falling, the fruit may stick up-

HOW MANGROVES BUILD ISLANDS

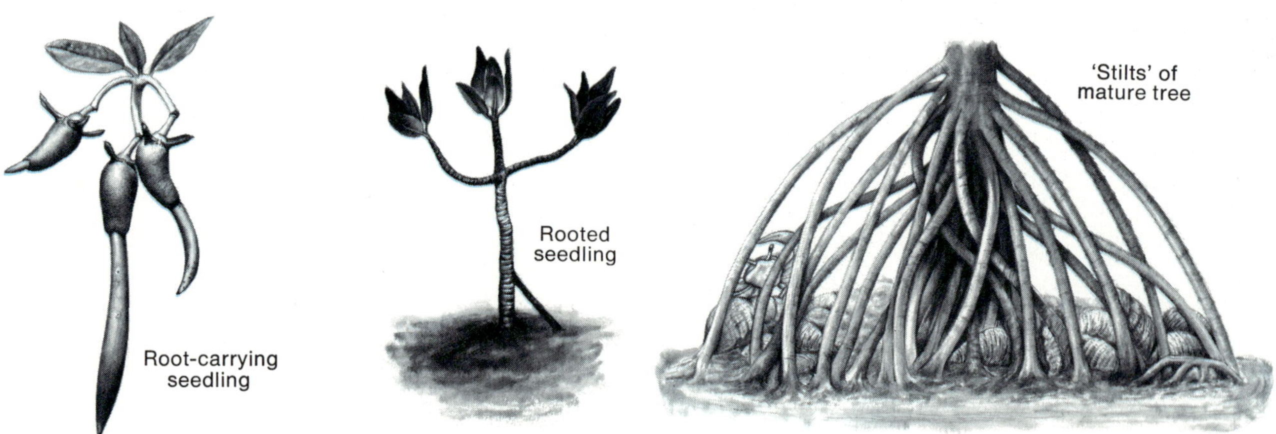

When the seedlings of the red mangrove fruit grow to about 6 in. long they drop from the tree and root in shallow water. Prop roots arch upwards from the growing seedlings to trap mud and debris and so start the slow creation of a plant-made island

Indonesian mangrove islet where they nest. Small animals in the swamp and fish in the nearby coastal waters ensure an abundant food supply

right in the soft mud at low tide, establishing a new plant within the protective radius of the parent. Other seedlings float away and take root elsewhere when they wash ashore, perhaps some distance from dry land if the water is shallow enough to expose the bottom at low tide. Before long, a thicket of trees is growing from the sea. Leaf litter and sediment collect in the tangle of the prop roots and, with continuing accumulations, eventually rise above the surface to form a soft, swampy island. In a few years, the island is so dense that it withstands even violent waves.

As new generations of red mangroves move outwards from the collected sediment, widening the island, another species, the black mangrove, begins to take hold in the vacated centre, by now perhaps elevated a foot or more above sea level. The process continues, and other plants, such as sea grape, appear and add to the structure of the island and to its diversity of plants. Flotsam and sediment, shell and coral rubble continue to wash in, and are retained in ever-increasing quantities.

Early in the island's growth animals arrive. As soon as the first young mangroves are securely rooted, oysters, barnacles, sponges and many other sedentary marine creatures adhere to stems and prop roots. The accumulated silt attracts the larvae of burrowing worms, clams and shrimps, which soon begin to establish large populations. Mobile mud crabs and small fish throng in the warm, turbid water, while fiddler crabs, land crabs, snails and insects swarm in the swamp and on the jungle of woody prop roots. Land hermit crabs climb to the tops of the mangroves. Seabirds, shore birds and even inland water birds visit mangrove islets.

A mangrove island is not easy to explore, for what appears in the distance to be a pleasantly forested shoreline turns out to be an incredibly tangled mass of woody stems and roots. If the supporting structures arch high enough, a small boat can pass under them; otherwise, the visitor must clamber over the rigid but often slippery stems and roots. In the interior of the islet, the soft, wet soil does not provide secure footing. Over a very long period, however, the soil packs, and mangrove islands may become firm and large enough to coalesce, or even join and extend the mainland shore. Such islands occur not only in the Caribbean but also in the Pacific and Indian oceans, especially near river mouths.

Long before man appeared, corals, coralline algae and mangroves were adding to the land mass of the earth. The efforts of these animals and plants seem minute compared with the vast geological forces that have shaped the world, but such reef and island constructions have established new environments for the development of plant and animal specialities, and important resting places for ocean migrants.

Islands and men

*With great daring, man long ago began making his way to islands. Now
his growing populations and technology threaten their devastation.
But, with wisdom and determination, the island heritage may yet be saved*

Profound impulses attract men to islands. For
the human spirit an island is a refuge as well as a
challenge. Islanders the world over are a recognisable breed: the fisherfolk of Shetland, the
Faeroes and Iceland; the whalers of Nantucket;
the families of Tristan da Cunha, exiled because
of a volcanic eruption, who preferred to return to
their bleak island home in the mid-Atlantic
rather than accept the relatively comfortable
routine of life in Britain. Tough, self-sufficient
and courageous, such people spend their days in
tune with the surrounding sea. To be an islander
is to belong to a proud fraternity, envied, roman-
ticised, but little understood by the city-dweller.

Anthropologists know little about how soon
early man was able to reach islands. Java man,
Homo erectus, who lived more than a million years
ago, was limited in intelligence and technical skill;
he was probably unable to cross straits of open
water, although he may have ridden on logs, at
least, on rivers and lakes. Much later, modern
man, *Homo sapiens*, built rafts and canoes with
which he successfully journeyed across sea-water

DESCENDANTS OF THE PIONEERS *Samoan fishermen
work from the harbour rocks near Pago-Pago. These
Polynesians are accomplished fishermen and navigators,
whose ancestors undertook great voyages into the
unknown Pacific. Over hundreds of years they settled
thousands of islands between New Zealand and Hawaii*

to offshore islands. He may have been seeking
food, avoiding enemies, or simply responding to
that irresistible urge common to all men, curiosity.
The distances traversed were not great at first, but
the barrier had been broken.

The Western world usually associates the
'discovery' of islands with the ancient Greeks and
Phoenicians, the Vikings, and the Portuguese,
Spanish, Dutch and British mariners of the last
few centuries. Most histories of European explora-
tions scarcely mention that many of the Asian and
Pacific islands visited, even those far removed
from continents, already had human inhabitants.
The references are usually disparaging comments
about 'savages'. Such history underestimates or
ignores previous ventures that – in the opinion of a
growing number of anthropologists and historians
– must rank among the greatest adventures since
modern man first emerged some 40,000 years ago.

The North Atlantic has no chains of islands like
those of the tropical Pacific: there is little to attract
and much to threaten the voyager, and European
man reached out slowly for the open sea. Trade in
Irish bronze and Cornish tin, over 2000 years ago,
followed the coasts. Brendan and Madoc, Irish
saint and Welsh prince, may have made ocean
voyages but we know of them only from legends.
It was the Norsemen who first crossed the North
Atlantic, and the Portuguese – by then equipped
with the astrolabe, a forerunner of the modern
navigator's sextant – who first ranged widely
across these expanses of forbidding Northern seas.

SAGAS OF THE FIRST ISLANDERS

The Phoenicians and Vikings, two of the most adventurous Western peoples, covered great cumulative distances by hugging coasts or by sailing relatively short passages from island to island. The Phoenicians reached both the British Isles and the Indian Ocean by following continental margins. The Vikings crossed wider expanses of stormy waters in sailing the open ocean between Scandinavia, Orkney, Shetland, the Faeroes, Iceland, Greenland and the North American mainland. Arabian and Indian explorers were also sailors, as were the Chinese, who arrived in Taiwan to find it already populated by a Malay people, and who landed in Japan to find it occupied by the Ainus.

Islands in the Atlantic, some only a few hundred miles from a continent, were not discovered by Europeans until about the 15th century, when the Portuguese and others came upon Madeira, the Azores, and the Cape Verde Islands and, later, Bermuda, Ascension, St Helena, Tristan da Cunha and the Falklands. In the Indian Ocean, only Madagascar was found to be inhabited – not predominantly by Africans, however, but by people of Asian derivation.

While Western scholars predicted that mariners would fall off the edge of the world (provided the fearsome monsters of the deep allowed them to get so far), other men were sailing across one-third of the earth's surface, exploring and settling thousands of Pacific islands. South-east Asia seems to have been the starting point for the migrations of peoples through the largest ocean.

The story of the settlement of the Pacific is known today in only small fragments, pieced together from legends of the people, comparisons of widely scattered cultures, and archaeology.

The earliest people to set sail may have used rafts. But as time went on, both the coastal mainland people and those who were already islanders learnt by experience to build more seaworthy craft, and they became incredibly skilful sailors. The migrations, especially those over great distances, almost certainly involved single canoes (including double-hulled ones), since keeping track of another canoe on a dark night would have been extremely difficult, if not impossible.

The pioneers of the vast Pacific came in three waves. Firstly, there were the apparently primitive people who settled in Australia, the Philip-pines and New Guinea. Secondly, there were the voyagers, possibly related to the present inhabitants of Malaysia, who settled in Madagascar and the Islands of Micronesia. And, thirdly, there were the Polynesians, the race of incomparable master mariners who stamped their personality on a large proportion of the watery surface of the world.

The original Australians

The earliest ancestors of the Australian Aborigines arrived on the island continent between 25,000 and 30,000 years ago. Nothing is known of their

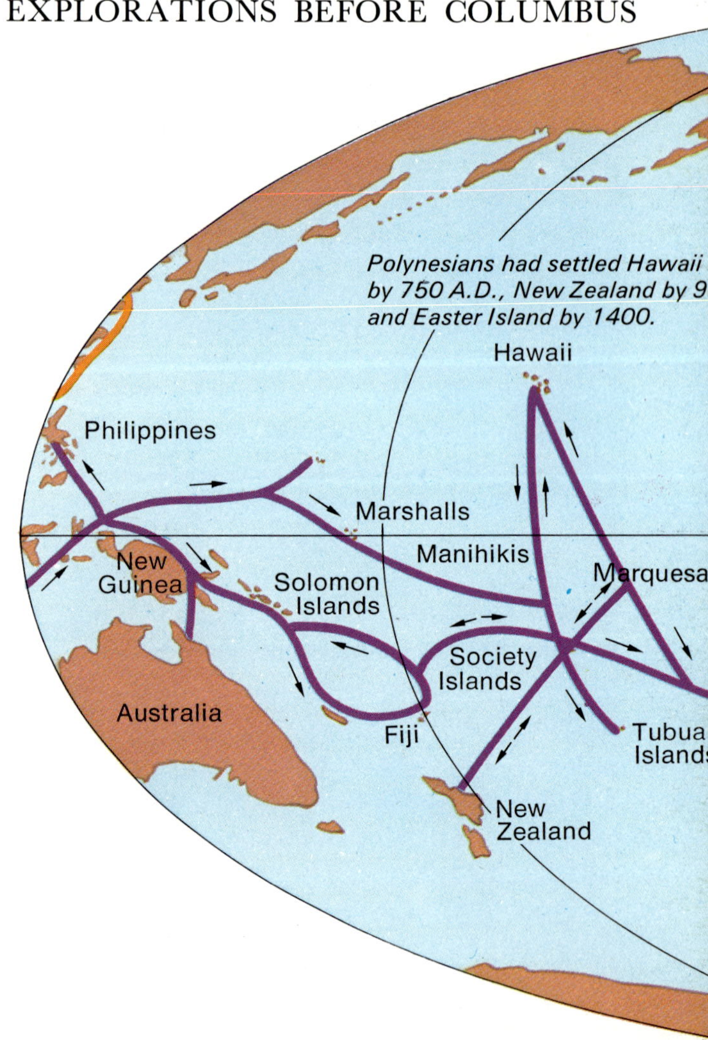

Early mariners set out from many lands to explore, trade and plunder.

origins or their methods of transport, but the way of life of their descendants is so simple and primitive that it is reasonable to suppose that it has changed very little over the succeeding centuries. Other voyagers, coming later, mingled with the first settlers.

Similar primitive peoples, with no knowledge of agriculture, settled in New Guinea and the the Philippine Islands. They lived by hunting and gathering fruits, and they carried their few possessions with them. They built no houses, domesticated no animals, and cultivated no plants until taught to do so by subsequent cultures. Later, a similar people, who were more experienced seamen with better boats, populated the islands of Melanesia, probably beginning with New Cale-

donia and the New Hebrides, and eventually getting as far as the Solomon Islands and Fiji.

The early Indonesians

The second wave of Pacific pioneers sailed west to become the dominant inhabitants of Madagascar, and east to populate the islands of Micronesia. They were apparently enough like the present-day people of Malaysia and Indonesia to be called Proto-Malays.

Their culture was more sophisticated than any established previously in the Pacific. For many centuries, it influenced all the islands from Madagascar to the central Pacific. There are few reliable dates in their history, but it appears that they were in the Philippines by 6000 BC.

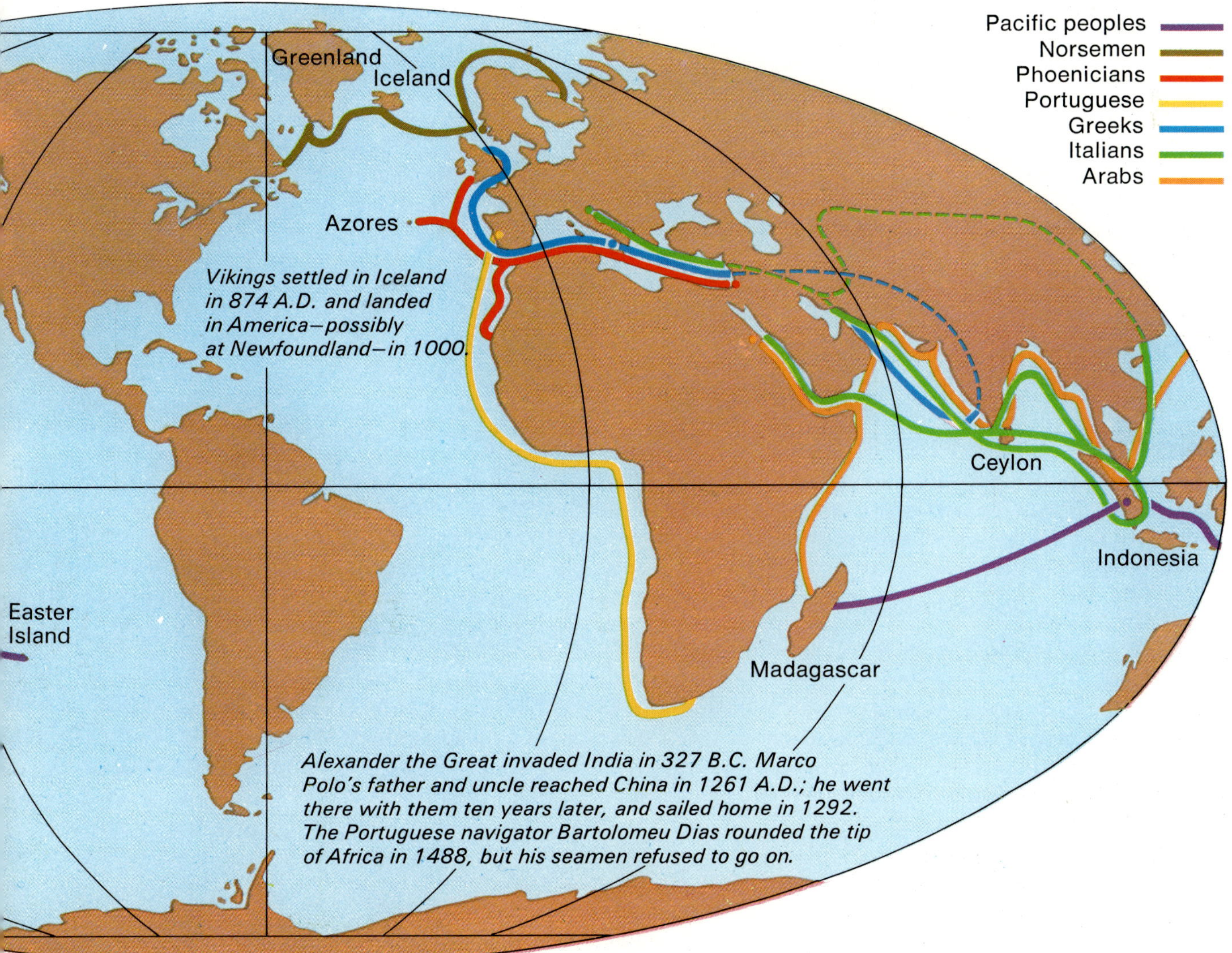

The Arab and European adventurers preferred to stay within sight of land, but the Pacific peoples sailed the open oceans

Several prehistoric races settled the islands of the Pacific. Although their origins are obscure and races intermingled during their migrations, they now form four major cultural groups – Polynesians, Micronesians, Melanesians and the Australian Aborigines

The Polynesians, intrepid masters of the vast Pacific

Finally came the Polynesians, whose origins, like those of their predecessors, are almost totally obscure. Their culture cannot be traced back to the Asian continent; the thread to their identity seems to be lost well out at sea on certain islands east of the New Hebrides. They were definitely not South American.

The Polynesians began their voyages several thousand years ago. They depended on tools of stone and shell (there are no metal ores on volcanic or coral islands), and they had no navigational instruments. Although winds, currents and stars changed so much en route as to make it appear that the Polynesians were in a different world, their sensitivity to the sea, unequalled by that of any other people in history, enabled them to go on. They gradually populated almost every island of habitable size in the tropical Pacific.

The travels of the Polynesians form a record of almost incredible achievement. Nearly 3000 years ago, they passed through Fiji in Melanesia, settled in Tonga, and gradually moved into Samoa. They sailed on to the Marquesas, where they were living 2000 years ago. This island group seems to have been the point of departure for their final and greatest voyages into the unknown. They went on to settle Easter Island in the east, the Hawaiian Islands in the north and New Zealand in the south – their furthest adventure into temperate regions. The huge triangular area they populated covers more than 7 million square miles.

Several centuries after the original voyages and settlements, other boatloads of Polynesians, primarily from Tahiti, reached Hawaii and New Zealand. The Polynesian culture – traditions, language and artifacts – spread with remarkable rapidity. A modern Maori visiting Hawaii can understand much of the language, and many geographical names of Polynesian origin occur, with only slight variation, all over the Pacific.

CENTURIES-OLD CRAFTSMANSHIP *Present-day Polynesians fashion a canoe of traditional design, but using modern tools and nails. The ancient mariners made dugouts with stone tools, or sewed hull planks together with coconut fibres*

LEARNING THROUGH PLAY *Polynesians begin to learn the arts of navigation at an early age – even with model canoes. The islanders' ocean-going canoes have as many as three masts, and an outrigger or smaller canoe for stability*

RIDDLE OF THE ROCKS *Stone heads up to 20 ft high are found on the slopes of Rano-Raraku volcano on Easter Island. No one knows the significance of these massive effigies, but it is believed they were connected with burial rites*

BOAT THAT BRAVED THE OCEAN *Powered by a sail of coconut fibres, the double-hulled canoe was typical of early Polynesian vessels. The hulls – as long as 80 ft – were lashed together, with a deckhouse for living-quarters between them*

EPIC VOYAGES LOST TO HISTORY

The story of the Polynesians is only partly known today. Without a written language, their history had to be transmitted in long and complex chants, which few of the modern islanders have ever heard. Some of the chanted histories were recorded by early Western explorers such as Captain James Cook, but of course they heard and comprehended only a few of them. Modern archaeology, linguistic comparisons and anthropology help to link groups of these great seafarers, but knowledge of them will always remain fragmentary and raise more questions than modern inquirers can ever hope to answer.

Polynesian trans-oceanic vessels, with or without sails, were huge, consisting usually of two canoes up to 80 ft long lashed together side by side with a deckhouse between them and held together by fibre ropes and wooden pegs.

The boats sailed before the wind beautifully.

They could travel upwind reasonably well, their deep, round hulls resisting drift, although the pounding they must have taken in sailing into the wind weakened the lashings and was probably hazardous if prolonged. In the 1000 or more years of trans-Pacific voyaging, some of the vessels must certainly have broken up. Others must have been carried out of control by storms, especially if the trip lasted beyond the summer months.

Populations can grow explosively in the restricted area of an island, and there is evidence that this may have been one factor behind intended migrations. When the Polynesians undertook such trips, women, children, plants and domestic animals went with the men. When landfall was successfully made, the colonisers were immediately ready to begin a productive new life.

But it was not always easy. Pacific islands vary widely in form and climate. Some are low atolls, others are high and mountainous; some have heavy rainfall, others are almost desert. After an exceptionally long voyage, the plants and animals brought by the colonisers, upon which they depended for food, might be too weak to survive. Even if they were healthy, the plants might not be able to grow in the soil (especially on coral islands), and the animals might not find the right food. Only rarely did an island offer the settlers a satisfactory staple food already growing wild.

Even a successful start might not assure the future. Islands in dry regions might suffer severe droughts. Plants would then wither, leaving the people dependent on fishing or on catching seabirds. If unsuccessful, the islanders would be left without provisions and possibly without enough strength to migrate further. Only the more recent tragedies are reliably documented, but archaeological evidence suggests that, earlier, whole island populations must have perished.

European explorers reported traces of former populations on a number of uninhabited islands. Fanning Island is a small coral island north of the Equator, between Tahiti and Hawaii. There Captain Cook found the remains of a rectangular building made of coral blocks, a few primitive tools, some fishhooks, ornaments and several graves. On Malden Island, dippers still lying in water-collecting hollows spoke eloquently of the hardships of drought.

Some groups of colonisers must indeed have been confronted with a serious shortage of drinkable water. In a few Pacific islands, the fresh water usually present under atolls and low islands is either absent or contaminated by salt water. Many present-day inhabitants of the Gilbert Islands have become accustomed to drinking brackish water that other people, even neighbouring islanders, find so obnoxious that it is almost impossible to swallow or keep down.

The further the Polynesians migrated across the ocean, the fewer familiar reef fish they found. Although reef fish populations throughout the Pacific are fairly similar to one another, the diversity of species decreases toward the east. In addition, a number of ordinarily harmless reef fish can become toxic, their toxicity varying from region to region according to the algae they eat. Polynesians therefore could not always rely on the same reef fish for sustenance.

Their many hardships did not deter the Polynesians. An adaptable people, they even found a way to 'rotate' islands as others rotate their crops. Manihiki and Rakahanga atolls are

THE POLYNESIANS' CARGO

Breadfruit

Taro

Coconut

Fowl

Sweet potato

Dog

Pig

South Sea islanders frequently carried seedlings, fruits and animals, including stowaways, in their canoes during long ocean voyages. Some of the cargo was used to feed the travellers, but food plants and farm animals were preserved to establish on their new island homes

25 miles apart. The Polynesians dwelt on only one island at a time, periodically moving back and forth. In this way the vegetation and fish populations of each atoll were given time to recover between periods of exploitation.

There is considerable evidence that the Polynesian settlers of the Hawaiian Islands, possibly for several centuries after their first arrival, made repeated voyages to Tahiti and back. In time, however, for unknown causes, the great double-hulled canoes were scrapped, the art of making them was lost, and all communication ceased. The Hawaiians continued to sail about their own archipelago, at times using canoes made from Oregon firs or California redwoods that had drifted across from North America. Except for these short voyages, the isolation of the Hawaiians was complete until the arrival of Europeans in the latter part of the 18th century.

MIMICKING THE BIRDS *New Guinea men, as resplendent as the birds of paradise whose feathers they wear for decoration, imitate*

PEOPLE OF THE PACIFIC

As colourful as their animal neighbours

The islands across the Pacific Ocean are inhabited by four peoples: the Aborigines from Asia who settled in Australia thousands of years ago; the Polynesians who are found from Samoa to Easter Island; the Melanesians in the south-west near Fiji and New Guinea, and the Micronesians to the north. All preserve many of the traditional skills and cultures of their ancestors.

ABORIGINE CELEBRATION *Australian Aborigines imitate the display dance of the brolga crane during their song and dance gatherings, called corroborees. They paint their bodies to represent the brolga feathers. Each Aborigine tribe has a mythology, dialect and culture of its own*

STONE MONEY *Large circular stones symbolise the wealth of a whole village on Yap Island, but they are not used in trading. The Micronesian inhabitants bring the stones from other islands by canoe*

the birds' courtship dances in their tribal displays. At such festivals the tribesmen compete to outdo one another in brilliance

BALANCING THE CANOE *A Society Islands sailor balances his canoe by standing on the outrigger. Polynesians and Micronesians still travel hundreds of miles across the ocean in such fragile craft*

ORDEAL BY FIRE *After a ceremony of prayer and purification, Fijian fire-walkers tread over red-hot stones in a rite of faith. Burning logs heat the stones in the firepits for ten hours before the walkers enter. Scientists believe that they may hypnotise themselves before the ordeal*

FOLLOWING THE BIRDS *Gulls and terns fly no further than about 20 miles from their nests to fish, and they are known always to return at night. Mariners watching a flock at dawn or dusk can therefore tell in which direction land lies*

AMAZING FEATS OF NAVIGATION

For Westerners, the science of navigation requires a high degree of technical skill and sophisticated instruments that keep track of position and course. The astrolabe, derived from an ancient Greek invention and used until it was superseded by the sextant in the middle of the 18th century, enabled a mariner to deduce time and latitude with remarkable precision. The Polynesians had nothing of this sort; they had no maps, written records or even the crudest instruments. They estimated distance only by the passage of time; the longer the voyage, of course, the greater was the chance of their making a navigational error.

Western man, in attempting to comprehend the feats of the Polynesians, has on occasion incorrectly described a calabash (a vessel for carrying fluids) with holes drilled in it and stick 'maps' (teaching aids) as navigational aids. He can only partly appreciate what these sailors were able to sense and understand. Even the instruments of the last century were not always able to locate an island. In 1837, a British ship set sail for Raratonga from an island less than 150 miles away. Despite knowledge of Raratonga's location and possession of compass, telescope and high crow's nest from which to keep watch, the captain gave up after a week of fruitless searching.

The Polynesians were well aware that the visible signs of a tropical island appear long before the island itself comes into view. Birds and floating plant debris (flotsam) indicate land, and often the direction in which it lies. The high islands of the Pacific are nearly always wreathed in clouds formed by the elevation and cooling of moisture-carrying trade winds. Such a stationary cloud formation on the horizon is a sure sign of an island's existence, although if the winds and ocean currents are adverse, a day or two may be required to sail close enough to see the mountains themselves. Thermal updrafts from an atoll may also cause characteristic clouds to develop and hang almost motionless. The undersides of these clouds often reflect the pale azure of lagoon waters, very different from the deep blue-black of the open sea. An atoll's presence may thus be discerned at great distances before its vegetation sketches the barest outline on the horizon.

Maps of familiar individual islands and groups

of islands were, of course, made for local use. Islands were shown on such maps by shells or pieces of coral fastened to sticks, and the best route of navigation might also be indicated.

How 'wave craft' is taught

One of the most remarkable techniques by which Polynesian mariners sensed the presence of islands was instilled in childhood into those trained to command the canoes. The famous *mattangs* that can be found in museums, or are now made in non-authentic form for tourists, were not maps at all, but Micronesian devices for instructing boys in the principles of wave motion. Such a stick chart could very well be used today as a visual aid in a physics laboratory, for it represents nearly every possible configuration of wave reflection and refraction.

Waves streaming across the ocean from a distant storm move in a regular pattern until they are steepened by a shoal or beach, or are bent out of their parallel array by the land. When waves hit an island, some are reflected back in the direction from which they came, while others are deflected at angles and warped around the island to continue in altered form on the other side. The primitive, yet sophisticated, stick chart indicates all the basic patterns of such wave action.

Each basic wave pattern had its own name and could be pointed out on a chart to others, no matter which way the chart was oriented. In this way, an experienced navigator could transmit his knowledge to youngsters, who then went to sea with him to practise and develop their uncanny sensitivity to the great ocean surrounding them.

From an aeroplane or a high cliff, wave effects are readily observed. But how can they be recognised at the surface, especially in a small boat that rises and falls not only with these waves but also with ocean swells of a different frequency? It is quite possible that three or more different sets of waves may affect a boat simultaneously.

The Polynesian navigator went to the bow of his canoe, crouched down in the hull, and literally *felt* every motion of the craft. Within minutes he was able to determine the position of the nearest island, of intervening reefs and of other islands near by. So accurate was the technique that he might not even need visual confirmation. No one knows precisely how effective wave navigation was over long distances, but it is now believed that these early Polynesians may have been able to locate accurately the position of islands and reefs from as far away as 100 miles.

Polynesians followed signs in the sea and the sky to find their destinations. Navigating between known islands, they observed the rising of certain stars or lined up with landmarks of the island they were leaving to head the right way

Towering cloud banks often hang motionless over islands

Birds in the sky and flotsam in the sea indicate land near by

Crossing wave patterns indicate disruption by a nearby island

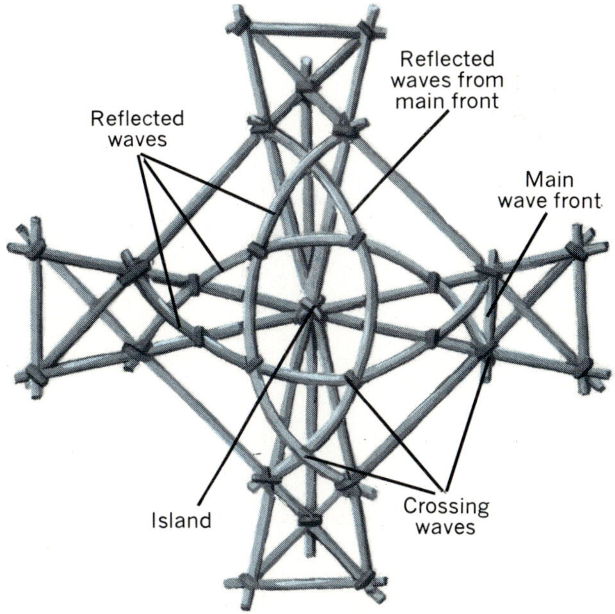

TEACHING AID *Young Polynesian sailors learnt how to interpret the wave patterns which a canoe would encounter in the neighbourhood of an island by studying stick charts made by experienced navigators. The charts, called mattangs, were shaped to illustrate a typical wave pattern*

How Polynesian mariners navigated by the night sky

Interpreting wave patterns was only a secondary navigational technique for the Polynesians. Like the later Westerners, they are primarily celestial mariners, using stars as fixed points, especially when they sail to known destinations.

Polaris, the North Star, is either invisible or too low in the sky to be a useful aid to navigation over most of the region covered by the Polynesian voyagers. The navigators therefore used two other major stars, Sirius and Arcturus, as their main guides. Sirius, the brightest star in the sky, passed directly over Tahiti. The red star Arcturus travelled over the island of Kauai in the Hawaiian chain (it now passes over Oahu). Since relying on only two stars would have been dangerous, a good navigator could recognise as many as 300 other stars. With such knowledge, he could take rear sights if the stars ahead were hidden by clouds. But Pacific islanders generally sailed during the most favourable season, when storms were rare and summer westerlies were almost constant, and so there was little likelihood that long-lasting

cloudy conditions would complicate their journey.

As a night of sailing wore on, each of the main guide stars appeared to dip into its own 'pocket' on the horizon. In other words, it set in the same place each night as long as the observer remained at the same latitude. To reach another latitude, a canoe sailed north or south until a given star dipped into the ocean in its known spot every night. Thus the passage of zenith stars over various parts of Polynesia and the setting locations of others were of immense importance to the early voyagers' inter-island navigation.

When a voyage began, the great canoes set sail at dusk, using island reference points to head in the proper direction, which, especially for the Hawaii-Tahiti trips, was often known from previous expeditions. An immediate fix was made on a star nearing the horizon; when it dipped below, the next descending star became the navigator's new guide star, and so on.

The Pacific sky was as familiar to a Polynesian as a road map of his own area is to a modern motorist. During the day, a mariner navigated chiefly by following the fairly dependable trade winds, but because these changed from time to time, he had to take corrective action at night. His only means of determining longitude was to estimate elapsed time. When he arrived at the proper latitude, he simply turned and sailed due east or west, estimating his longitudinal progress until the destination was reached.

At least one Westerner has proved the effectiveness of the Polynesian method of stellar navigation by sailing from Tahiti to New Zealand. Lacking the confidence of the old Polynesians, however, he was accompanied by a skilled scientific navigator whose readings were kept secret from him – but available in case of emergency.

Even today islanders in the more remote parts of the Pacific think nothing of sailing off to another island cluster a few hundred miles away. A generation or two ago, they did not consider voyages of 1500 miles extraordinary.

For more than 1000 years, the Polynesians remained supreme seafarers, even though their voyages, according to available accounts, could be of incredible hardship and danger. They spanned the whole Pacific island world long before Western man ventured away from his continents. They knew well their world's northern and southern limits, which coincided with the edges of that region in which the sun's rays seem to shine directly down to earth during the course of a year and which correspond closely to the Tropic of Cancer and the Tropic of Capricorn.

It is possible, too, that they journeyed far enough east to reach South or Central America. The sweet potato, which the first European explorers found to be an important part of Polynesian diet, is believed to have originated on the American continent.

THE NIGHT FISHERMEN *Flashing torches to lure fish, Tahitians circle a placid lagoon in the dusk. The islanders go on long night journeys, navigating by the stars, and on these voyages they use torches to keep in touch with other boats*

MAN—DESTROYER OF ISLAND LIFE

Undoubtedly the Polynesians and their predecessors altered the ecological system of every island they landed on, no matter how brief their stay. One large double-hulled canoe could transport not only several dozen human beings, but perhaps 20 tons of provisions – breadfruit, bananas, coconuts, taros, sweet potatoes, sugarcane, various seeds, pigs, dogs and chickens – as well as stowaway rats. Once an island was reached, the pigs, and possibly the chickens, were set free to forage for themselves, and were allowed to go wild to be hunted later. The destruction of vegetation by such grazing and rooting land animals was incalculable.

In more recent times, European mariners released goats on islands in various parts of the world to assure themselves a source of fresh food on their return or in case of shipwreck. Close-cropping goats are even more destructive than wild Asian pigs. The barrenness of several of the Antilles in the southern Caribbean is in part accounted for by the large herds of wild goats that roam the hillsides.

The Polynesians were not great hunters and often used fire to drive animals towards traps or confined areas. In New Zealand, the first human inhabitants based their economy on the hunting of that huge flightless bird, the moa. They used fire to concentrate their prey, and in only a little over 500 years the moa disappeared for ever. Later New Zealanders, the Maoris, found that bracken, whose starchy root was a useful staple food, flourished best after fire, and this led to further destruction.

Finally, European settlers cleared land for pasture, introduced sheep, pigs, deer and a wide range of birds including starlings and sparrows, and made further inroads into the forests of native trees. Because of extensive destruction of their environment and new competition, much of the rich and varied bird life which the islands had supported disappeared.

On many smaller islands across the Pacific, forests were destroyed, and grasses grew in their stead. If an adequate grass cover was too slow in developing, whole regions became eroded, revealing the many-hued volcanic soils beneath. Scrub forests on atolls were replaced by cultivated

EROSION FOR CENTURIES *Polynesians, hunting across New Zealand, burnt many of the islands' forests; hundreds of years later, European settlers caused serious erosion by clearing forests for pasture. Now 60 million sheep graze on the islands*

groves of coconut palms and by other planted crops. Cultivation of food plants was especially important on the smaller islands where burgeoning human populations required an increasing supply of food. The Polynesians often excavated, terraced and enriched the soil, and by their farming increased the diversity of plant species present on the islands.

Among the few works of ancient Pacific islanders still extant are the famous rice terraces north of Baguio on the island of Luzon in the Philippines. The terraces, which are so arranged that water can spill from each paddy to the one below, are among the greatest works of man. They were built by the Ifugao people, descendants of a wave of Proto-Malays long preceding both the Philippine lowlanders and the Micronesians elsewhere. The Ifugao are highly efficient hydraulic engineers and have established an unusual degree of harmony with nature.

The isolation of an island is shattered when the first human being sets foot on its shores. As the previously uninhabited Atlantic islands of Madeira, St Helena and the Canaries were subjected to deforestation and grazing, their appearance and their ecology changed markedly. Charles Darwin commented on the barren and devastated island of Ascension when he visited it in 1836.

IMPORTED DESTROYERS *Introduced species, such as goats, have overrun many native plants and animals on the smaller Leeward Islands in the Caribbean. Particularly serious erosion occurs when the goats uproot the ground covering*

PADDY FIELDS IN THE MOUNTAINS *Built over a period of 2000 years, these rice terraces are 6000 ft up in the Luzon mountains in the Philippines. They provide 100,000 acres of level ground for cultivation and farmers work constantly to keep them in repair*

IN THE WAKE OF MAN

H-BOMB TEST SITE *A water-filled crater is a grim reminder of the atomic and hydrogen bombs that were exploded on Eniwetok atoll near Bikini in the 1950's. As radioactivity fades, plants and animals have begun to return to both atolls*

A CHANGED WORLD *The Hawaiian environment has been permanently changed by sugar-cane plantations and enormous sugar refineries. Intensive farming produces a yearly crop of more than 1 million tons on 221,300 acres of plantations*

In the Pacific, the changes brought about by the arrival of the Polynesians had a more profound effect on the wild life and plants of the islands than did the advent of Western agriculture centuries later. When the first Western explorers entered the ocean, it is doubtful that a single island still retained its original community of plants and animals. But despite the new life introduced by the Polynesians, the islands were still paradises. They had changed, but they were unspoilt.

In the two centuries since Western man arrived, the story has been an increasingly gloomy one. Forests have been razed to the ground and obliterated; seals, turtles and other huge marine animals have been hunted into extinction; reef fish and shell-fish, when not poisoned by pollution, have been harvested to near oblivion. A great many plants – weeds as well as ornamental and food species – have been introduced, as have insects, European rats, mongooses, South American toads, Malaysian birds and a host of other creatures. Because they were unable to compete against these invaders, many native island plants and animals now exist only on inaccessible mountains and on the few islands that have remained uninhabited.

Lost to extinction

The Pacific islands have lost more species to extinction than any other place known, and possibly have received more imported species as well. Millions of sheep feed on New Zealand's pastures, which themselves are composed of imported grass, sometimes causing erosion by over-grazing. Phosphate mining has almost obliterated the vegetation of some islands.

On Midway Island, heavily populated by albatrosses, attempts have been made to remove or exterminate the birds because the island is also an American naval base. Copra plantations cover many atolls where native vegetation once grew. On Hawaii, a road was cut through native trees on a high mountain to enable tourists to see an untouched tropical island forest. It is untouched no longer; for every car bringing sightseers also brings seeds, and the plants introduced in this way are spreading like a disease through the older vegetation.

SCARRING THE EARTH'S FACE *Opencast mines scar the flanks of New Caledonia's mountain ranges, which are rich in nickel – the island is third in world production – and other minerals. Even when mines are abandoned, animals and plants return very slowly*

Why islands are especially vulnerable

Probably nowhere are the unsightly and destructive effects of civilisation more evident than on the little islands of the South Seas that continue to be romantically presented in travel brochures. For a long time it seemed that their distant guardian nations were unaware of the devastation of the land and of the poverty and frustration of the local human populations; or, if they were aware of the problems, they were unable to solve them.

Islands are vulnerable to complete devastation by intruders for the very reason that makes their native plant and animal life unique. The species that are found on an unspoilt island have been developed from the very few that were able, many thousands of years ago, to survive the long migrations across great expanses of sea. Protected since then by ocean barriers from competition against dominant mainland forms, they gradually developed alone to become unique species. If they are allowed to remain in isolation, without having to face threats and possible extinction from introduced species, they continue to develop in ways that add much to our knowledge of life on earth. But Western man creates instability and change, rapidly converting something unique into just another bit of mainland: an isolated copy of the commonplace.

A FUTURE FOR THE ISLANDS

Ever since Charles Darwin landed on the Galapagos Islands in 1835, scientists have been discovering that the surprising life forms of remote islands provide clues to many of the unsolved mysteries of nature. Now these same islands are playing a part in providing material for a new and vital field of study: the science of ecology; for, unless we learn the lessons of the Galapagos – and other islands that have been far more spoilt – our own existence on the planet may soon be endangered.

Of course, islands as much as continents are for people, and it would be wrong to prevent the development of island groups, such as the West Indies, which are already highly populated. Yet there are still islands which are uninhabited – and which can be kept so – and others where plan-ning and restraint can prevent further damage.

The governments of the larger islands are now actively engaged in their own programmes of conservation. In New Zealand naturalists are determined that the takahe, the bird which for fifty years was believed to be extinct and then found again, should not vanish a second time into oblivion. Efforts are also being made to guard the country's own symbol, the kiwi, and the other vulnerable flightless birds. Only naturalists, carrying special passes, are allowed to set foot on the offshore islands which have become the last refuge of the tuatara. In Australia the state governments are at last turning their attention to the kangaroos which are being slaughtered to make food for pets and provide leather. In Madagascar, forest wildlife parks are being

FORESTS DESTROYED *Before European colonisation in the 17th century, forests of ebony covered the mountains of Mauritius in the Indian Ocean. Now some 40 per cent of the island is occupied by sugar-cane fields*

SET IN A TURQUOISE SEA *The turquoise waters of the Pacific wash over the coral fringing the island of Moorea. Like almost all the world's islands Moorea has been altered by man; but it remains one of the most beautiful*

established in an eleventh-hour attempt to save the hard-pressed families of lemurs.

But the small remote islands present problems which can only be solved by concerted international efforts, and there are now encouraging signs that the world's governments are determined to prevent further depredations. On the Galapagos Islands and on Aldabra, saved a few years ago from becoming an airfield, there are scientific research stations and active conservation programmes which should ensure that the giant tortoises do not follow the dodo into extinction. A plan to protect all the unspoilt islands of the Pacific has been developed under the International Biological Programme and agreements on island conservation are being prepared.

If these national and international programmes for conservation can prevail against short-term greed, there is still hope that the wonders of island life – evolved over so many millions of years – may be preserved for our descendants.

THE VANISHED DODO *Evolving in a world without predators, the dodo of Mauritius became large and fat and lost the ability to fly. Sailors killed many dodos, and they became extinct when pigs and monkeys were introduced by man*

Bibliography

and recommended reading

The publishers wish to acknowledge their indebtedness to the following books and periodicals, which were consulted for reference or as sources of illustration. The fields of island ecology and oceanography are fascinating and fast-moving. We hope that reading *The Fascinating Secrets of Oceans and Islands* will encourage you to delve into other publications and periodicals

ABYSS by C. P. Idyll (Constable)

ALDABRA ALONE by Tony Beamish (Allen & Unwin)

ANCIENT VOYAGERS IN POLYNESIA by Andrew Sharp (Angus & Robertson)

ANIMAL SPECIES AND EVOLUTION by Ernst Mayr (Harvard University Press)

ANIMALS IN MIGRATION by Robert T. Orr (Macmillan)

ANIMALS OF AUSTRALIA by A. Poignant (Dodd, Mead)

ANIMALS WITHOUT BACKBONES by Ralph Buchsbaum (Penguins)

ASPECTS OF DEEP SEA BIOLOGY by N. B. Marshall (Hutchinson)

ATLANTIC CORAL REEFS by F. G. W. Smith (University of Miami Press)

ATLANTIC SHORE, THE by John Hay and Peter Farb (Harper & Row)

ATOLL ENVIRONMENT AND ECOLOGY by Herold J. Wiens (Yale University Press)

AUSTRALIA (Australian News and Information Bureau)

AUSTRALIA AND THE PACIFIC ISLANDS: A NATURAL HISTORY by Allen Keast (Hamish Hamilton)

AUSTRALIAN BIRDS by Robin Hill (Nelson)

AUSTRALIAN BIRDS IN COLOUR by Keith Hindwood (East-West Center Press)

AUSTRALIAN GREAT BARRIER REEF, THE by Keith Gillett (A. H. & A. W. Reed)

AUSTRALIAN REPTILES IN COLOUR by Harold Cogger (East-West Center Press)

AUSTRALIAN SEASHORES by William J. Dakin and others (Angus & Robertson)

BETWEEN PACIFIC TIDES by Edward F. Ricketts and Jack Calvin, revised by Joel W. Hedgpeth (Stanford University Press)

BIOGRAPHY OF THE SEA, A by Richard Carrington (Chatto & Windus)

BIOLOGY OF MARINE MAMMALS, THE by Harald T. Andersen (Academic Press)

BIRD WONDERS OF AUSTRALIA by Alec H. Chisholm (Angus & Robertson)

BIRDS OF HAWAII by George C. Munro (Tuttle)

BIRDS OF THE ATLANTIC OCEAN by Ted Stokes (Country Life)

BIRDS OF THE OCEAN by W. B. Alexander (Putnam)

BIRDS OF THE SOUTHWEST PACIFIC by Ernst Mayr (Macmillan)

BIRDS OF THE WORLD by Oliver L. Austin Jr. (Hamlyn)

BLUE WHALE, THE by George L. Small (Columbia University Press)

BRIDGE TO THE PAST by David Attenborough (Harpers)

BUSH DWELLERS OF AUSTRALIA Australian News and Information Service (Rigby)

CARIBBEAN REEF FISHES by John E. Randall (T.F.H. Publications)

CIRCLE OF FIRE, THE by Carl Heintze (Meredith Press)

CORAL ISLAND by Marston Bates and Donald Abbott (Scribners)

DANGEROUS MARINE ANIMALS by Bruce W. Halstead (Cornell Maritime Press)

DARWIN AND THE GALAPAGOS by John Livingston and Lister Sinclair (Canadian Broadcasting Corp.)

DARWIN'S FINCHES by David Lack (Peter Smith)

DIVING FOR SUNKEN TREASURE by Jacques-Yves Cousteau & Philippe Diolé (Cassell)

EARTH BENEATH THE SEA, THE by Francis P. Shepard (Johns Hopkins)

ECOLOGICAL ANIMAL GEOGRAPHY by W. C. Allee and K. P. Schmidt (John Wiley)

ECOLOGY by Peter Farb and the Editors of *Life* (Time Life)

ECOLOGY AND FIELD BIOLOGY by Robert L. Smith (Harper & Row)

ECOLOGY OF INVASIONS BY ANIMALS AND PLANTS, THE by Charles S. Elton (Methuen)

EDGE OF THE SEA, THE by Rachel Carson (MacGibbon & Kee)

EXPLORING THE OCEAN WORLD by C. P. Idyll (Crowell)

EXPLORING THE SECRETS OF THE SEA by William J. Cromie (Allen & Unwin)

FIELD BOOK OF SEASHORE LIFE by Roy Waldo Miner (Putnam)

FIELD GUIDE TO THE BIRDS OF NEW ZEALAND, A by R. A. Falla, R. B. Sibson and E. G. Turbott (Collins)

FISHES, THE by Francis Downes Ommanney (Time Life)

FOUR-LEGGED AUSTRALIANS by Bernhard Grzimek (Collins)

FRAIL OCEAN, THE by Wesley Marx (Ballantine)

FRESHWATER TORTOISES OF AUSTRALIA AND NEW GUINEA by John Goode (Lansdowne Press)

FRONTIERS OF THE SEA by Robert C. Cowen (Gollancz)

FUNDAMENTALS OF ECOLOGY by Eugene P. Odum (W. B. Saunders)

FURRED ANIMALS OF AUSTRALIA by Ellis Troughton (Angus & Robertson)

GALAPAGOS: ISLANDS OF BIRDS by Bryan Nelson (Longmans)

GALAPAGOS, NOAH'S ARK OF THE PACIFIC, THE by Irenaus Eibl-Eibesfeldt (Doubleday)

GALAPAGOS: THE FLOW OF WILDNESS by Eliot Porter (Sierra Club)

GALAPAGOS, WORLD'S END by William Beebe (Putnam)

GALATHEA DEEP SEA EXPEDITION, THE by Anton F. Bruun (Allen & Unwin)

GREAT RED ISLAND, THE by Arthur Stratton (Macmillan)

GUIDE TO MARINE FISHES by Alfred Perlmutter (Bailey Bros.)

HALF MILE DOWN by William Beebe (Lane)

HANDBOOK OF NORTH AMERICAN BIRDS, *Vol. 1* by Ralph S. Palmer (Yale University Press)

HARVEST OF THE SEA by John Bardach (Allen & Unwin)

HAWAII: A NATURAL HISTORY by Sherwin Carlquist (Natural History Press)

HAWAII: A PICTORIAL HISTORY by Joseph Feher (British Museum Press)

HIDDEN SEA, THE by Douglas Faulkner and C. Lavett Smith (Viking Press)

HISTORY OF FISHES, A by J. R. Norman (Hill & Wang)

ISLAND LIFE by Sherwin Carlquist (Natural History Press)

ISLAND LIFE by Alfred Russell Wallace (Macmillan)

ISLES OF THE CARIBBEES by Carleton Mitchell (National Geographic Society)

ISLES OF THE SOUTH PACIFIC by Maurice Shadbolt and Olaf Ruhen (National Geographic Society)

KINGDOM OF THE OCTOPUS by Frank W. Lane (Jarrold)

LAND OF A THOUSAND ATOLLS by Irenaus Eibl-Eibesfeldt (MacGibbon & Kee)

LEMUR BEHAVIOUR: A MADAGASCAR FIELD STUDY by Alison Jolly (University of Chicago Press)

LIFE IN MICRONESIA by E. H. Bryan Jr. (Hourglass)

LIFE OF FISHES, THE by N. B. Marshall (Weidenfeld & Nicolson)

LIFE OF SEA ISLANDS, THE by N. J. Berrill and Michael Berrill (McGraw)

LIFE OF THE OCEAN, THE by N. J. Berrill (McGraw)

LIFE OF THE SEASHORE, THE by William H. Amos (McGraw)

LIFE ON THE SEA-SHORE, THE by A. J. Southward (Heinemann)

LIVING FISHES OF THE WORLD by Earl S. Herald (Hamish Hamilton)

LIVING TIDE, THE by N. J. Berrill (Gollancz)

LIVING WORLD OF THE SEA, THE by William J. Cromie (Prentice-Hall)

LOWER ANIMALS, THE by Ralph Buchsbaum and Lorus J. Milne (Doubleday)

MAMMALS IN HAWAII by P. Quentin Tomich (Bishop Museum Press)

MAMMALS OF THE WORLD by Ernest P. Walker (Johns Hopkins)

MAN AND DOLPHIN by John C. Lilly (Gollancz)

MAN'S PLACE IN THE ISLAND ECOSYSTEM by F. R. Fosberg (Bishop Museum Press)

MARINE ANIMALS: PARTNERSHIPS AND OTHER ASSOCIATIONS by R. V. Gotto (English Universities Press)

MARINE GEOLOGY OF THE PACIFIC by H. W. Menard (McGraw)

NATURAL HISTORY OF MARINE ANIMALS by George E. Macginitie and Nettie Macginitie (McGraw)

NATURAL HISTORY OF SHARKS, THE by Thomas H. Lineweaver III and Richard H. Backus (Deutsch)

NEW DICTIONARY OF BIRDS, A by S. Landsborough Thomson (Nelson)

NEW ZEALAND SEASHORE, THE by J. Morton and N. Miller (Collins)

OCEAN, THE by W. H. Freeman (Scientific American)

OCEAN ISLAND, THE by Gilbert C. Klincel (Doubleday)

OCEANOGRAPHY by Peter K. Weyl (Wiley)

OCEANS, THE by H. U. Sverdrup and others (Prentice-Hall)

OPEN SEA: ITS NATURAL HISTORY, THE by Sir Alister Hardy (Fontana)

PACIFIC BASIN BIOGEOGRAPHY: A SYMPOSIUM by J. Linsley Gressitt (Bishop Museum Press)

PACIFIC SEASHELLS by Spencer W. Tinker (Tuttle)

PENGUINS by John Sparks and Tony Soper (David & Charles)

PLANKTON OF THE SEA, THE by R. S. Wimpenny (Faber)

PLANTS AND THE MIGRATIONS OF PACIFIC PEOPLES by Jacques Barrau (Bishop Museum Press)

POLYNESIAN CULTURE HISTORY by Genevieve A. Highland and others (Bishop Museum Press)

RED BOOK, THE by James Fisher, Noel Simon and Jack Vincent (Collins)

SEA, THE by Leonard Encel (Time Life)

SEA, THE by Robert C. Miller (Nelson)

SEA AROUND US, THE by Rachel Carson (MacGibbon & Kee)

SEA BIRDS OF THE SOUTH PACIFIC OCEAN by P. P. O. Harrison (Princeton University Press)

SEA SHORE, THE by C. M. Yonge (Collins)

SEABIRDS by J. Fisher and R. M. Lockley (Collins)

SEALS OF THE WORLD by Judith E. King (Natural History Museum)

SEALS, SEA LIONS AND WALRUSES by Victor B. Scheffer (Oxford University Press)

SEAS, THE by Frederick S. Russell and C. M. Yonge (Warne)

SEASHELLS OF NORTH AMERICA by R. Tucker Abbott (Dean)

SECRET ISLANDS, THE by Franklin Russell (Hodder & Stoughton)

SHARK! KILLER OF THE SEA by Thomas Helm (Hale)

SHARK: SPLENDID SAVAGE OF THE SEA, THE by Jacques-Yves and Philippe Cousteau (Cassell)

SHOALS OF CAPRICORN, THE by F. D. Ommanney (Longmans)

SHOREBIRDS OF NORTH AMERICA, THE by Peter Matthiessen, edited by Gardner D. Stout (Viking)

SILENT WORLD, THE by Jacques-Yves Cousteau (Hamish Hamilton)

STOCKS OF WHALES, THE by N. A. Mackintosh (Fishing News (Books) Ltd)

SURTSEY: NEW ISLAND IN THE NORTH ATLANTIC by Sigurdur Thorarinsson (Cassell)

THIS GREAT AND WIDE SEA by R. E. Coker (University of North Carolina Press)

TREASURY OF AUSTRALIAN WILDLIFE, A by D. F. McMichael (Ure Smith)

TROPICAL PACIFIC, THE by N. V. Harris (University of London Press)

TROPICAL QUEENSLAND by Stanley & Kay Breeden (Collins)

TUATARA, LIZARDS, AND FROGS OF NEW ZEALAND, THE by Richard Sharell (Collins)

TURTLE by Archie Carr (Cassell)

UNDERSEA FRONTIERS by Gardner Soule (Rand McNally)

UNDERWATER GUIDE TO MARINE LIFE, THE by Carleton Ray and Ciampi Elgin (Kaye & Ward)

VERTEBRAE STORY, THE by Alfred S. Romer (University of Chicago Press)

VIKINGS IN THE PACIFIC by Peter H. Buck (Phoenix Books)

VOLCANOES by Cliff Ollier (MIT Press)

VOLCANOES by Kent H. Wilcoxson (Cassell)

WHALE, THE by Leonard Harrison Mathews and others (Allen & Unwin)

WHALES by E. J. Slijper (Hutchinson)

WHALES, DOLPHINS AND PORPOISES by Kenneth S. Norris (University of California Press)

WILD AUSTRALIA by Michael K. Morcombe (Hamlyn)

WILD FLOWERS OF AUSTRALIA by Thistle Y. Harris (Angus & Robertson)

WILD MAMMALS OF MALAYA, THE by Lord Medway (Oxford University Press)

WILDLIFE OF THE SOUTH SEAS by F. A. Roedelberger, Vera I. Groschoff and Peter J. Whitehead (Constable)

WILDLIFE OF WESTERN AUSTRALIA by Peter Slater and Eric Lindgren (Western Australian Newspapers Ltd)

WONDROUS WORLD OF FISHES (National Geographic Society)

WORLD BENEATH THE OCEANS by T. F. Gaskell (Aldus)

WORLD BENEATH THE SEA by James Dugan and others (National Geographic Society)

WORLD OF BIRDS, THE by James Fisher and Roger Tory Peterson (Macdonald)

WORLD OF THE POLAR BEAR, THE by Richard Perry (Cassell)

WORLD OF THE WALRUS, THE by Richard Perry (Cassell)

YEAR OF THE SEAL, THE by Victor B. Scheffer (Charles Scribner's)

YEAR OF THE WHALE, THE by Victor B. Scheffer (Souvenir Press)

ZOOGEOGRAPHY by P. J. Darlington Jr. (Wiley)

Periodicals

ANIMAL KINGDOM, New York

AUDUBON, New York

BIOSCIENCE, Washington, DC

BIRDS OF THE WORLD, London

INTERNATIONAL WILDLIFE, Milwaukee, Wisconsin

NATIONAL GEOGRAPHIC, Washington, DC

NATURAL HISTORY, New York

NATURE, London

OCEANOLOGY INTERNATIONAL, Houston, Texas

OCEANS, San Diego, California

OCEANUS, Woods Hole, Massachusetts

PURNELL'S ENCYCLOPEDIA OF ANIMAL LIFE, London

SCIENCE, Washington, DC

SCIENTIFIC AMERICAN, New York

SEA FRONTIERS, Miami, Florida

SMITHSONIAN, Washington, DC

Index

Figures in italic type refer to illustrations, diagrams or maps and sometimes, in addition, to text on the same page.

Index

Figures in italic type refer to illustrations, diagrams or maps and sometimes, in addition, to text on the same page.

Index

Figures in italic type refer to illustrations, diagrams or maps and sometimes, in addition, to text on the same page.

Index

Index

Figures in italic type refer to illustrations, diagrams or maps and sometimes, in addition, to text on the same page.

Figures in italic type refer to illustrations, diagrams or maps and sometimes, in addition, to text on the same page.

Index

Figures in italic type refer to illustrations, diagrams or maps and sometimes, in addition, to text on the same page.

Index

Figures in italic type refer to illustrations, diagrams or maps and sometimes, in addition, to text on the same page.

Index

Figures in italic type refer to illustrations, diagrams or maps and sometimes, in addition, to text on the same page.

Index

Picture credits

The publishers would like to express their thanks to the photographers whose work appears in this book.

1–5 Josef Muench. 6–7 Heather Angel. 10–11 Neville Fox-Davies. 12 David Muench. 14 D. Muench. 15 Edward Malsberg. 16 Dennis Brokaw. 17 (left) John W. Evans; (right) H. Angel. 18 Douglas P. Wilson. 19 Jack Dermid. 21 (top) Anthony Mercieca; (bottom) Walter Dawn. 22 (top) Betty Randall; (bottom) Anthony Bannister. 23 (top left) Gary G. Gibson; (top right) H. Angel; (bottom) D. Brokaw. 24 E. Malsberg. 25 William H. Amos. 27 (top left) John S. Flannery; (bottom left; right) J. Dermid. 28 (top) W. Dawn; (bottom) Helen Cruickshank. 29 Caulion Singletary. 30 From *The Life of an Estuary*, by Robert M. Ingle, © May 1954, by Scientific American, Inc. All rights reserved. 31 Laurence Lowry/ Rapho Guillumette. 32 Douglas Faulkner. 34 George Kelvin. 35 Kellner Associates (from *The Continental Shelves*, by K. O. Emery, © September 1969, by Scientific American, Inc. All rights reserved). 36 (left) D. P. Wilson; (right) W. Dawn. 37 (top left) Kellner Associates; (top right; centre right) D. P. Wilson; (bottom right) Harry McNaught. 38 (top) W. Dawn; (bottom) D. P. Wilson. 39 (top left) W. Dawn; (top right; bottom) D. P. Wilson. 40–41 Darrell Sweet. 42 D. Faulkner. 43 (top) D. Faulkner; (bottom) Don Wobber. 44 (top) Allan Power/National Audubon Society; (bottom) D. Faulkner. 45 Ron Church. 46 Kellner Associates. 47 Robert C. Murphy. 48 *American Fish Farmers* Magazine. 49 (left) Jack McKenney; (right) Tom Myers. 50 Keith Gillett. 52 (top) K. Gillett; (bottom) G. Kelvin. 53 Anthony Petrocelli. 54 (top) D. Faulkner; (bottom) Valerie Taylor. 55 (left; top right) D. Faulkner; (bottom right) Eva Cellini. 56 David Doubilet. 57 A. Bannister. 58–59 Wayne & Karen Lukas/Group IV/National Audubon Society. 60 (top) D. Faulkner; (bottom) Walter Deas. 61 K. Gillett. 62 D. Faulkner. 63 (top) D. Faulkner; (bottom) E. Cellini (from R. V. Gotto, after F. A. Potts). 64 (top) D. Faulkner; (bottom) A. Bannister. 65 (left) E. Cellini; (right) D. Faulkner. 66 T. R. Beaven. 67 (top) E. Cellini; (bottom) D. Faulkner. 68 Ron Taylor. 69 E. Cellini. 70–71 (top left) Ben Cropp; (top centre; bottom right) D. Faulkner. 72 Edmund Hobson. 73 D. Faulkner. 74 (top) E. Cellini; (bottom) D. Faulkner. 75 D. Faulkner. 76 (top; bottom right) D. Faulkner; (bottom left) E. Cellini. 77 (top left; bottom left; bottom right) D. Faulkner; (top right) E. Cellini. 78 E. Hobson. 79 (left) D. Faulkner; (right) R. Church. 80 A. Power. 81 (top) R. Church; (bottom) D. Faulkner. 82 D. Faulkner. 83 (top) B. Cropp; (bottom) D. Faulkner. 84 (top) Graham Pizzey; (bottom) D. Faulkner. 85 Jen & Des Bartlett. 86–87 A. Petrocelli. 88 Flip Schulke/Black Star. 90 (top) William M. Stephens; (bottom) E. Malsberg. 91 (top) D. Faulkner; (bottom) Robert C. Hermes/National Audubon Society. 92 (left) D. Faulkner; (right) D. P. Wilson. 93 Margiocco/Popperfoto. 94 R. Taylor. 95 (top) D. Faulkner; (bottom) K. Gillett. 96–97 D. Sweet. 98 J. & D. Bartlett. 99 (left) James W. LaTourrette; (top right) R. Church; (bottom right) D. P. Wilson. 100–1 Bernie Pertchik. 102 (top) H. McNaught; (bottom) George X. Sand. 103 E. J. Zimbelman. 104–5 H. McNaught. 105 (right) R. Taylor. 106 (top) B. Cropp. 106–7 Eva Cropp. 108 W. H. Amos. 110 W. H. Amos. 111 (top) Dennis M. Opresko; (bottom) E. Malsberg. 112 G. Kelvin (after N. B. Marshall and C. P. Idyll). 113 Peter David/Photo Researchers. 114 (top) E. Cellini; (bottom) P. David/Photo Researchers. 115 W. H. Amos. 116 E. Malsberg (from N. B. Marshall, after V. V. Tchernavin). 117 W. H. Amos. 118–19 Davis Meltzer. 120 (top left) R. Church; (top right) Walter Jahn/U.S. Naval Oceanographic Office. 120–1 (bottom) R. Church. 122 (top) William R. Curtsinger; (bottom) Photo L.R.A./Groupe Cousteau. 123 (top) Eric Barham; (bottom) Kellner Associates; (bottom right) R. Church/E. Barham. 124 E. Malsberg. 125 Elkan J. Morris. 126 Jessie O'Connell Gibbs. 128 G. Kelvin (after F. S. Russell and C. M. Yonge). 129 (top) P. David/ Photo Researchers; (bottom) U.S. Naval Oceanographic Office. 130 Robert E. Schroeder. 131 (left) E. Cellini; (right) H. Angel. 132 (top) Karl Kenyon/National Audubon Society; (bottom left) George Holton/Photo Researchers; (bottom right) Fred Bruemmer. 133 (top) Francisco Erize; (bottom) Cordell Hicks. 134–5 Steve Wilson/DPI. 136 G. G. Gibson. 137 Kellner Associates. 138 H. Angel. 139 Kellner Associates. 140 (left) D. Faulkner; (right) B. Cropp. 141 (top) E. Malsberg; (bottom left) Douglas Baglin; (bottom right) A. Power. 142 (top) Robert Potts; (bottom) Eric Hosking/Bruce Coleman Inc. 143 (top left) G. Kelvin; (top right) David & Katie Urry/Bruce Coleman Inc; (bottom) W. R. Curtsinger. 144 E. Malsberg. 145 Kellner Associates. 146 Chuck Nicklin. 148 (top) Kellner Associates; (bottom) H. Angel. 149 David C. Stager. 151 D. Baglin. 152 (top) D. Faulkner; (bottom) D. Doubilet. 153 (left) Kellner Associates; (right) American Museum of Natural History. 154 Thase Daniel. 155 (top) F. Erize; (bottom) Lancelot Tickell. 156 W. R. Curtsinger. 157 Michael C. T. Smith. 158 W. R. Curtsinger. 159 (top; bottom) W. R. Curtsinger; (bottom right) Katia Bojiloba/DPI. 160 Alvin E. Staffan. 161 (left) Kellner Associates; (right) Kenneth W. Fink; (bottom right) E. Cellini. 162 F. Erize. 163 T. Daniel. 164 (top) Frederick Baldwin/National Audubon Society; (bottom) Kellner Associates. 165 F. Bruemmer. 166 William F. Bryan. 167 (top) Herbert Clarke; (bottom) W. F. Bryan. 168 (top) Russ Kinne/Photo Researchers; (bottom) Steve & Dolores McCutcheon. 169 T. Daniel. 170 Gerald Ferguson. 172 Alfred L. Pentis. 173 (top) Bucky Reeves/National Audubon Society; (bottom) R. Church. 174–5 H. McNaught. 176–7 Theodore Walker. 178–9 H. McNaught. 180 D. Faulkner. 181 J. & D. Bartlett/Bruce Coleman Inc. 182 (left) John G. Ross/Photo Researchers; (top right) J. W. LaTourrette/Miami Seaquarium. 182–3

(bottom) J. & D. Bartlett/Bruce Coleman Inc. 184 (top) J. McKenney; (bottom) Kellner Associates. 185 Peter L. Vila. 186 George Musil. 187 Kellner Associates. 188 (top) Bob Brooks; (bottom) N. Merrit/Photo Aquatics. 189 (top) Adam Wolfitt/Susan Griggs; (bottom left) N. Merritt/Photo Aquatics; (bottom right) Ralph Pinto. 190–1 Sven Gillsäter. 191 (top right) Chuck Diven. 192–3 J. Muench. 194 Robert L. Bowman. 196–7 H. C. Berann, based on material by Prof. Bruce Heezen (Lamont Geological Observatory) and Marie Tharpe (U.S. Oceanographic Office)/© Verlag DAS BESTE GmbH, Stuttgart. 196 (bottom) G. Kelvin. 198 R. Kinne/Photo Researchers. 199 (top) T. Iwago/Photo Trends; (bottom) S. Gillsäter. 201 S. Gillsäter. 202 (top left) James A. Kern; (top right) Edward F. Anderson; (bottom) Photographic Library of Australia. 203 (top) J. A. Kern; (bottom left) E. S. Ross; (bottom right) E. Malsberg. 204 (left) Sherwin Carlquist; (right) G. Kelvin. 205 E. Cellini. 206 (top) Karl Weidmann/National Audubon Society; (bottom) G. Kelvin. 207 (top) Fred Ward/Black Star; (top right) Roy Woodbury; (bottom left) T. Daniel; (bottom right) William J. Bolte. 208 N. Smythe. 209 Tom McHugh/Photo Researchers/Courtesy of the Chicago Zoological Park. 210–11 L. Tickell. 212 (top) Leonard Lee Rue III; (bottom) Tee Balog. 213 (left) Raul Mina Mora; (right) L. Tickell. 214 Alan Root. 216 G. Kelvin. 217 Orion Press/FPG. 219 Eiji Miyazawa/Black Star. 220 (left) George Laycock; (right) M. Kellner. 221 Grant Haist. 222 Charles Fracé. 223 David Cavagnaro. 224 A. Root. 225 (top; bottom) A. Root; (middle) Okapia. 226 (top) G. Haist; (bottom) T. Daniel. 227 M. Kellner. 228 G. Kelvin. 229 William H. Sager. 230 W. H. Sager. 231 (top) T. Daniel; (bottom) Werner Stoy/Camera Hawaii. 232 (top; bottom left) Richard E. Warner; (bottom right) G. Laycock/Photo Researchers. 233 R. E. Warner. 234 Quentin Keynes. 236 M. Kellner. 237 Roland C. Clement. 238 Howard Uible. 239 Peter Johnson. 240 E. Cellini. 241 H. Uible. 242 H. Uible. 244 (top) Q. Keynes; (bottom) Edwin Gould. 245 (top) P. Johnson; (bottom) George Holton/Photo Researchers. 246 M. Kellner. 247 (top) G. R. Roberts; (right) E. Cellini. 248 (top) K. W. Fink; (bottom) M. F. Soper. 249 M. C. T. Smith. 250 John Warham. 251 New Zealand National Publicity. 253 M. F. Soper. 254 Stan & Kay Breeden. 256–7 M. K. Morcombe. 258 (top) D. Baglin; (bottom) Harold J. Pollock. 259 Adaptation from Life Nature Library: *The land and wildlife of Australasia* © 1964 Time Inc. All rights reserved; and *Atlas of Australian Resources*, Dept. of National Development, Canberra, 1970. 260 M. K. Morcombe. 261 (top left; bottom) M. K. Morcombe; (top right) E. Cellini. 262 (left) M. K. Morcombe; (right) Okapia. 263 G. Kelvin. 264 (top) S. & K. Breeden. 265 (top) G. Kelvin; (bottom left) S. & K. Breeden; (bottom right) D. Baglin. 266 (top) Steve Pearson; (bottom) Anthony Healey. 267 H. McNaught. 268 (left) S. & K. Breeden; (right) G. Kelvin. 269 (top) A. Healey; (bottom) S. & K. Breeden. 270 (left) A. Root/Okapia; (right) E. Cellini. 271 S. & K. Breeden. 272 Okapia. 273 G. Kelvin. 274 (left) A. Healey; (right) E. Cellini. 275 Len Robinson. 276 M. K. Morcombe. 278 (left) A. Root; (right) John Carnemolla. 279 (top) John Markham; (bottom) H. J. Pollock. 280 L. Robinson. 281 (top) A. Root; (bottom left) M. K. Morcombe; (bottom right) D. Baglin. 282 David Corke. 283 A. Root. 284 (top) William S. Peckover; (right) E. Malsberg. 285 A. Root. 286 Photographic Library of Australia. 287 (left) A. Root; (top right; bottom right) L. Robinson. 288 (top) Warren Garst/Van Cleve Photography; (middle) A. Root; (bottom right) E. Malsberg. 289 Okapia. 290–1 H. McNaught. 292 M. K. Morcombe. 293 (top) Okapia; (bottom) National Zoological Park. 294 M. K. Morcombe. 295 (left) Jack Fields; (right) Okapia. 296 E. Malsberg. 297 M. K. Morcombe. 298 (left) H. J. Pollock/FPG; (right) J. Carnemolla. 299 (top) John Kaufmann; (bottom left) M. T. Tanton; (bottom right) A. Root. 300 S. & K. Breeden. 301 R. Kinne/Photo Researchers. 302 Wallace A. Cole. 304 Aevar Johannesson. 305 Richard Frohs. 306 (top) A. Johannesson; (bottom) Q. Keynes. 307 (top) John E. Randall; (bottom) T. Stell Newman. 308 (top left) G. Kelvin; (bottom) Camera Hawaii. 309 A. Johannesson. 310 (top) G. R. Roberts; (right) Harrison Forman. 311 G. Tomsich/M. Grimoldi. 312 (left) M. Kellner; (right) From *Annales du Jardin Botanique de Buitenzorg*, Vol. XLVI–XLVII. 313 Ivan Polunin. 314 W. H. Amos. 315 Peter Sanchez. 316 (top) W. Stoy/Camera Hawaii; (bottom) D. Muench. 317 (top left) D. Muench; (bottom left) W. Stoy/Camera Hawaii. 318 (top) Hawaii Volcanoes National Park; (bottom) M. Kellner. 319 W. H. Amos. 320 P. Sanchez. 321 W. H. Amos. 322 D. Faulkner. 324–5 H. McNaught. 327 Photographic Library of Australia. 328 G. Kelvin. 329 Cameron B. Kepler. 330 (top) C. B. Kepler; (bottom) D. Baglin. 331 (top; bottom right) C. B. Kepler; (bottom left) G. Holton/Photo Researchers. 332 (top left) G. R. Roberts; (bottom) E. Cellini. 332–3 M. Philip Kahl. 334 David Moore/Black Star. 336–7 G. Kelvin. 338 M. Kellner. 339 (top; bottom left) D. Moore/Black Star; (bottom right) J. E. Randall. 340 Paul Rockwood/National Park Service. 341 E. Cellini. 342–3 (top) J. Fields. 342 (bottom left) J. Fields/Photo Researchers; (bottom right) H. J. Pollock. 343 (bottom left) J. Fields/Photo Researchers; (bottom right) D. Moore/Black Star. 344 J. Carnemolla. 345 M. Kellner. 346–7 Robert Lebeck/Black Star. 348 G. R. Roberts. 349 (top) Bill Atkinson; (bottom) International Foto File. 350 (top) S. Arthur Reed; (bottom) Van Bucher/Photo Researchers. 351 J. Fields. 352 (left) Frederick Ayer/Photo Researchers. 352–3 (bottom) J. Fields. 353 (top right) E. Cellini.

Acknowledgments

The publishers wish to acknowledge the generous assistance and advice of:

Andrew Berger, University of Hawaii
Joseph P. Branham, University of Hawaii
Alec Chisholm, Cremorne Point, New South Wales, Australia
Harold Cogger, Australian Museum, Sydney
Maxwell S. Doty, University of Hawaii
R. Hammond, Division of Marine Biology, Commonwealth Scientific and
Industrial Research Organization, Cronulla, New South Wales, Australia
Allan Keith, New York City
Gordon A. Macdonald, University of Hawaii
Basil Marlow, Australian Museum, Sydney
Michael Morcombe, Armadale, Western Australia
Dieter Mueller-Dombois, University of Hawaii
Arthur Reed, University of Hawaii
Sidney Townsley, University of Hawaii

® 'The Reader's Digest' is a registered
trademark of The Reader's Digest Association
Inc. of Pleasantville, New York, U.S.A.

Published by
The Reader's Digest Association Limited
25 Berkeley Square, London W1X 6AB
The Reader's Digest Association (Canada) Limited
215 Redfern Avenue, Montreal 215, Quebec
The Reader's Digest Association Pty Limited
Reader's Digest Building, 26–32 Waterloo Street,
Surry Hills, Sydney, N.S.W. 2010
The Reader's Digest Association Limited
Nedbank Centre, Strand Street, Cape Town

Printing by Petty & Sons Limited, Leeds
Binding by Hazell Watson & Viney Limited, Aylesbury